D1143370

Essentials of Bacterial Physiology

The Class of '75
Douglass College

Charles C. Amilo
Richard A. Bronen
Alexandria Cabrera
John Chu
Kathleen G. Clodius
Gail Dargenzio
Elaine M. Ehlers
Virginia Glaser
Edwin R. Guzman
Carolyn M. Hirst
Janice R. Hulbert
Vidyut Jhaveri
Gary J. Jones
Patricia E. Kasenchar
Lynda A. Kiefer
Carol M. Krause
Steve Mamber
Beatrice F. Minassian
Anubha R. Mukherjee
Jenny Ng
John O'Keefe
Arlene Potts
Betsy S. Powell
Russell Rawlings
Jo-Ann Sciacchitano
Gail P. Simon
Ramona M. Slepetis
Vera Vasquez-Ortegon
Bridget A. Walsh
Susanne Weiser

Essentials of Bacterial Physiology

W. W. Umbreit
Rutgers University

and

The Class of '75
Douglass College

Distributed by

BURGESS PUBLISHING COMPANY
MINNEAPOLIS, MINNESOTA

Library of Congress Catalog Card Number: 76-13615
ISBN: 0-87933-221-2

78 77 76 1 2 3 4 5
Manufactured in the United States of America.

LIBRARY OF CONGRESS CATALOGING IN PUBLICATION DATA
Umbreit, Wayne William, 1913–
Essentials of bacterial physiology.
A revision of An introduction to bacterial physiology, by E. L. Oginsky and W. W. Umbreit.
Includes index.
1. Bacteria—Physiology. I. Rutgers University, New Brunswick, N. J. Douglass College. Class of 1975. II. Oginsky, Evelyn Lenore, 1919– An introduction to bacterial physiology. III. Title.
[DNLM: 1. Bacteria—Physiology. QW52 U49e]
QR84.05 1976 589.9'01 76-13615
ISBN 0-87933-221-2

Exclusive Distributor:
BURGESS PUBLISHING COMPANY

ISBN: 8087-2107-0

Preface

A preface is a place where an author can thank those who assisted him, explain the reasons why he wrote the book (if only to ease his conscience), and warn the readers of what pitfalls they might encounter—all of these in the relative security of knowing that, with the exception of yourself, very few will read or pay much attention to the little gems of wisdom (or eccentricity) revealed herein. An author of an introductory work, even an advanced introduction, is at a serious disadvantage. The work will be reviewed by people who have an advanced knowledge—either of the entire field or more likely of some specialized area of it. And they will be disappointed to find that their pet areas are frequently treated with something of a lick-and-a-promise, presumably in favor of those areas (clearly of less importance) which happen to be the favorites of the author. Many such reviewers have had little contact with the student, who has yet no such knowledge, and for whom the book was written. This is not a book for the advanced microbial physiologist. It is for the beginners—whom you hope will not be misled too far by it.

Since the publication of the last edition of Oginsky and Umbreit in 1959, Dr. Oginsky and I have recognized that a revision was necessary and we tried to write one. Indeed, we had three versions somewhat completed when we were out run by events, beset with problems not related to physiology (one of which was the 3,000 miles separating our laboratories), and simply could not find time to complete the work. But we did have a slightly out of date manuscript when I again was called upon, after a lapse of a few years, to teach Bacterial Physiology to a group of seniors. I found that they had no place to go for any reasonably coherent discussion of the subjects I intended to cover; reviews were useful but too detached and too scattered. The texts available at the time were either too specialized or too old. The class and I, with perhaps a little pressure on my part, agreed to write such a book. I would write out my lectures and have them put on stencils (and mimeographed), a chore done by Ms. Linda Eveleigh and Ms. Eleanor Filoramo. Each individual in the class would summarize a review of some importance and these, after editing, were also reproduced. Naturally, I leaned fairly heavily on the earlier manuscript but by the summer of 1975 we had a fairly up to date, somewhat composite manuscript, that required only "minor" revision. This minor revision proved to be a formidable task and took a great deal of time until the spring of 1976. I am sure that there is much still wrong—but we must stop somewhere. My major debt, then, is to Evelyn Oginsky and to the class of '75 whose names appear at the front of the book. I am also indebted to Mrs. Elain Derry who tried to get most of the errors of case, tense, and agreement out of the manuscript while typing the camera-ready copy. Mona Scherage prepared several of the drawings. Other figures were taken from the literature and we are grateful to the authors and publishers for permission to reproduce them, as acknowledged in the respective figures.

A note of thanks to Dr. Christine Reilly, Dr. Elizabeth Cook, and Ms. Eleanor Boshko of the Douglass faculty for helpful suggestions while teaching the course. Somewhat to our mutual surprise we discovered that Dr. Albert Moat was also writing a book on bacterial physiology, but by exchanging manuscripts, in a somewhat crude state, of course, we found that the two books were complementary and we both decided to go ahead anyway.

I sometimes wonder how this book will be received. Are we still able to touch the next generation or has the toll of time taken too much and the advance of the art, the methodology, and the knowledge been too swift so that what I, and perhaps my generation, still regard with wonder and acclaim is the obvious commonplace to the next generation. The only way we can tell is to see what happens to this volume. Wish it luck.

WAYNE W. UMBREIT

Contents

Essentials of Bacterial Physiology

1 Introduction: The Outer Surface of Bacteria

The purpose of bacterial physiology is to understand the functions of bacterial processes from the viewpoint of what good (or harm) such processes do for the bacteria themselves. It is an attempt to "grasp and to understand the essential character of life" (as Aristotle said about biology in general). Physiology is the study of function, the function of the organism itself.

To approach this understanding, bacterial physiology uses whatever tools may be available—the tools of biochemistry, molecular biology, enzyme chemistry, genetics, cultural bacteriology, and even other fields; but the focus of the endeavor is to understand the way in which the cell can live, grow, survive, and reproduce. It is futile to try to set up fine distinctions between bacterial physiology and other related fields; there exists an area of knowledge and a philosophical approach that is bacterial physiology.

Some years ago E. L. Oginsky and I wrote a book called *Introduction to Bacterial Physiology*, now out of print. Today, there is much more knowledge, a vastly increased scientific literature, and several volumes of advanced texts; yet an "introduction" is needed even more, partly because the knowledge to be surveyed is so vast and partly because that knowledge, as presently organized, is so specialized.

Yet an "introduction" is also somewhat unnecessary since you have already been introduced to physiology in your courses in general bacteriology or general microbiology. What is needed is a summary, the essentials as it were, of a vast and complex field. We shall therefore attempt to outline a series of concepts that have proved useful in the understanding of the life of a bacterial cell, and which have, on the basis of experimental work, at least a reasonable possibility of being correct. The use of these concepts permits a beginning toward understanding of the life of the cell.

We have organized these chapters to provide brief

discussions of the major concepts of bacterial physiology. Occasional references are inserted in the text for two purposes: (1) to provide evidence for the concept discussed, and (2) to serve as a source of further information in case you wish to examine a given concept more thoroughly. So as not to interfere with the continuity of the discussion, we have merely given them a number and listed them as "papers" at the end of the chapter. At the end of each chapter we have also provided a list of reviews or books on the subject, including those that we have found especially useful. But in no case do we cover the literature with any degree of completeness. Indeed, in your advanced courses you may find our treatment superficial. But our purpose is to provide an outline of concepts, an intellectual bridge between the introductory treatment and the advanced treatise. We attempt to delineate the essentials.

The intent of the references following each chapter is to provide more information on the specific topics discussed. However, the literature is so vast that we have restricted our references to recent reviews that have appeared in *Bacteriological Reviews*, *Annual Review of Microbiology*, *Annual Review of Biochemistry*, and *Advances in Microbial Physiology*. An occasional book is also cited. In addition, we refer to the Benchmark Papers in Microbiology, each volume of which collects the classic and critical papers on the subject. These are published by Dowden, Hutchinson & Ross, Inc., and are distributed by Halsted Press (605 Third Ave., New York, N.Y.). For those reviews summarized individually in the text, a footnote indicates which review the summary is based on.

In this text we assume that you have had an adequate course in general or beginning microbiology, some organic chemistry, and sufficient contact with biochemistry so that words like histidine, gene, DNA, polymer, and protein are familiar and that you have a conception of what they are. We assume that you are looking for an introduction, not an advanced treatise, and that you want this adequately founded and with adequate detail, but that you are concerned with broad principles and concepts.

Bacteria, of course, are procaryotic cells, which means that their nuclear structure is not bounded by a physical barrier or membrane, that most of their energy-yielding enzymes are located in the cell membrane and not dispersed in the cytoplasm, that their flagella are chemically distinct from the membrane and do not originate from it, and that they contain a substance in their cell walls clearly unique in nature, that is, muramic acid. These four characteristics constitute the definition and the justification of the separation of these organisms from the eucaryotic plants and animals. They comprise a separate kingdom, indeed a unique way of life.

BACTERIAL SIZE

Bacteria so small that it is difficult for us to readily grasp their smallness in any kind of concrete conception. One normal bacterium weighs 10^{-12} g, or 10^{-9} mg, or 1/10,000,000 μg, and even 80 percent of that weight is water. One microgram of wet bacteria contains 1 million individual cells. One microgram of dried bacteria contains from 3 to 7 million cells. These cells have an enormous surface in relation to their total volume, and it is at their surface that many of their most important reactions take place.

BACTERIAL SURFACE

Gram-positive cells appear to have a relatively uniform surface when looked at in the electron microscope, that is, after having been dried. But the surface of gram-negative organisms may be variegated and of varied depth. Cells may be smooth or resemble a butternut, walnut, or Brazil nut. Some seem to have a fibrous and even wooly-dog appearance. At times such photographs, especially in the scanning electron microscope, seem to show connections between cells, and in many cases the surface may be smooth enough, yet patchy, with areas of different depths, and even with channels and holes boring deep into the surface (1). One can see little of the surface in the light microscope, so we are entirely dependent upon the appearance after drying. We can only speculate as to how an undried cell looks.

CAPSULES

Many cells have capsules or a slime layer about them. Capsule is probably a better word; it implies some degree of organized structure. By capsule we mean something of relatively low rigidity outside and beyond the well-staining part of the cell. It stains poorly and is probably not a vital part of the cell, although such terms can lead to tricky definitions. But it is clearly there. The capsular area may or may not be uniform in depth. It usually appears uniform, but this may be due to shrinkage during staining or drying. So without trying to make fine distinctions, the capsule is a material, relatively uniform in composition, out beyond the cell wall although this itself may be difficult to precisely specify in the gram-negatives. We tend to think of it as dispensable, something rather loosely attached, which does not need to be present for the cell to live. Its presence tends to provide "smooth" colonies, its absence, "rough."

There are three major types of capsules. First are those that are primarily carbohydrate but contain nitrogen, usually in the form of glucosamine, or, if lacking in nitrogen, will tend to have uronic or glucuronic acids rather than simple sugar polymers; these capsules are under close genetic control (as examples one might cite capsules of the pneumococcus, which we shall consider later). Second are capsules, essentially without nitrogen, that are simple polymers of sugars, rarely containing uronic acids and the like. These are primarily dextrans (glucans), levans (fructose polymers), galactosans, and so on, but there is one case, *Acetobacter*, where the polymer is cellulose. Third are capsules consisting of polypeptides that contain one amino acid; to my knowledge, polymers of D-glutamic acid are typical.

Whether or not a capsule is present depends upon the genetic constitution (as in pneumococcus), the environment (low temperatures favor capsule formation in *Klebsiella* and *Alkaligenes*), and nutrition, especially the presence of certain carbohydrates in the medium.

Functions of Capsules

The function of capsules is not entirely clear. In the pneumococcus it can be considered a device for evading previously generated antibody, and, in general, one finds capsulated strains in hostile environments. For example, *Nitrosomonas*, organisms for whom organic matter is toxic, evidently grow in sewage as capsulated forms. Thus one function may well be protection. A second function might be that of a reserve storage material. In times of plenty the organism may convert some of its nutrient into a form readily utilized by it but not generally available to its competitors. Indeed, in some cases, *Rhizobium*, for example, the capsular material, synthesized in times of plenty, seems to be utilized under starvation conditions. But there are other cases where the capsule does not seem to be utilized and may indeed be a waste product. For example, certain streptococci can ferment sucrose, utilizing the fructose but piling up a large capsule of dextran,which is evidently not utilizable later. Thus capsules seem to function as protection, reserve materials, or waste products. When one considers some of the very elaborate mechanisms of synthesis, this seems a little unsatisfactory. But insight or experiment have not indicated more specific functions.

Several interesting facts about capsules may be mentioned. These apply to individual organisms and specific kinds of capsules and may not apply to other organisms and capsules. For example, *Bacillus anthracis* forms a capsule of polyglutamic acid. This is not formed by avirulent strains, but only by virulent ones (2). As a first hypothesis, one might assume some protective factors associated with the capsulation, but this is not necessarily so. Carbon dioxide is necessary for the formation, not of the capsule, but of the enzymes that make the capsule. And carbon dioxide is also necessary for the formation of the toxin that is an important factor in virulence. The capsule will not form in nutrient broth or even on nutrient agar, but will form after those media are treated with charcoal, which removes long-chain fatty acids that evidently inhibit capsule formation. Capsule formation requires carbon dioxide as bicarbonate, and at pH 7.4 an atmosphere containing 5–10 percent carbon dioxide is required. Mutants can be obtained that do not require CO_2. The capsule is composed solely of D-glutamic acid, which is made, starting at the stationary phase, directly from L-glutamic acid by a system that does not involve ribosomes. On transfer of the culture for some years, the ability to make a glutamate capsule may be lost, but it may be replaced by a polysaccharide capsule.

In *Azotobacter* the capsule consists of a sort of hair-like fiber (3). Some strains are infected with a phage with a very short tail, so short that it cannot get through the capsule, an example of capsular protection. But when the phage gets through the capsule in an occasional cell, it directs the synthesis of a capsule-destroying enzyme, which strips off the capsules of the remaining cells, thus rendering them susceptible to the phage (4).

Some organisms do not form capsules. Of 242 strains of *Pseudomonas aeruginosa* studied, only two were capable of forming capsules (5). In certain of the *Leuconostoc* and *Streptococcus (bovis, salivarius)*, the organisms are not capsulated when grown on glucose but are heavily capsulated when grown on sucrose. They contain a dextransucrase that liberates fructose and transfers the glucose moiety of the sucrose to the nonreducing end of a linear chain (1–6) to form a glucose polymer of 10^6 to 10^7 molecular weight. In *Leuconostoc mesenteroides* one also obtains branched polymers at the C_3 hydroxyl. *Bacillus subtilis*, a few pseudomonads, and some *Enterobacter* produce a levan from sucrose. It is thought by many that the dextrans and levans so produced contribute importantly to plaque formation on teeth and thus are an important factor in dental caries. Some *Neisseria* form an amylopectin from sucrose that is a linear polymer with 1–4 links with branches at 1–6. One should point out that these are polymers produced at the cell surface, not inside the cell. Indeed, sometimes the enzyme is actually secreted into the medium

In *Alkaligenes*, the capsule is polysaccharide yet the organisms do not use glucose (6). They form the polysaccharide material entirely from amino acids, especially from proline or tyrosine. If the culture is shaken, little capsule is produced, and a lower temperature (perhaps 20°C) favors capsule formation. In *Alkaligenes faecalis* the polysaccharide capsule is mostly glucose with a little galactose; the glucose moiety may be esterified with succinyl groups. In *Arthrobacter viscosus*, the polysaccharide is composed of glucose, galactose, and mannuronic acid in equal portions, and about two thirds of the free hydroxyl groups are acetylated. This polysaccharide has the very unusual property of forming a gel.

REFERENCES

Reviews

Sutherland, I. w. 1972. Bacterial exopolysaccharides. *Advan. Microbial Physiol. 8*:143—214.

Papers

1. *Ann. N. Y. Acad. Sci. 236*:63 (1974).
2. *J. Gen. Microbiol. 43*:119 (1966).
3. *J. Bacteriol. 76*:119 (1958).
4. *J. Bacteriol. 84*:1209 (1962).
5. *J. Bacteriol. 89*:1432 (1965).
6. *J. Bacteriol. 89*:1521 (1965).

The following are typical textbooks used in modern courses in bacterial physiology.

Gunsalus, I. C., and R. Y. Stanier, eds. 1960—1964. *The Bacteria*, 5 vols. Academic Press, Inc., New York.
Lamanna, C., and M. F. Mallette. 1965. *Basic Bacteriology*, 3rd ed. Williams & Wilkins, Baltimore, Md.
Moat, A. G. 1976. *Microbial Physiology*. Ronald Press, New York.
Rose, A. H. 1975. *Chemical Microbiology*, 3rd ed. Butterworth, London.
Sokatch, J. R. 1969. *Bacterial Physiology and Metabolism*. Academic Press, Inc., New York.

2
Cell Wall of Gram-Positive Bacteria

CELL WALL STRUCTURE

Beneath the capsule is the cell wall; its structure differs profoundly between gram-positive and gram-negative bacteria. Undoubtedly, this difference in structure accounts not only for the distinct staining characteristic, but also for most, if not all, of the other properties associated with gram-positive or gram-negative bacteria. All bacterial cell walls (with a very few exceptions) contain *murein*, the basic unit of microbial cell wall structure; it is also known as muramic peptide complex, mucopeptide, peptidoglycan, and a few other similar names. These terms have been used so loosely that they no longer refer to specific parts of the bacterial cell wall structure, but to the structure as a whole or to some large "chunk" of it. Murein is a substance that occurs only in procaryotic cells (except if eucaryotic cells are invaded by procaryotes, as in Rickettsial infected animal cells). It is associated with and is almost the identifying fingermark of the procaryote. In gram-positives it may comprise two thirds of the weight of the cell wall, and in gram-negatives it may be less than 1 percent; yet in both cases it is responsible for the rigidity of the cell wall (with a few exceptions). When it is removed, cylindrical cells usually become spherical.

At the core of the murein structure is a derivative of glucosamine, actually the three lactyl ether of glucosamine called by the trivial name muramic acid (see Figure 2-1). In essentially all cases, muramic acid exists as part of a chain, usually in the *N*-acetylated form, the chain consisting of (acetyl muramic) (acetyl glucosamine) (acetyl muramic) (acetyl glucosamine), and so on; the chain extends for 4 to 40 units, possibly longer. The muramic acid has a free carboxyl group associated with the ether-linked lactyl, and this carboxyl group is associated with a series of peptides, the most

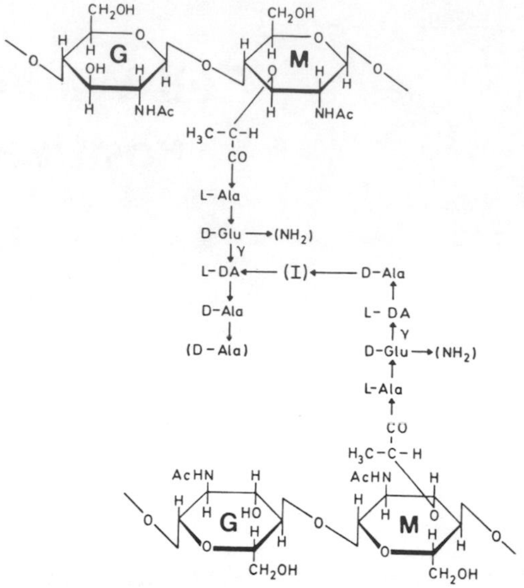

FIGURE 2-1. *Structure of a typical peptidoglycan showing cross-links. G = glucosamine, M = muramic acid (lactyl ether of glucosamine), Ac = acetyl, L-DA = diamino acid (lysine, diaminopimelic), I = interpeptide bridge. After Schleifer, K. H., Zeit. Immun. Forsch.* 149*:104–117 (1975). Used with permission.*

common of which is muramic to L-alanine to D-glutamic to L-lysine (or diaminopimelic) to D-alanine (diaminopimelic, known as DAP, is "carboxylated lysine").

In all of the hundreds of bacteria studied (gram-positive and gram-negative), this quadripeptide is rather constant. Figure 2-2 shows the kinds of variation that have been observed, but it is usually L-alanine—D-glutamic—L-lysine (or meso-DAP), and D-alanine, and essentially always in this order.

Two muramic peptides in two separated (acetyl glucosamine)—(acetyl muramic) chains may or may not be (but generally are) cross linked, as indicated in Figure 2-1. There are two major kinds of such cross links. Type A (the more common) links the lysine or DAP of one chain (position 3) to the D-alanine of the second peptide (position 4); these are called, not inappropriately, 3—4 links. The type B cross links are constructed from 2—4 links. Type A exists in several variations. One might have (A1), a direct cross link, the amino group of the lysine reacting with the free carboxyl of the D-alanine. This forms a very tight structure. One also finds (A2) that the link between them is a muramic-type peptide, that is, D-alanine (attached to lysine position 3), L-lysine—D-glutamic—L-alanine attached to position 4 of

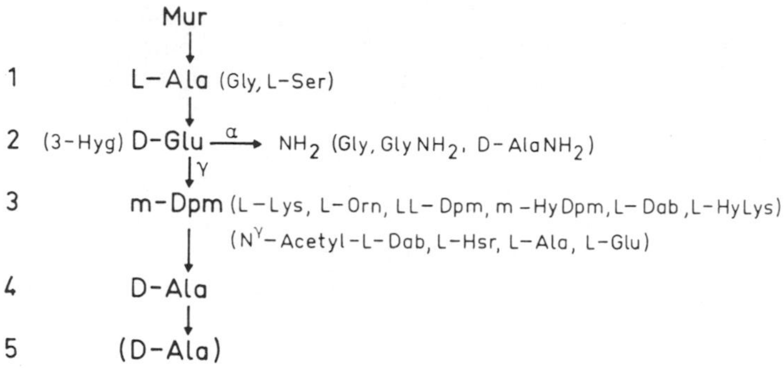

Mur
↓
1 L-Ala (Gly, L-Ser)
↓
2 (3-Hyg) D-Glu —α→ NH_2 (Gly, Gly NH_2, D-AlaNH_2)
↓γ
3 m-Dpm (L-Lys, L-Orn, LL-Dpm, m-HyDpm, L-Dab, L-HyLys)
↓ (N^{γ}-Acetyl-L-Dab, L-Hsr, L-Ala, L-Glu)
4 D-Ala
↓
5 (D-Ala)

FIGURE 2-2. *Variations in the peptide subunit of the peptidoglycan. The pentapeptide is shown. Abbreviations are: Hyg = hydroxiglutamic acid; gly NH_2 indicates that the α carboxyl of glutamic is combined with the amino group of glycine whose carboxyl group in turn is an amide; mDpm = meso-diaminopimelic; HyDpm = hydroxidiaminopimelic; Dab = diaminobutyric; the remainder are conventional abbreviations. From Schleifer, K. H. and O. Kandler, Bacteriol. Reviews 36:407-477 (1972). Used with permission.*

the next peptide. The most common cross link (A3) is that composed of monocarboxy amino acids, all of the same kind. *Staphylococcus aureus*, for example, has five glycines as the cross link. *Micrococcus roseus* has four L-alanines as the cross link, *Streptococcus pyogenes* has L-threonine, and *Lactobacillus acidophilus* has D-isoasparagine. The next type of cross link (A4) is also composed of monocarboxyamino acids, but different ones: glycine—serine—glycine, or glycine (2 to 5) plus alanine, and so on. Finally, (A5) the linking peptide may have dicarboxy acids as well, for example, asparticalanine. The type B cross link seems to be less common, but since the link via D-glutamic (position 2) and D-alanine (position 4) is between two carboxyl groups there needs to be a diamino acid in the bridge. This is usually lysine, but on rare occasions, it can be ornithine.

CELL WALL BIOSYNTHESIS

The biosynthesis of cell walls is of some considerable interest. We shall concentrate (as have most others) on the synthesis of the murein as represented in the cell wall of *Staphylococcus aureus*. The synthetic pathway is, for purposes of discussion, conveniently divided into three parts, each of which is somewhat different than the others, and each of which introduces a somewhat new insight as to how synthesis may occur in the living cell. We shall look first at the synthesis of muramic acid itself (illustrated in Figure 2-3). Glucose is converted

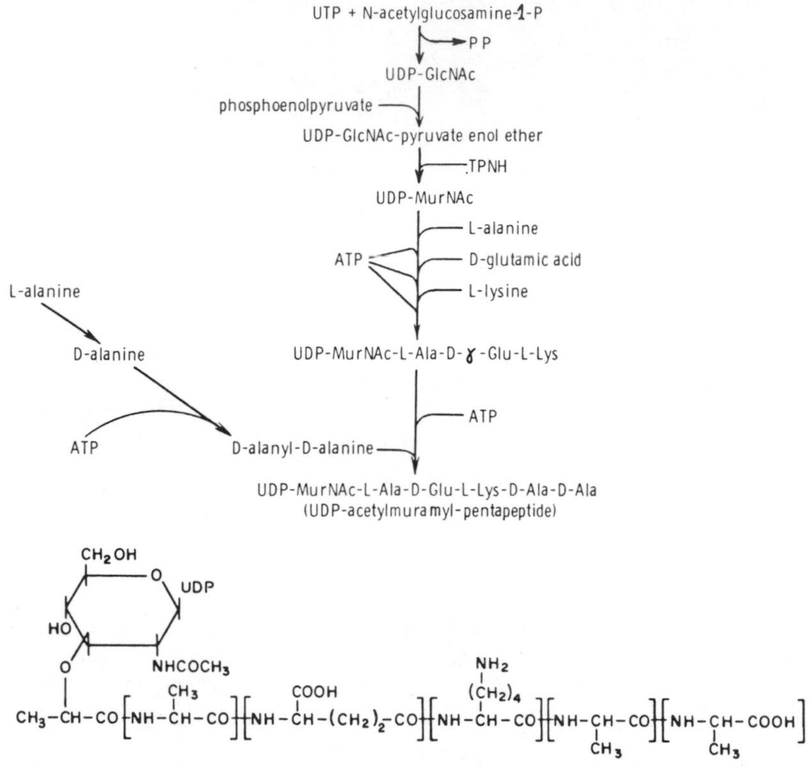

FIGURE 2-3. *The first stage of cell wall synthesis; the formation of UDP-N acetyl muramic pentapeptide. The detailed structure of the amino acid portion of the pentapeptide is given at the bottom. The α carboxyl of the glutamic acid and the ε amino group of lysine are free and available for cross linking. From Blumberg, P. M. and J. L. Strominger, Bacteriol. Reviews* ***38****:291-335 (1974). Used with permission.*

into glucosamine, then into acetyl glucosamine-l-phosphate, which unites with its obligatory carrier. This substance condenses with phosphopyruvate to form UDP-acetyl-glucosamyl-pyruvyl-ether. This is then reduced, via $NADPH^+$, to the lactyl ether, a UDP-acetyl-muramic acid, which first adds L-alanine, then D-glutamine, and then L-lysine (or DAP), each being added as the free amino acid, that is, not attached to a transfer RNA or anything of the sort. The last addition is somewhat more interesting; two D-alanines are condensed and added as a dipeptide. The end product of this series is thus UDP-acetyl-muramic pentapeptide.

The second stage begins when the pentapeptide undergoes an exchange with a lipid carrier, whose nature we shall discuss later (Figure 2-4). Suffice it to say here that it is a C_{55} isoprenyl alcohol. It is of some interest that this "translocase" is membrane bound, but has recently been removed from the membrane by use of Triton X-100 (a detergent); in solution it is inactive unless any of a number of phospholipids are added (1).

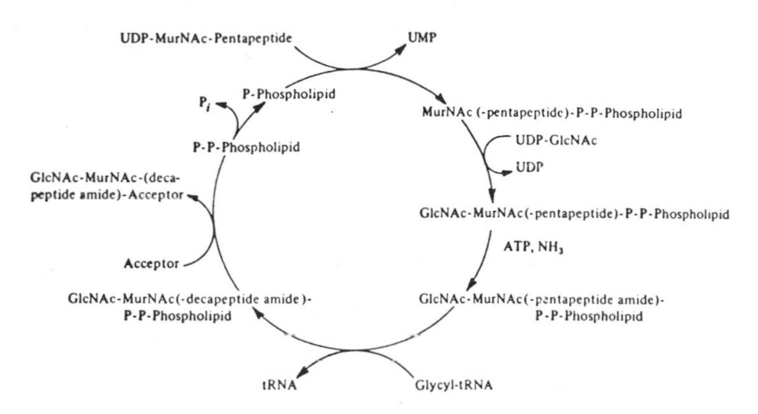

FIGURE 2-4. *The second stage of cell wall synthesis; the addition of cross linking amino acids. The case illustrated is that of Staphylococcus aureus. Five glycines are added to form a decapeptide. From Blumberg, P. M., and J. L. Strominger, Bacteriol. Rev. 38:291-335 (1974). Used with permission.*

What appears to be happening is that, to forge the peptide, the enzymes are held on the sugar (by way of the UDP carrier), but once the peptide is formed, the cell now wishes to tinker with the sugar end, so the molecule is held by way of lipid at the peptide end. At this point, acetyl glucosamine is added, and, for reasons not quite clear, the free carboxyl of the fifth amino acid (D-alanine) and the carboxyl group of the glutamic acid are tied up by forming the amide,which presumably protects them. At this point, glycine is added to the lysine, the carboxyl group of the glycine uniting with the ε-amino group of the lysine to form a peptide; then four more glycines are added until there are five altogether. The glycines, however, are carried in an activated form on a glycyl tRNA, and there is a special tRNA to carry glycine into the cell wall, in addition to the glycyl tRNA that carries glycine into protein, although both glycyl tRNAs seem to be active in protein synthesis.

After these reactions, one has acetyl glucosamine, acetyl-muramic decapeptide, the five amino acids from the muramic chain (L-alanine, D-isoglutamine (the amidated glutamic), L-lysine, D-alanine, D-alanine) plus the five glycines attached to the lysine. The next stage is to incorporate it into cell wall. In the normal cell wall approximately 10 percent of the acetyl-muramic peptides have an extra D-alanine residue and are pentapeptides rather than tetrapeptides. Furthermore, most of these bear open pentaglycine chains attached to lysine. That is, while the acetyl glucosamine—acetyl muramic part is already condensed into the long polysaccharide chain, the peptide parts are not, and these serve as the growing points of the cell wall. These growing points now serve as acceptors of lipid decapeptide and, by a process of transpeptidation, the extra D-alanine is removed and the glycine cross-link attached (Figure 2-5). There is also an enzyme, called D-alanine carboxypeptidase, that removes this extra (the fifth) D-alanine (and thus terminates chain building). Both the transpeptidase and the D-alanine carboxypeptidase are sensitive to penicillin and may bear an important relation to its action. The D-alanine carboxypeptidase has recently been removed from the membranes of *Bacillus subtilis* (2), but for it to be active, a lipid or a proper detergent must be added. In short, such membrane proteins are active only in the form of lipid or detergent micelles, either in the membrane or in solution.

THE ACTION OF PENICILLIN[1]

The central feature of penicillin is the highly strained β lactam ring, which may be that part of the molecule which reacts with the penicillin target. The side chains attached to the 6-aminopenicillanic acid have a profound effect upon its action and influence resistance to penicillinase tolerance of acidity, and ability to penetrate bacterial cells. The low toxicity of penicillin for the animal suggests that the drug inhibits some bacterial function which has no similar function in higher cells, such as the bacterial cell wall. It was observed very early that low penicillin concentrations caused the formation of deformed cells, and it was suggested that penicillin interfered with the formation of cell wall. Indeed, the formation of protoplasts and sphaeroplasts lacking the muramic acid complex by growth in low levels of penicillin was based upon this action. Additionally, UDP acetylmuramic peptides accumulated in cells inhibited by penicillin, which suggests that the penicillin acted at a stage subsequent to the nucleotide precursor.

When the enzymes involved in cell wall synthesis were studied, it was found that two were sensitive to penicillin. One, called transpeptidase, is the enzyme causing cross-linking of the muramic peptide chains. In this

[1]Blumberg, P. M., and J. L. Strominger. 1974. Interaction of penicillin with the bacterial cell: penicillin-binding proteins and penicillin-sensitive enzymes. *Bacteriol. Reviews 38*:291—335. (Reviewed by Vidyut Jhaveri.)

GlcNAc-MurNAc · L-Ala · D-Glu · *meso*—DAP · D-Ala · D-Ala

\+

GlcNAc—MurNAc · L-Ala · D-Glu · *meso*—DAP · D-Ala · D-Ala

GlcNAc—MurNAc · L-Ala · D-Glu · *meso*—DAP · D-Ala

GlcNAc—MurNAc · L-Ala · D-Glu · *meso*—DAP · D-Ala · D-Ala + D-Ala

A

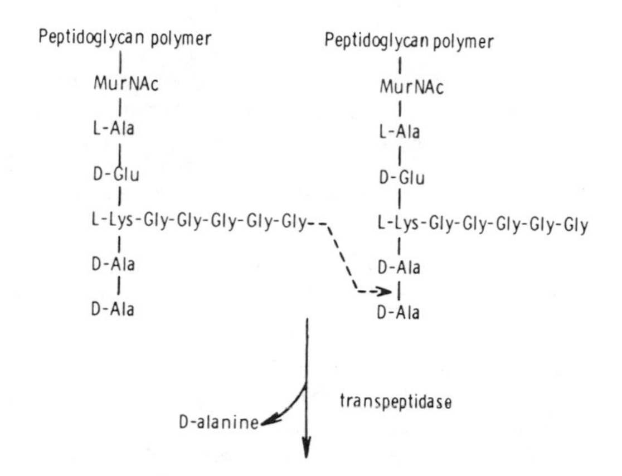

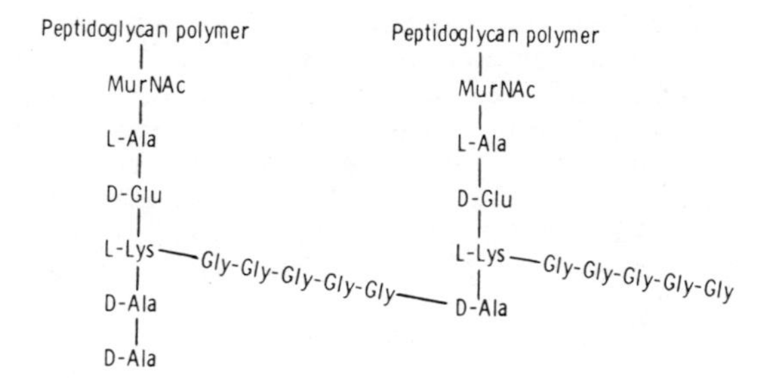

B

FIGURE 2-5. *Final stage in cell wall synthesis; cross-linking of peptidoglycan polymers by transpeptidation. A is for Escherichia coli, B for Staphylococcus aureus. From Blumberg, P. M. and J. L. Strominger, Bacteriol. Reviews 38:291-335 (1974). Used with permission.*

case it was believed that the penicillin was an analogue of the terminal D-alanyl alanine of the pentapeptide and that the penicillin reacted with the enzyme to form a penicilloyl enzyme which was inactive but stable.

A second penicillin-sensitive enzyme is carboxypeptidase, which removes the second D-alanine in the muramic pentapeptide and thus stops cross-linking. In *B.subtilis* it is membrane bound. However, this is probably not the enzyme responsible for the death of the cell since it can be inhibited 95 percent and the cell still grows well.

A third penicillin-sensitive enzyme is found primarily in *Escherica coli*, called endopeptidase, it cleaves off the last two D-alanines in the pentapeptide between D- and DAP.

If radioactive penicillin is supplied to cells, it binds to certain materials in the cell called penicillin binding components (PBC). In *B. subtilis*, for example, there are five PBC's. One is carboxypeptidase and another probably the transpeptidase; the remaining three are unknown. It is not clear which if any of these PBC's are actually responsible for killing the cell; clearly, inhibition of carboxypeptidase is not. Other organisms have differing numbers of PBC's. Although it is clear that penicillin does indeed inhibit cell wall synthesis, it is not certain that it does not also do something else to kill the cell.

TEICHOIC ACIDS

The amino sugar chains with their peptide tails and cross links constitute the backbone of the procaryote cell wall. In gram-positive bacteria such murein complexes may be as much as 60—80 percent of the cell wall. But many gram positives have only 50 percent or so of murein, and the other 50 percent is either teichoic acid, uronic acid polysaccharide, or both. It turns out that there are two types of teichoic acids. One is based on ribitol (Figure 2-6A) and appears to be associated with the cell wall structure itself and to be largely confined to gram-positive organisms. Note the D-alanine and the phosphate. The cell wall of *B. subtilis*, for example, has 5.3 percent phosphorus, owing to its content of teichoic acid. The sugar group may vary; in *Staph. aureus* it is acetyl glucosamine; in *B. subtilis* it is glucose.

The other kind of teichoic acid is based on glycerol (Figure 2-6B), although it is somewhat like the ribitol except for being two glycerols, one piled on top of the other. It too has D-alanine and sugar constituents. This type of teichoic acid is found associated with the membranes, and is found in both gram-positive and gram-negative cells, although evidently it may also be a constituent of the cell wall proper.

The teichoic acids are attached, when in the cell wall, to the amino sugar chain. Their function is not clear; they are extractable with trichloroacetic acid and can be

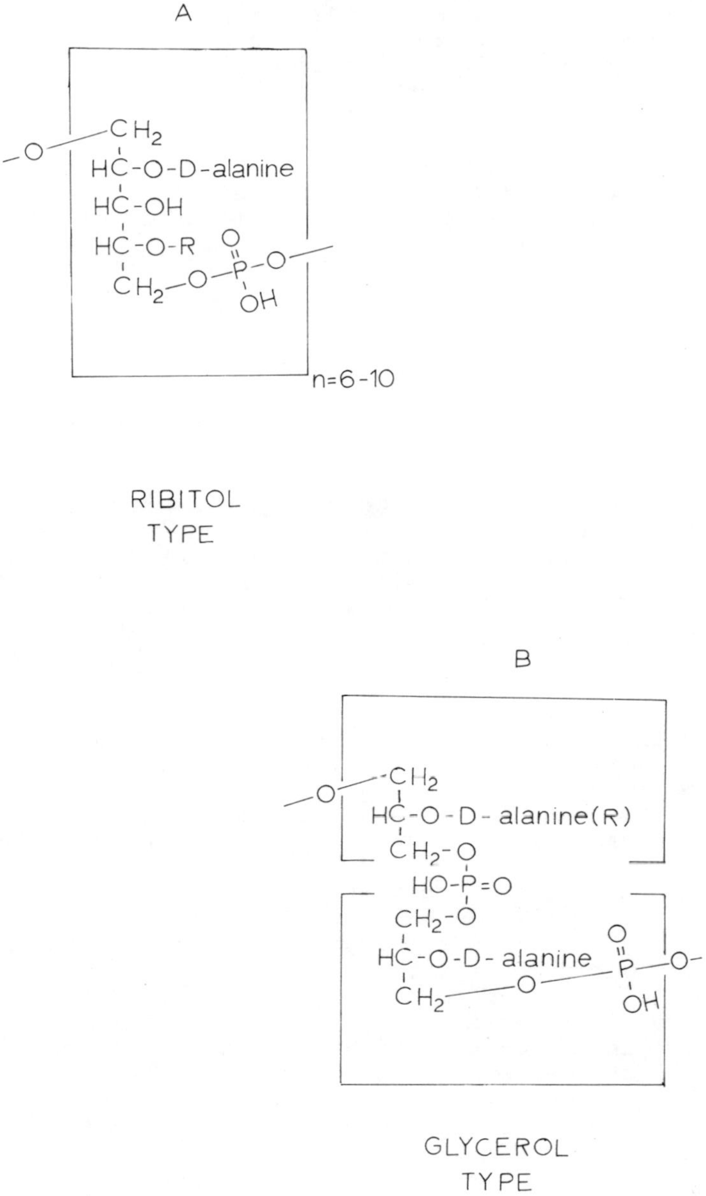

A

n=6-10

RIBITOL
TYPE

B

GLYCEROL
TYPE

FIGURE 2-6. *Techoic Acids. Part A represents the ribitol type, frequently called "cell wall" techoic acids. The D-alanine may be in positions 2 or 3. The R group may be acetyl glucosamine or glucose. Part B represents the glycerol type, frequently called "membrane" techoic acids. D-alanine may be replaced in some organisms by an "R" group. In* <u>*Streptococcus*</u> <u>*faecalis*</u> *the "R" is kojibiose with the D-alanine esterified to either carbon 3 or 4, and constitutes the "D" antigen.*

removed, yet the rigidity of the cell wall is retained. If cells are grown in very limited phosphate, teichoic acid can be replaced by uronic acid polymers. They seem to be involved in the synthesis or at least attachment of some of the pneumococcus capsules (types 6, 18, 19, and 34 especially). It is becoming apparent, however, that the immunological specificity of the cell surfaces is a function of the type of sugar attached to the teichoic acid, which depends upon the actual location of the complex on the surface.

IMMUNOLOGICAL PROPERTIES OF TEICHOIC ACIDS[1]

Teichoic acids are found in probably all gram-positive cells but are not usually synthesized by gram-negative cells. Conventionally, teichoic acids are regarded as either cell wall or cell membrane teichoic acids. If linked to peptidoglycan, they are considered cell wall teichoic acids; if associated with the membrane, they are considered membrane teichoic acids. The occurrence of wall teichoic acids varies among genera, but membrane teichoic acids appear in all gram-positive cells. Membrane teichoic acids all contain the following:

1. Linear backbone of polyglycerol phosphate.
2. A linkage of that backbone through a phosphodiester group using a 1,3 attachment.
3. All the structural variation is confined to the nature and extent of the sugar substitution on the -OH-groups at position 2 of the glycerol units.
4. D-alanine is the usual substituent found.
5. All membrane teichoic acids are linked covalently to the membrane glycolipid; this complex is called lipoteichoic.

The wall teichoic acids have structural diversity. They can:

1. Be either glycerol teichoic acid similar to the membrane teichoic acid or ribitol teichoic acid in which ribitol replaces glycerol as the backbone.
2. Have a restricted array of sugar substituents.
3. Vary in structure, which can change the overall shape or conformation of the molecule and its serological specificity.
4. Bond covalently to the peptidoglycan of the wall components.

Results indicate that teichoic acids free of wall or membrane components do not induce antibody formation but will react with antibodies already formed against a cell or cell fraction. This classifies the teichoic acid as a hapten. It is not clear whether antigenic specificity resides in the location, the amount, substituent groups, or linkage of the teichoic acid.

[1]Knox, K. W., and A. J. Wicken. 1973. Immunological properties of teichoic acids. *Bacteriol. Reviews 37*:215–257. (Reviewed by Susanne Weiser.)

Experiments have shown that antibodies specific for membrane teichoic acid have agglutinated whole organisms. How is this possible, if the membrane teichoic acid is located in a membrane beneath the cell wall? Is it possible that antibodies can penetrate the cell wall to react with membrane, or do they react with membrane components protruding through the cell wall, or do they do both?

Electron micrographs show that some cell walls have a highly porous network capable of passing molecules the size of IgM, but others have pores too small for any antibody to pass. It is possible to have the membrane teichoic acid stick out from the membrane and thus come near enough to the outer surface of the cell wall to act as a surface antigen.

Where and how do the antibodies bind to the teichoic acid? In general, the teichoic acid will have a terminal group determinant, and specificity depends on the carbohydrate located at that terminal site. In other cases the specific determinant site is an integral part of the backbone and is a nonterminal site. Substituent carbohydrates are sometimes the immunodominant component of the teichoic acid. For instance, a cross reaction will occur between antibodies to wall ribitol teichoic acid and membrane glycerol teichoic acid of the same organism, owing to a common carbohydrate substituent. It is possible for antibodies to form with specificity to the D-alanine substituents but these instances are few. Antibodies have also been demonstrated for the backbone of the teichoic acid, but these were phosphodiester bonds involving 1,3 linkage; when a teichoic acid with a 1,2 linkage was tested, it did not react. Thus almost all the teichoic acids can cause antibody production in some organisms. Teichoic acids as group antigens in the classification of bacteria are of limited usefulness.

OTHER CELL WALL CONSTITUENTS (GRAM-POSITIVE)

The cell walls of staphylococci and streptomyces seem to be mostly muramic teichoic acid complexes, but other gram-positives may contain other materials, usually polysaccharides. Micrococci and bacilli have polysaccharides degradable to glucose, galactose, and mannose. In addition to these three sugars, rhamnose is found is cell wall polysaccharides from streptococci, lactobacilli, clostridia, and propionic acid bacteria. Corynebacteria, mycobacteria, and nocardia seem to have a similar polysaccharide containing arabinose. Group A streptococci have a protein (the "M" protein) which seems to be the principal virulence factor in this group.

The cell walls of various gram-positive organisms differ in composition, but they are still closely enough related to be recognized as a biological entity. When the cell is plasmolyzed, the wall holds the cell firm. Although the wall may shrink when dried or fixed, it is still able to maintain the cell shape (else we would not see rods). It is not stained by the usual methods, yet is easily discerned on the electron microscope. Clearly, the cell wall serves as a container, a corset, for the cell, but it also serves as a recognition surface, especially for antibody-forming cells.

PROTOPLASTS AND SPHEROPLASTS

The protoplasm of several species of gram-positive bacteria can be freed from the cell wall and still remain a relatively stable structural unit (the *protoplast*), although it is much more sensitive to external influence than is the intact cell. The two primary methods of producing protoplasts are (1) by enzymatically digesting materials of the cell wall (with lysozyme treatment, for example), or (2) by preventing the growth of the cell wall murein with penicillin, without preventing the growth of that portion of the cell beneath the cell wall. A third, somewhat more limited method is to grow organisms (especially certain streptococci) that use lysine (or DAP) in forming their cell walls in the presence of only minimal amounts of lysine (or DAP), so that they are unable to make complete cell walls.

Since there is no cell wall to prevent the continual expansion of the protoplast due to water intake, protoplasts well and burst in solution of low osmotic pressure. Therefore, under most circumstances protoplasts will not survive unless a *stabilizing agent* is present. This agent is frequently 0.2—0.6 *M* sucrose, sometimes with a balanced salt solution; its function is to provide a stable osmotic environment. Osmotic pressure is not the entire story, however, since different materials (sucrose, sodium chloride, and cellobiose, for example) in solutions of equal osmotic pressure produce different results (i.e., sucrose is generally a good stabilizer, cellobiose generally poor). The exact composition of the solution that permits the protoplast to survive seems to vary with each strain of bacterium.

The protoplast possesses the permeability properties of the intact cell and carries out (when suitably protected) many of the processes that can be carried out by the cell. Protoplasts have been shown to synthesize proteins and nucleic acids, to support the development of bacteriophage or spores, and, in fact, to carry out all the cell's functions that are not dependent upon the cell wall. Although early studies indicated that protoplasts were not capable of synthesizing cell wall structures, which suggested that a partly intact murein chain was necessary as a base upon which to build further structures, later studies have shown "reversion" to cellular forms by protoplasts. Ability to resynthesize cell wall is dependent upon how the protoplast was made.

Similar preparations of cells lacking the rigid layer may be made from gram-negative organisms. In this case, however, the cell is still surrounded by the lipoprotein and lipopolysaccharide materials and is not therefore entirely free of cell wall materials. These structures are called *spheroplasts*. They are analogous to protoplasts, but are less sensitive to osmotic shock and still retain the nonrigid portion of the cell wall.

RELATION BETWEEN CELL WALL COMPOSITION AND CELLULAR SHAPE

Protoplasts and most spheroplasts are spherical (or sometimes pleomorphic) and are invariably gram-negative, indicating that in the gram-positive organisms it is the rather thick murein layer that is responsible for the gram-positive staining reaction. Some spheroplasts from gram-negative organisms may be rod-shaped even though they are osmotically sensitive, indicating that rigidity is not always solely dependent upon the murein network, but that on occasion the lipoprotein and lipopolysaccharide may be rigid enough to sustain the cylindrical form.

The function of the cell wall is clearly that of a "corset" and is necessary for a rod-shaped bacterium to remain in the form of a cylinder. The molecular conformations of the rigid layer that result in the production of either a cylinder (straight as in the bacilli, or curved as in the vibrios) or a sphere as in the cocci are still unknown.

There is preliminary evidence that the glycan strands of cocci may be shorter than those of rods. Indeed, in *Strep. pyogenes* the evidence seems to show that the strands are only five acetyl hexosamines in length. Some *Arthrobacter* undergo a sphere-to-rod transition during growth. It has been found that the chain length of the glycan units in the coccus form is from 14 to 62 hexosamines, whereas that from the rods varies between 114 and 135 hexosamines. In addition, the murein from the cocci contains an L-alanyl-glycyl-glycyl bridge, whereas the rod form has a single L-alanyl bridge, about two thirds of the peptide tails being cross linked in either form. Possibly, the short chain length combined with the longer bridge provides for a more readily bendable rigid structure, and after all the coccus must have more curvature to its cell wall than the rod. One might even suppose that the rod-shaped form developed because the cell wall was easier to build.

CONTRASTS BETWEEN GRAM-POSITIVE AND GRAM-NEGATIVE CELL WALLS

It is convenient to emphasize at this point the contrast between the cell walls of gram-positive and gram-negative bacteria. A convenient model of each is given in Figure 2-7. The wall in gram-positive organisms is composed of layer after layer of muramic complex, cross linked, evidently in both directions by way of the peptide bridges. Although other substances may be located in this wall, its

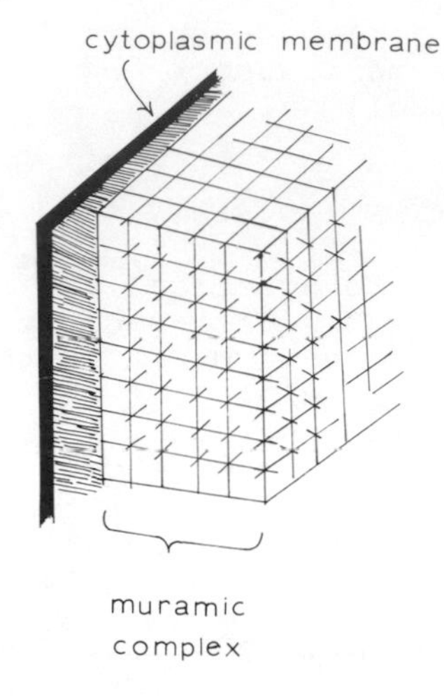

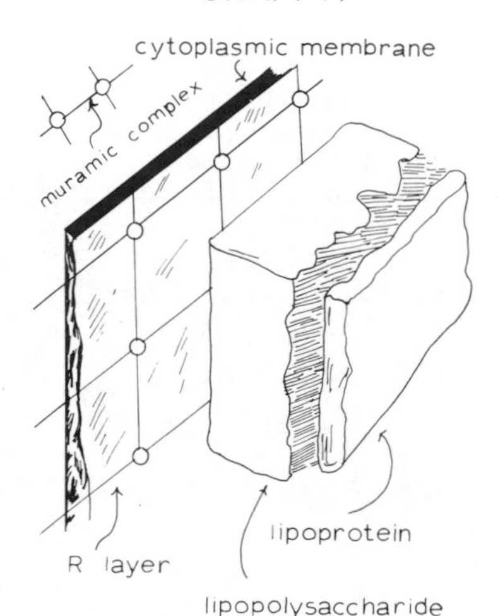

FIGURE 2-7. *Diagram of cell walls. The cell wall of gram-positive cells is composed of layer after layer of muramic complex appropriately cross linked. That of the gram-negative cell is composed of an "R" layer containing muramic peptides that cross link protein globules, outside are layers of lipopolysaccharide and lipoprotein. These differences are reflected in the composition of the cell wall.*

Material	*Gram-positive*	*Gram-negative*
Protein	*small*	*20% or more*
Lipids	*<2%*	*10—20%*
*Amino sugars**	*15—20%*	*2—3%*

**A measure of the muramic complex.*

main characteristic is that of the murein complex. The gram-negative cell has a relatively thin R layer composed of short strands of muramic complex linking together rather large protein units, as we shall describe in Chapter 3. Beyond this there is an outer layer of lipopolysaccharide and lipoprotein, perhaps not as uniformly distributed as in the model, but certainly with some relation to each other.

In both gram-negative and gram-positive bacteria the murein is rigid, and its function is to offer protection against water inflow. It is a "corset" that prevents the cell from expanding too far by taking in more water than is needed. This kind of a rigid structure offers certain difficulties in cell growth and cell division. How does a cell expand, for example? How, in a more practical sense, does it add more material and thus grow, while still maintaining its rigid and enclosed nature? We shall discuss these problems in Chapter 7.

THE CONSEQUENCES OF CELL WALL STRUCTURE: THE GRAM STAIN

We mostly examine bacteria in the light microscope after staining. So far as one can tell, the staining of bacteria is dependent upon the adsorption of the dye to the protein of the cell. Since virtually all stained cells can be decolorized by alcohol or acid alcohol (except those that are acid fasts), one does not see any great evidence for chemical combination between dye and protein. But there is one stain, so different from the others and used so widely in microbiology, that it warrants special consideration. This is the gram stain, so named because it was first described in 1884 by a Danish bacteriologist, H. C. I. Gram. It has become so widely used that it no longer requires capitalization and is written gram stain. The story of how Gram happened to come across this unusual procedure would be of some interest, but apparently this has been lost to history. Gram describes what he did but not why he did it.

Gram's paper is, however, interesting enough to quote in part. Gram was working at this time in Friedlander's laboratory. In the following direct translation (provided by M. Solotorovsky) from Gram's paper, the Koch—Ehrlich method referred to in the first sentence is the acid fast stain.

> A good differential stain for the tubercle bacillus has been achieved with the Koch—Ehrlich method. The need of a comparable method for other bacteria may now be fulfilled by a differential staining method for pneumococci that I have developed in collaboration with Friedlander. The present report extends the method to other bacteria.
>
> As already described by Friedlander, the sections are treated with absolute alcohol and then immersed in Ehrlich's aniline gential violet solution for 1—3 minutes. The stained slides are transferred without rinsing or rapid rinsing in alcohol to an iodine—potassium iodide solution (1 gram iodine, 2 grams potassium iodide, 300 grams water) for 1—3 minutes. The blue-violet color

> of the section changes to blackish purple-red. The sections are treated with absolute alcohol until completely decolorized and clarified in cedar oil. Bacteria are stained intensely blue and the background tissue is light yellow. The background tissue may be counterstained with Bismarck brown. Such preparations retain their stain for 4 months. The method is rapid, convenient and also applicable to smear preparations.
>
> Other aniline stains such as fuchsin aniline and aqueous gentian violet are not satisfactory. Tincture of iodine or aqueous potassium iodide cannot be used because they decolorize the bacteria. Rinsing the stained sections with dilute alcohol or water produces variable staining reactions. The stain is partially resistant to decolorization with 3 percent solutions of hydrochloric acid or sulfuric acid in alcohol.
>
> The method was applied to organisms associated with croupous pneumonia, pyemia, suppurative nephritis, suppurative arthritis following scarlet fever, multiple brain abscesses, osteomyelitis, typhoid fever, liver abscesses, erysipelas, tuberculosis, anthrax, and gas gangrene. [Here follow several pages of observations.]
>
> In closing, I should like to reemphasize that the aniline gentian violet, unlike other aniline dyes, stains bacteria in preference to tissue cell nucleic.
>
> This method appears to be an advance in bacteriological technique.
>
> [There is then a curious testimonial, as follows.] Addendum of the Editor (C. Friedlander): I should like to note that the Gram method is an outstanding, indeed in many cases, the best available method for staining bacteria.

There is a further item of interest. Gram had looked at lung sections of many cases of pneumonia and always seemed to find a gram-positive diplococcus. Evidently, they decided to isolate the organism, and with the next case that came to autopsy they found a gram-negative capsulated rod, which was clearly the cause of pneumonia and which they isolated as "Friedlander's bacillus" known today as *Klebsiella*. In the days when they worked, no distinction was made between bacterial lobar pneumonia (caused predominantly by *Diplococcus*, which were clearly described by Gram) and bacterial bronchial pneumonia caused predominantly by *Klebsiella*.

Several important *experimental* facts about the gram stain (sometimes still disputed) include the following:

1. Crystal violet (sometimes called methyl violet; the older term is gentian violet) is absolutely essential. No other dye will work. Other dyes have been claimed, but none acts nearly as well or gives the same separation of organisms.
2. Iodine is essential. Again, a wide variety of other agents have been tried. None works as readily or as consistently as iodine.
3. The reaction is influenced by pH. At a pH below 5.5 all cells stain gram negatively. At a pH of above 11 all cells remain violet, but they do so whether iodine is there or not; hence this is not really a gram stain.
4. If the cell wall is crushed, broken, or removed (protoplast), the cells are gram-negative.
5. The concentration of alcohol used in the decolorization process is important. It should be 90 percent or greater. The nature of the alcohol, that is, ethyl, propyl, and so on (even acetone), does not matter, within certain limits of course.
6. The counterstain (usually safranin) is not specific, and a wide array of dyes can be used. The counterstain does serve the function of displacing a certain amount of adsorbed crystal violet, which might otherwise confuse the observations made without counterstain.

Clearly, the widespread use of the gram stain long before there was any real knowledge of cell wall structure attests to the usefulness of the stain as an index of the physiological characteristics of the organisms. But since we now know that it is the profound difference in cell wall structure which is reflected in the gram stain, we also know that many of the physiological properties of the cells thus arise because of the different nature of the cell wall.

Today we know that the basis of the gram stain is the following. In gram-positive organisms, crystal violet enters. If one washes after this step, the only crystal violet left is that adsorbed to cell protein. Iodine then enters. Iodine combines with crystal violet, forming a complex that is still adsorbed to the protein and is not extractable by water. The complex is soluble in alcohol, but if 90 percent or more is used, a shrinkage of the muramic layers occurs, and the complex is retained inside the cell (3). The muramic layer in the gram negatives is so differently constructed that such shrinkage does not occur and the violet complex is extracted. The counterstain is used to displace any residual adsorbed crystal violet in the gram negatives and thus gives a clearer differentiation. Two changes in technique may be used to improve the differentiation in some cases. First, one stains with crystal violet, drains the slide but does not wash, then adds iodine, drains, and blots dry. One obtains a slide with a somewhat messier background but with a more definitive staining of gram positives in doubtful cases. In the other technique, after the iodine has been removed, the slide is blotted dry before the alcohol is added so that one does not expose the slide to less than 90 percent alcohol.

In the past there have been many theories concerning the gram stain, but the one described here seems most likely to be correct. However, one theory had clearly reproducible supporting observations and pointed in a completely different direction. It was found that, if one took gram-positive organisms and treated them with "bile salts," one could extract a "material" and the cells were then gram-negative. If this material was added to the cells from which it had been removed, they became gram-positive again. The material was identified as magnesium ribonucleate. Indeed, ribonuclease could be used to render gram-positive cells gram-negative, and probably the "bile salts" were active because they contained this enzyme. It appeared that gram-positive bacteria had a layer of magnesium ribonucleate which was responsible for their gram reaction. Isolated and purified cell walls do not contain ribonucleic acid, but of course they are gram-negative. A few samples of ribonuclease did not render gram-positive cells gram-negative; that is, the actual enzyme carrying out the process is not ribonuclease but a contaminant of such enzyme preparations. Finally, it has been shown that gram-positive cells rendered gram-negative in a variety of ways can indeed be reconverted into the gram-positive state by magnesium ribonucleate. However, this can be done only when the magnesium ribonucleate is precipitated *in situ*, and its action appears to "plug up holes" in the cell wall created by previous treatments (4). It is now abundantly clear that it is the unique nature of the cell wall of gram-positive bacteria which is responsible for their staining properties.

REFERENCES

Reviews

Archibald, A. R. 1974. The structure, biosynthesis and function of teichoic acid. *Advan. Microbial Physiol.* *11*:53—96.

Braun, V., and K. Hantke. 1974. Biochemistry of bacterial cell envelopes. *Ann. Rev. Biochem.* *43*:89—121.

Cole, R. M. 1965. Bacterial cell-wall replication followed by immunofluorescence. Symposium on the Fine Structure and Replication of Bacteria and Their Parts, *Bacteriol. Reviews* *29*:326—344.

Gander, J. E. 1974. Fungal cell wall glycoproteins and peptido-polysaccharides. *Ann. Rev. Microbiol.* *28*:103—191.

Glauert, A. M., and M. J. Thornley. 1969. The topography of the bacterial cell wall. *Ann. Rev. Microbiol. 23*: 159—198.

Necas, C. 1971. Cell wall synthesis in yeast protoplasts. *Bacteriol. Reviews 35*:149—170.

Rogers, H. J. 1970. Bacterial growth and the cell envelope. *Bacteriol. Reviews 34*:194—214.

Salton, M. R. J. 1960. *Microbial Cell Walls*. Wiley, New York.

Schleifer, K. H., and O. Kandler. 1972. Peptidoglycan types of bacterial cell walls and their taxonomic implications. *Bacteriol. Reviews 36*:407—477.

Shockman, G. D. 1965. Unbalanced cell-wall synthesis: autolysis and cell-wall thickness. Symposium on the Fine Structure and Replication of Bacteria and Their Parts, *Bacteriol. Reviews 29*:345—358.

Papers

1. *Proc. Natl. Acad. Sci. 69*:1972 (1972).
2. *J. Biol. Chem. 248*:6759 (1973).
3. *J. Gen. Microbiol. 30*:233 (1963).
4. *J. Bacteriol. 90*:1500 (1965).

3
Cell Wall of Gram-Negative Bacteria

The present concept of the structure of the cell wall of gram-negative bacteria was shown in Figure 2-7. A more correct, but of course more complex, representation is given in Figure 3-1. Outside of the osmotic membrane (the cytoplasmic membrane) is a rigid layer, relatively thin, consisting of peptiglycan, being rigid, it is called the R layer. Immediately beyond this is a periplasmic space filled with numerous substances, as we shall see later. Beyond this is a thicker layer composed largely of lipopolysaccharide, with some lipoprotein (which may not be uniformly distributed), which acts as a molecular sieve or filter, and which many call the "outer membrane." The choice of words is unfortunate, since "membrane" has some connotations that are inappropriate to this structure. A better name for it, which we shall use, is "package," although the terms "envelope" and "outer barrier" have also been used.

To start from the outside, not much is known about the lipoprotein. In *Escherichia coli* the major lipoprotein seems to be of relatively low molecular weight (7,200) and exists in two forms (1). About one third is bound *via* the carboxy terminal end to the peptidoglycan, presumably projecting into or through the periplasmic space. About two thirds is not so bound and exists in the package layer itself.

LIPOPOLYSACCHARIDES

The lipopolysaccharides are surprisingly well known and those most thoroughly examined are the "O" antigens (which are also the endotoxins) of *Salmonella* and *E. coli*. The lipopolysaccharides consist of three parts: the *backbone*, the *core*, and the *side chain*. The backbone (see Figure 3-2) consists of lipid A, a curious 8-carbon acid (KDO: 2-keto-3-deoxyoctonic acid), some heptose phos-

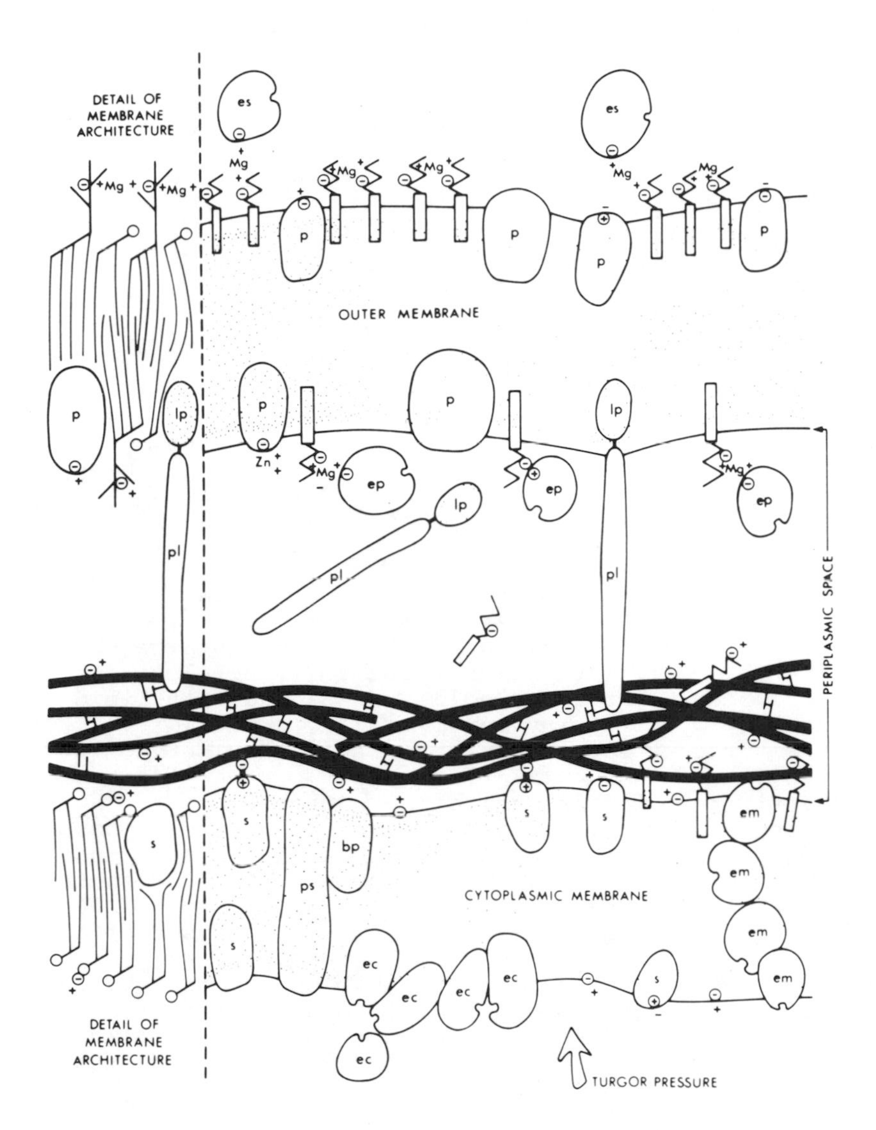

FIGURE 3-1. *Detailed representation of the cell gram-negative bacteria. The thick solid lines represent peptidoglycan, p = protein, bp = binding protein; ec, em, ep, es = enzymes, associated with cytoplasmic membrane, with cell wall synthesis, in periplasmic space and a cell surface. Lp and pl are lipoprotein, ps = permease, s = structural protein. From Costerton, J. W., J. M. Ingraham and K. J. Cheng, Bacteriol. Reviews 38:87-110 (1974). Used with permission.*

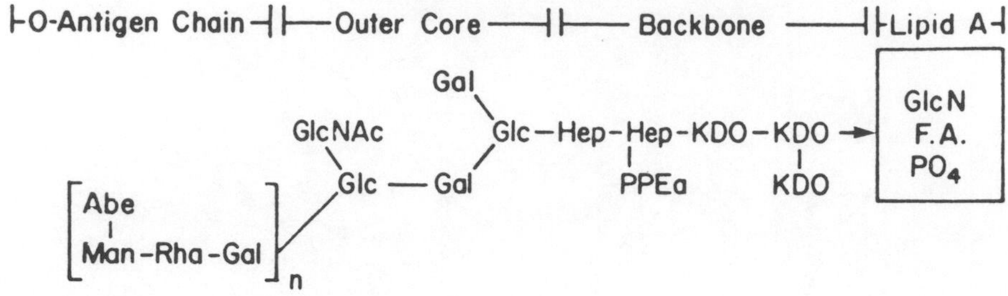

O-Antigen Chain | Outer Core | Backbone | Lipid A

Gal
GlcNAc
Glc – Hep – Hep – KDO – KDO → GlcN F.A. PO_4
Glc — Gal
PPEa
KDO
[Abe | Man-Rha-Gal]n

FIGURE 3-2. *Structure of the lipopolysaccharide of Salmonella typhimurium. The abbreviations are: Abe = abequose (3,6-dideoxy-galactose), Ea - ethanolamine, F.A. = fatty acid, Gal = galactose, Glc = glucose, GlcN = glucosamine, Ac = its acetyl derivative, Hep = glycero-D-manno-heptose, KDO = 2-keto-3-deoxyoctonate (3-deoxy-D-manno-octulosonate), Man = mannose and Rha = rhamnose. From Osborn, M. J., P. D. Rick, V. Lehmann, E. Rupprecht, and M. Singh, Ann. N. Y. Acad. Sci. 235:52-65 (1974). Used with permission.*

phates, and *O*-phosphorylethanolamine. The backbone is essentially similar in most gram-negatives, the core varies a little from one to another, and the side chain differs in detail (but is constant in principle) with each organism. The fatty acids associated with lipid A are lauric, myristic, hydroxymyristic, and palmitic. The length of the heptose chain appears to vary between two and ten from one organism to another. The heptose most frequently reported is L-glycero-D-mannoheptose. The KDO is synthesized *via* a condensation of arabinose-5-phosphate and phosphoenol pyruvate (as shown in Figure 3-3) and transferred to CPT before attachment to the glucosamine component of lipid A.

The core in the group B *Salmonella* is composed of a glucose-galactose-glucose-acetyl glucosamine sequence with the first glucose attached both to the heptose and a second unsubstituted galactose, as illustrated in the final stage of synthesis shown in Figure 3-4. In *E. coli* the core is glucose—galactose—acetyl glucosamine and colitose (3,6-dideoxy-L-galactose).

If one looks at this sequence of monosaccharides in the core and side chain regions of *S. typhimurium* lipopolysaccharide, one can see that if a mutant were unable to add glucose to the heptose backbone, none of the other sugars could be added, whereas in a mutant unable to add galactose, the first glucose could be added but not the second, and therefore no subsequent sugars. A mutant lacking the ability to add N-acetyl glucosamine could still add the first three sugars, but not the side chain. It was by the study of such mutants, which form rough or R-type colonies, instead of the smooth or S-type colonies of the wild-type parents, that the biosynthetic pathway and the structure of the core were determined (Figure 3-5). The mutants were found to have a defect either in

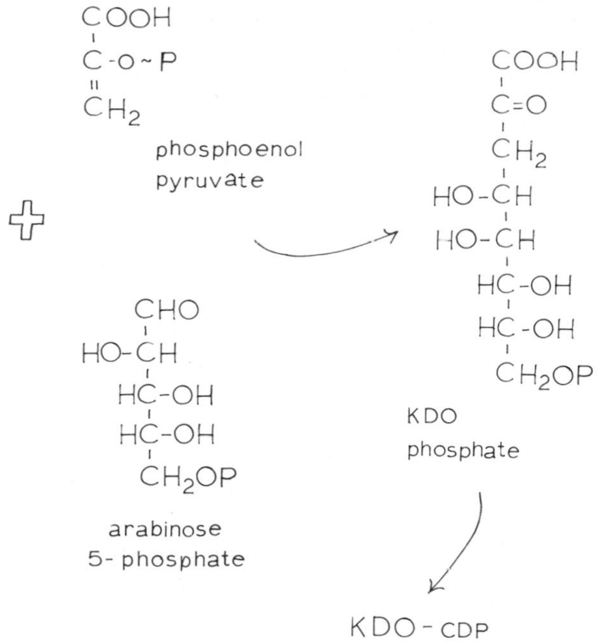

COOH
|
C-O~P
||
CH_2

phosphoenol pyruvate

+

CHO
|
HO-CH
|
HC-OH
|
HC-OH
|
CH_2OP

arabinose 5-phosphate

→

COOH
|
C=O
|
CH_2
|
HO-CH
|
HO-CH
|
HC-OH
|
HC-OH
|
CH_2OP

KDO phosphate

↓

KDO-CDP

FIGURE 3-3. *The synthesis of KDO.*

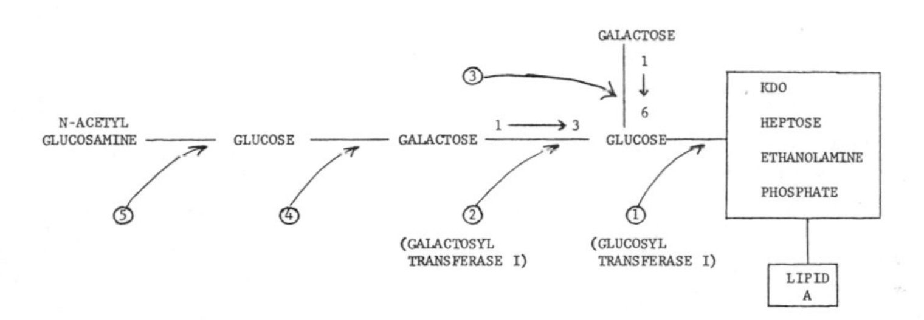

FIGURE 3-4. *Steps in the biosynthesis of the "core" of the lipopolysaccharide of* *Salmonella typhimurium.* *From Rothfield, L., and M. Takeshita, Ann. N. Y. Acad. Sci.* *133:385—390 (1966). Used with permission.*

a step leading to the synthesis of the UDP derivative of the sugar (e.g., epimerase, pyrophosphorylase, etc.), or in the transfer of the UDP-sugar for chain elongation. By studies of the composition of the core of such mutants, the sequence of core biosynthesis shown in Figures 3-4 and 3-5 was determined. Each sugar of the core is added in sequence from the UDP-sugar by a specific transferase. The enzymes that carry out these syntheses are located in the cytoplasmic membrane. If membrane fractions possessing this synthesizing activity are extracted with a lipid solvent, activity is lost, but can be restored by the addition of phosphatidyl ethanolamine. The phospholipid

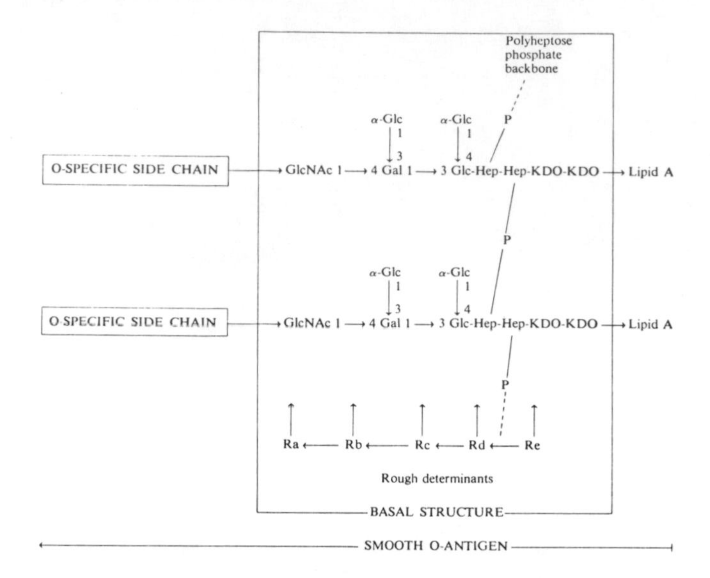

FIGURE 3-5. *The use of mutants in the determination of core structure (Shigella flexneri). From Simmons, D.A.R., Bacteriol. Reviews 35:117-148 (1971). Used with permission.*

must have ethanolamine in the phosphorylated position, and both R_1 and R_2 of the glycerol portion must be present and contain either unsaturated or cyclic (cyclopropane) fatty acids. The synthesis occurs in five steps, as shown in Figure 3-4 and from "mono layer experiments" it appears that at least four of the enzymes (labeled 1 through 4) are lined up in that order on the membrane.

The side chain, or O antigen (somatic or cellular antigen in contrast to H or flagellar antigen) is illustrated in Figure 3-6. A glucosyl carrier lipid (a C_{55} alcohol) unites with UDP-galactose, similar to the union of UDP-acetyl-muramic pentapeptide, which then accepts the various sugars from their carrier nucleotides in a fashion determined by the enzymes. When it has reached its necessary length and composition (the tetrasaccharide), one of the tetrasaccharides adds to another and then to another, each step freeing one glycosyl carrier lipid pyrophosphate. The number of tetrasaccharide units to be added to the carrier lipid evidently varies, but once the appropriate number is reached, the completed side chain [i.e., $(\text{tetrasaccharide})_n$] is transferred from the lipid carrier to the core terminal sugar (in the case of *Salmonella* B acetyl glucosamine), thus completing the package layer. Rough mutants that lack side chains but have a complete core are defective only in the enzyme for side chain attachment.

We need now to say a word about the C_{55} isoprenoid alcohol phosphate that has appeared as a lipid carrier in

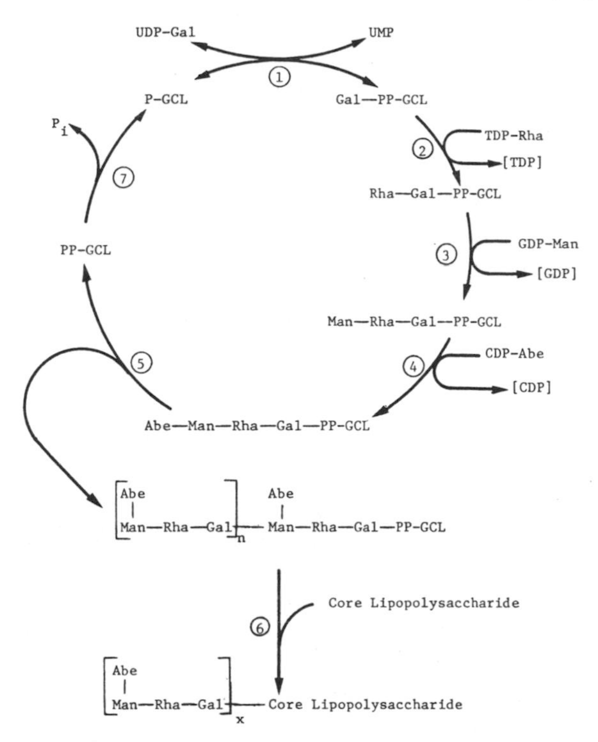

FIGURE 3-6. *Pathway of biosynthesis of the side chain (the "O" antigen) in Salmonella typhimurium. From Osborn, M. J. et al. as in Figure 3-2. Used with permission.*

capsule formation (Chapter 1), in peptidoglycan formation (Chapter 2), and now here in side chain formation (2-5). Its structure is shown in Figure 3-7. It contains 10 isoprene units. The monophosphate is involved in mannolipid synthesis in yeast and, in general, whenever a sugar must be added. The pyrophosphate is the one more commonly employed when more than one sugar must be added. All these substances are synthesized in the cytoplasmic membrane, yet eventually appear in the package. Indeed, the side chains (O antigens) are responsible for the antigenic differences among *Salmonella* species as well as for antigenic similarities; for example, group B *Salmonella* all contain abequose (3,6-dideoxy-D-galactose) in the side chain, while group D contains tyvelose (3,6-dideoxy-D-mannose). There are, of course, cross reactions within each group.

What is the sense, what is the purpose of all this variation in the side chains? We have earlier seen in gram-positive bacteria, that the nature and location of the teichoic acids determined serologic specificity. The O antigens and the teichoics specify the surface of the cell; these are the sites where antigenicity and antibody

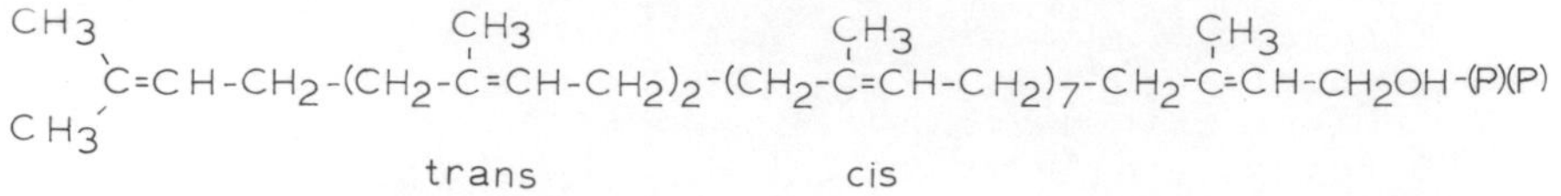

$$(CH_3)_2C{=}CH\text{-}CH_2\text{-}(CH_2\text{-}\overset{CH_3}{C}{=}CH\text{-}CH_2)_2\text{-}(CH_2\text{-}\overset{CH_3}{C}{=}CH\text{-}CH_2)_7\text{-}CH_2\text{-}\overset{CH_3}{C}{=}CH\text{-}CH_2OH\text{-}(P)(P)$$

trans cis

FIGURE 3-7. *The C_{55} isoprenoid alcohol whose phosphates serve as the lipid carrier in the biosynthesis of capsule, peptidoglycan and "package" components of microorganisms. Its monophosphate is abbreviated CLP; the pyrophosphate, CLPP.*

reaction occur. They are marks of distinction. They serve to differentiate one organism from another. Distinctive thy surface as thine enzymes can make (to paraphrase Shakespeare), and thou wilt be known as a distinct individual. These are insignia of rank; they signal, I am *Salmonella typhimurium*, not just any *Salmonella*.

IMMUNOCHEMISTRY OF THE SHIGELLA "O" ANTIGENS[1]

As might be imagined, the situation is somewhat more complicated than that outlined in the previous paragraphs. There are structural and biological similarities between the O antigens of the various members of the Enterobacteriaceae. The O antigens are polysaccharide—lipid A—protein—lipid B complexes bound noncovalently to the cell wall mucopeptide. The lipid A component has not been completely characterized. It has been generally accepted that it has a backbone of hydroxy-myristol-glucosamine phosphate. All the hydroxyl groups in the hydroxy acid and glucosamine are esterified with long-chain fatty acids. The lipid A component has been associated with the endotoxic property, but lipid-A-poor lipopolysaccharides have been isolated that are highly toxic.

The lipid B component consists primarily of palmitic and oleic acids, which may act as a cofactor in the biosynthesis of the basal structure of the O antigen lipopolysaccharide component. Lipid-B-like material has been associated with the protein moiety.

The protein component (poorly characterized) has two important functions. It confers antigenicity on the macromolecular complex and determines the colicin activity of the organism. The protein component probably repre-

[1]Simmons, D. A. R. 1971. Immunochemistry of *Shigella flexneri* O antigens: a study of structural and genetic aspects of the biosynthesis of cell-surface antigens. *Bacteriol. Reviews 35*:117—148. (Reviewed by Russell Rawlings.)

sents the enzymes involved in the biosynthesis of the polysaccharide.

The polysaccharide component consists of heteropolymers that determine the serological specificity of the O antigen, with antibody response directed toward specific structural sequences in the polysaccharide O specific side chains. This component is nonantigenic, but reacts *in vitro* against antibody prepared against the whole antigen, and is properly termed a hapten.

Two methods used to isolate the O specific material are trichloroacetic acid and phenol—water extractions. The former has a lower yield, but it is more free of other cellular material such as nucleic acid. The trichloroacetic acid extract is highly toxic. Upon acid hydrolysis it can be dissociated into a nontoxic, nonantigenic polysaccharide (hapten), a phospholipid (lipid B), and a toxic, antigenic protein, which is a lipid A protein. By extracting with the phenol—water method and purifying in the ultracentrifuge, one can prepare a large quantity of toxic, nonantigenic lipid A polysaccharide.

There are eight major serotypes of *Sh. flexneri*. Analysis of the O polysaccharides showed that about 60—75% can be accounted for as carbohydrate, the other 30 percent being the lipid component. The carbohydrate portion is comprised of seven hexoses: L-rhamnose, *N*-acetyl-D-glucosamine, D-glucose, D-galactose, aldoheptose phosphate, 3-deoxy-2-oxyoctonate, and *O*-phosphorylethanolamine. Except for L-rhamnose, these hexoses are thought to be found in all members of the Enterobacteriaceae. They have been demonstrated in all the *Salmonella*, *E. coli*, and *Arizona*. These six hexoses are thought to comprise the basal structure of the O polysaccharide. Furthermore, this basal structure is the region containing the rough antigenic determinants. The O specific side chains comprise the other region which is responsible for the specificity and cross-reactivity. In *Sh. flexneri*, these O specific side chains are repeating units of D-glucose, *N*-acetyl-D-glucosamine, and L-rhamnose. The basal structure plus these O specific side chains comprise the smooth O polysaccharides.

By studying rough mutants isolated from smooth serotypes, it has been found that there are two types of smooth-to-rough mutation. One type results from a loss of a single side chain synthetase or transferase, resulting in a total absence of O specific side chains, but a complete basal structure. The other involves the loss of a single basal synthetase or transferase. In this case, there is a normal biosynthesis of the O specific side chains, but they cannot be assembled on the incomplete basal structure as the required acceptor is missing. These side chains give some indication as to the nature of the mutants, as their presence can be determined in the ultracentrifuge supernatant of the extracted rough polysaccharides. (Figure 3-5 shows the basal structure and where rough enzyme defects occur, termed Ra, Rb, Rc, Rd and Re mutants.) By studying the sugar content of the polysaccharides of such mutants, the structure and biosynthesis of the basal structure were determined. The basal structure is built from the lipid A end. For example, rough serotype 3a was found to have a lipopolysaccharide sugar content of 3-deoxy-2-oxo-octonate, *O*-phosphorylethanolamine, and aldoheptose phosphate, but lacked D-glucose, D-galactose, *N*-acetyl-D-glucosamine, and L-rhamnose. This has been termed an Rd mutant and cannot incorporate the first glucose residue into the basal structure. These mutants lack either a uridine diphosphate glucose synthetase or a uridine diphosphate glucose transferase.

The fine structure of *Sh. flexneri* O specific side chains has been determined in detail. The sequence of hexoses in the primary chain is the same (and is, in fact, for all of *Sh. flexneri* serotypes). The serotypes vary in the linkages and in the position and number (two to four) of the glucosyl side chains. In serotype 5b, the glucosyl groups are acetylated. The group-factor determinants are based around the rhamnosyl—(1—4)—rhamnose sequence. The type-specific factors center around the glucosyl side chains and their linkages to the primary O specific side chains.

THE R LAYER

There is still a rigid layer nearer to the cytoplasmic membrane. The original conception was that it had the structure suggested in Figure 2-7; the relatively short structure of murein served to connect the protein "balls," which one can see, on occasion, in the rigid layers obtained from gram-negative bacteria. The protein itself was thought to be a very special kind of protein, low in aromatic amino acids. Further studies show that it is indeed of a very special kind. The protein lacks aromatic amino acids and sulfhydryl groups, and has a greater than usual overlapping sequence, which suggests a unique and unusual function (1).

DERIVATIVES OF GRAM-NEGATIVE CELLS

About 30 years ago somewhat pleomorphic forms were discovered (called L forms, from Lister Institute) that were a stable growth of soft protoplasmic elements without defined morphology and which do not revert to the rigid bacterial form. It was then found that if spheroplasts from *Proteus* and some other organisms were carried in a series of subcultures, L forms could be obtained. In a sense, the protoplasts had lost the ability to make the murein complex, and indeed L forms are devoid of muramic acid. Their nutrient requirements were those of the strain from which they originated, but they no longer required the same degree of osmotic protection as did the original protoplasts. Clearly, L forms came from gram-negative organisms.

There is a still more disorganized protoplasm, the *Mycoplasma* or PPLO (Chapter 14), that possesses no cell wall structure or muramic acid, and may be multinuclear, with the whole blob of protoplasm dividing in a random fashion such that some cells may have so little of the mother cytoplasm that they are small enough to be able to pass through millipore filters. There are similarities between L forms and *Mycoplasma* with respect to colony form, cellular appearance, resistance to cell wall antibiotics, and requirement for a balanced salt environment. There are differences, however, in that the *Mycoplasma* show much poorer growth, have more exacting nutritive requirements, and have some cells so small that they can pass through filters. It appears that the *Mycoplasma* could well have originated from L forms by a further loss of genetic elements, presumably the same as the L forms originated from gram-negative bacteria.

SURFACE CHARGE

Bacteria have a surface charge. As such they can be made to move in an electric field. At neutral pH the charge is negative, and bacteria will move toward the anode. But the actual charge and thus the rate of movement are so variable as to have little significance.

REFERENCES

Reviews

Costerton, J. W., J. M. Ingram, and K. A. J. Cheng. 1974. Structure and function of the cell envelope of gram-negative bacteria. *Bacteriol. Reviews 38*:87—110.

Ellwood, D. C., and V. W. Tempest. 1972. Effects of environment on bacterial wall content and composition. *Advan. Microbial Physiol. 7*:83—118.

Fox, E. N. 1974. M proteins of group A streptococci. *Bacteriol. Reviews 38*:57—86.

Glaser, L. 1973. Bacterial cell surface polysaccharides. *Ann. Rev. Biochem. 42*:91—112.

Horecker, B. L. 1966. The biosynthesis of bacterial polysaccharides. *Ann. Rev. Microbiol. 20*:253—290.

Nowotny, A. 1969. Molecular aspects of endotoxic reactions. *Bacteriol. Reviews 33*:72—98.

Oseroff, A. R., P. W. Robbins, and M. M. Burger. 1973. The cell surface membrane: biochemical aspects and biophysical probes. *Ann. Rev. Biochem. 42*:647—682.

Reaveley, D. A., and R. E. Burge. 1972. Walls and membranes in bacteria. *Advan. Microbial Physiol. 7*: 2—82.

Richmond, D. V., and D. J. Fisher. 1972. The electrophoretic mobility of microorganisms. *Advan. Microbial Physiol. 9*:1—30.

Roantree, R. J. 1967. Salmonella O antigens and virulence. *Ann. Rev. Microbiol. 21*:443—466.
Rothfield, L., and D. Romeo. 1971. Role of lipids in the biosynthesis of the bacterial cell envelope. *Bacteriol. Reviews 35*:14—38.
Shaw, N. 1970. Bacterial glycolipids. *Bacteriol. Reviews 34*:365—377.

Papers

1. *Ann. N. Y. Acad. Sci. 235*:68 (1974).
2. *Proc. Natl. Acad. Sci. 57*:1878 (1967).
3. *Proc. Natl. Acad. Sci. 57*:1798 (1967).
4. *J. Biol. Chem. 247*:5107 (1972).
5. *J. Bacteriol. 112*:1306 (1972).

4
Permeation

BETWEEN CELL WALL AND MEMBRANE (PERIPLASMIC SPACE)

When one removes the cell wall, as in the preparation of protoplasts and spheroplasts, certain degradative enzymes appear in the medium as though they had existed in the area between the cell wall and the membrane. Similar enzymes are released by osmotic shock. Usually this is accomplished by incubation with high concentrations of sucrose (sometimes with EDTA), followed by exposure to water or dilute salt solutions. This procedure does not seem to damage the rigid cell wall, yet it allows enzymes to escape. Ribonuclease, alkaline phosphatase, other phosphatases, a type of DNAse, thymidine phosphorylase, and some other enzymes have been found to be released into the medium under these conditions. Presumably, they are loosely bound to the cell membrane or somehow held in the space between the cell wall and the cytoplasmic membrane. This is the *periplasmic space.*

By means of episomic transfer it is possible to obtain the synthesis of the *Escherichia coli* alkaline phosphatase in *Salmonella typhimurium*, an organism that does not possess alkaline phosphatase. In the *Salmonella* so treated the alkaline phosphatase is also released into the medium on spheroplast formation. Spheroplasts of *E. coli* (in contrast to whole cells) were unable to form alkaline phosphatase, but they did synthesize and secrete into the medium a protein antigenically related to the enzyme. There is evidently an important functional factor involved in the space between the cell wall and membrane, which further study may clarify.

The two observations just mentioned suggest, first, that cell organization helps determine where and in what condition an enzyme can occur. After all, the instructions carried to the *Salmonella* via the episome are capable of merely specifying that a protein of a certain length and sequence is to be made. Where the cell should put it, could not, by way of our present knowledge, be specified in the episomal DNA. Yet the cell places it in the periplasmic space. The fact that spheroplasts of *E. coli* made a precursor of the enzyme, not the enzyme itself, suggests that the enzymes of the periplasmic space (which are absent in the spheroplast) have some function in processing the protein as specified by the DNA to convert it to a functional enzyme. This might occur by removing one or more amino acids from the end of the chain, for example; but the important point is that this occurred in the periplasmic space. We usually think of the periplasmic enzymes as degradative, but *Desulfovibrio* contains hydrogenase in the periplasmic space (1), and possibly a more extensive search will reveal enzymes in addition to the degradative types. Furthermore, there is evidence for other materials in addition to enzymes in the periplasmic space.

Periplasmic spaces seem to be typical of gram-negative cells, but it seems possible that gram-positive cells have a similar phenomenon. Even in the absence of a visible space, one can conceive of a protein with one end at least loosely attached to the membrane, which projects out into the interstices of the cell wall, perhaps even between the spaces in the gram-positive murein, and which might be released from its membrane binding by heat shock or similar treatments.

THE CELL MEMBRANE

Beneath the cell wall, but apparently not necessarily attached to it, is the cell membrane. In the light microscope, the membrane is invisible as a distinct entity, but can be readily seen in the electron microscope. In gram-positive organisms the membrane is made of three layers, easily visible in electron micrographs of thin sections of bacteria. It is thought that the outer layers are protein and the central layer is lipid. The membrane may be a "considerable" distance from the cell wall; at least a space as wide as itself may separate it from the wall. The membrane is still evident on the surface of the protoplast. In gram-negative organisms the membrane is not as clearly separable into three layers and appears in most photographs to be rather tightly pressed against the cell wall; in others, it seems to be torn away from it.

It has been possible to prepare membranes of gram-positive bacteria in a pure state by diluting osmotically protected protoplasts until they burst, and then centrifuging down the membranes. In *Bacillus megaterium*, membranes prepared in this fashion constituted about 10 percent of the dry weight of the cell and contained essentially all the cell lipid. About one fourth of the membrane was found to be phosphatidic acid. The membranes of gram-positive bacteria are generally high in phosphatidyl glycerol and diphosphatidyl glycerol.

Membranes from gram-negative organisms are more difficult to prepare in a pure state since the spheroplast still contains the major lipid constituents of the cell wall, and the fragments formed on osmotic lysis consist not only of membrane but of various portions of the package as well. Yet it is clear that in both gram-positive and gram-negative microorganisms the membrane is the site of both differential permeability and the osmotically active barrier of the cells.

Note that we make a distinction between these. The osmotically active function is evident when cells are put into a solution of higher osmotic pressure than the cells (hypertonic solution). The cytoplasmic contents enclosed by the membrane condense and shrink, and plasmolysis results. This means that water passes freely through the membrane, but other materials do not pass freely. This is distinct from differential permeability in which certain things pass into the cells, whereas others do not, but water is not pulled in or out in the process.

Before we go on we should point out the membrane has many activities. We have seen that it is involved in the synthesis of capsules, murein complexes, the package constituents of the gram-negatives, and even the C_{55} alcohols, which are the lipid carrier components; to this we now add osmotic character and differential permeability. The membrane in bacteria has still further functions, but it is already clear that the membrane cannot carry out all these functions in the same space, and that on a molecular or functional basis the membrane must be a mosaic of different functions — here C_{55} synthesis, there permeation, further along muramic peptide synthesis, and so on.

Membranes can be isolated and their nature studied (for a review, see (2)). In the gram-positives it consists of 25 percent lipid, a surprisingly high 40 percent protein content, and all the phospholipid of the cell (for a review, see (3)). In gram-negatives, the membranes include various portions of the package, have small amounts of RNA and DNA associated with them, and contain at least 92 percent of the phospholipid of the cell (4). The composition of the membrane may vary greatly. For example, the fatty acids (for a review, see (5)) may vary depending on strain, media, rate of growth, and growth temperature (6) and very much the same can be said for the phospholipids (for a review, see (7)). These variations seem to have little effect upon the permeability properties of the organisms, from which one may conclude, at least on a very rough basis, that neither the phospholipids nor the fatty acids are *specifically* involved in permeability per se, or that, if they have such a function, one type easily replaces another. Lecithin (phosphatidylcholine) is missing in most bacteria except among the gram-positive spore formers. It was hoped that mutants which lacked the ability to make one or more phospholipids might show such striking alterations in permeability that one might be able to associate particular phospholipids with particular types of permeation, but this has not proved to be the case.

Because of a lack of good methods for approaching permeability, because some methods employed proved to be misleading, and, possibly, because not all organisms are alike in the way they handle permeation of different substances, there is at present a considerable amount of contradiction and confusion. It therefore seems best to begin with a picture of permeability that we have found satisfactory in that it is capable of describing most permeation phenomena. We recognize that in a few cases this may not be the explanation. These exceptions, however, do not invalidate the working hypothesis, which proves satisfactory, we suggest, in the majority of cases examined. We do not see any advantage to be gained by fancy diagrams and sketches of membranes. We simply do not know how a membrane looks at the molecular level, but two things are true about it:

1. Water passes freely back and forth.
2. *All other things do not pass* through the membrane, either in or out, except if the membrane possesses a specific mechanism for their penetration called a *permease*.

Some people have found the word permease offensive in that it implies a material similar to an enzyme. An enzyme has a product different from its substrate, whereas in the permease, substrate and product are the same. The term "carrier" is used by some. Nevertheless, we shall use the term "permease" because it emphasizes the specificity of the substance transported.

There are three theories of permeation, not necessarily mutually exclusive, each maintained with considerable emotion by its proponents. We shall describe the evidence, which we think favors the permease concept, and later consider other explanations.

We should also say that the permeation and exchange of inorganic ions may be somewhat different from that of organic materials. In animal cells, especially in the salt-excreting glands of sea birds, there are definite sodium pumps that remove excess sodium from the cell. This sort of thing does not appear to occur in bacteria. In *E. coli*, for example, although potassium is diffusable, sodium does not seem to penetrate, and so it does not have to be pumped out. Thus penetration of inorganic ions may be somewhat different from that of organic substances, but we will concentrate upon the permeation of organic materials.

Organic permeation is based upon the permease concept (for a review, see (8)) which is as follows:

1. In *E. coli*, galactose can be assimilated against a concentration gradient.
2. The agent carrying out this process is called a permease and resembles an enzyme as follows:
    a. It is stereospecific.
    b. It is inducible.
    c. Its development is inhibited by agents that inhibit protein synthesis.
    d. It shows typical Michaelis relations between the initial velocity of accumulation and the

external concentration of galactose.

3. The permease enzyme is different from the enzymes that act on galactose metabolically based on the following evidence:
   a. Permease and metabolic enzymes are inhibited by different inhibitors.
   b. They show different enzyme kinetics.
   c. They show different genetics. There are cryptic mutants, that is, organisms which have the enzymes to attack galactose but cannot do so because the galactose does not penetrate.
   d. One can obtain substrates for the permease which are not substrates for the metabolic enzymes; that is, they will penetrate but not be metabolized.

In the case of the galactose permease of *E. coli*, the galactose inside the cell is free galactose and the galactose has not been structurally altered by transport into the cell. Furthermore, in cells into which galactose is penetrating, glucose and amino acids might be penetrating at the same time, but they do not interfere with each other; that is, each permease is separate from another. Even in the case of the penetration of D-alanine and L-alanine into whole cells of *Bacillus subtilis*, two separate permeases are involved (9).

ISOLATION OF PERMEASE

It would seem, off hand, to be rather difficult to isolate a permease, since its function is to transport things across a membrane; one could hardly remove it from the membrane to isolate it without destroying its function, and it would therefore be rather difficult to know when one had it. Yet the very galactose permease we have been discussing has been so isolated by a very ingenious method. *N*-ethylmaleimide, which combines with sulfhydryl groups, was found to combine with and irreversibly inactivate the galactose permease, but could be prevented from acting if the enzyme were loaded with substrate (galactose or thiodigalactoside). Therefore, *E. coli* was suspended in a great excess of the thiodigalactoside and treated with excess *N*-ethylmaleimide. This reacted with all the available sulfhydryl groups except those of the permease, which were protected by their substrate. The cells were then removed and suspended in a buffer containing no thiodigalactoside (to now expose the permease enzyme) and were treated with ^{14}C *N*-ethylmaleimide, which now combined only with the permease, since this was all that was left for it to react with. Then one isolated the protein containing the radioactivity as the permease. It was found to be a lipoprotein, as might be expected (10,11).

EXIT PERMEASES

We postulate that the cell membrane is not penetratable except if there is a specific permease for this purpose. It would seem possible that if things went into a cell by a permease they might come out by the same permease; that is, permeases are reversible. The evidence, however, is quite the contrary. Exit permeases differ from enter permeases. Evidence for exit permeases as distinct from enter permeases is the the following: first, entrance and exit show different degrees of inhibition with the same and with different inhibitors. Second, if a mere reversal (or a mere diffusion, a leak) were responsible for exit, one would expect exit to be highly temperature dependent (which it is not) and highly dependent upon the physical state of the membrane. In *E. coli* there is a galactose in permease and a glucose in permease, and the entry of galactose is not influenced by the presence of glucose; the two permeases are independent. However, if cells are preloaded with galactose, adding glucose markedly increases the rate of galactose exit. Just how this is brought about is not clear, but it does indicate that the galactose in and out permeases are different. Third, the genetics of in and out permeases are different. It therefore seems clear that there exist different permeases for penetration into the cell and for excretion of the substance from the cell, and if these are not present there will be no movement of substances across the membrane (12).

NATURE OF ACTIVE TRANSPORT

Galactose in *E. coli* is transported into the cell against a concentration gradient, that is, a greater concentration in the cell than outside. Such transport is called *active* transport and requires energy. We believe that the transport of most organic materials is active, and that there are very few transported that do not require energy. Incidentally, one may distinguish between them by carrying out the reaction at temperatures where energy generation is negligible, at 0°C, for example. For our purpose, we shall regard most if not all organic transport as active. Highly specific transport systems which do not involve energy (called *facilitated diffusion*) can be regarded as "permeases" which are not linked to an energy source or, indeed, in some cases these may be the "active" permeases acting in the absence of the normal energy supply.

One would suppose that if energy is required ATP could well be involved. But the surprising thing about active transport is that the substance needed to provide the energy is highly specific. Two basic active transport systems are known, with some indication that others exist: (1) the phosphopyruvate system, and (2) the D-lactate system.

Phosphoenolpyruvate System

In the case of *E. coli*, lactose, maltose, galactose, and fructose are each transported by a distinctly different permease. Yet mutants have been obtained that simultaneously lose the ability to grow on all four sugars. Since it was unlikely that they would lose all four permeases, it seemed probable that the mutants were deficient in some factor that was common to all four permeases, presumably the thing that made them "active." These car$^-$ (carbohydrate minus) mutants then would lack either a way of generating the transport energy or a way of connecting it to the permease. It was subsequently found that such mutants could transport and utilize the four sugars if they were provided with phosphoenolpyruvate (not replaceable by ATP, GTP, etc.), and it turned out that there was a histidine protein common to all the four permeases. This protein has been isolated; it is released by shock (suggesting that it is in the periplasmic space), has a molecular weight of only about 9,500, and contains no detectable tyrosine, tryptophan, or cysteine (it does contain phenylalanine) (13). What evidently happens is that the histidine protein reacts with phosphoenolpyruvate to form pyruvate and phosphohistidine protein, which reacts with the sugar permease and the sugar to deliver the phosphorylated sugar to the permease, which then transports it. The transport requires phosphoenolpyruvate, and nothing else works nearly as well. The transport system is easily lost, presumably because the histidine protein is either on the outer surface of the membrane or in the periplasmic space.

D-Lactate System

If one prepares protoplasts and spheroplasts and lets them swell and burst,and then centrifuges down the membranes, one obtains hollow spheres, called *vesicles*. The burst of the spheroplast is not, in most cases, an explosive burst that disintegrates the whole membrane, but rather more like popping a cork out of a small hole; when the cell contents have flowed away, the membrane, no longer subject to such disruptive osmotic forces, can go back together again. Such preparations prepared from *E. coli* will transport galactose, arabinose, glucose-6-phosphate, gluconate, glucuronate, and 15 amino acids, providing D-lactate is supplied. No other natural substance will work nearly as well (although some artificial materials, such as ascorbic phenazines, show some activity (14)).

Mutants may be obtained that lack D-lactic dehydrogenase, and these do not transport the above materials in spite of the fact that they have normal respiration-linked pathways. Vesicles prepared from these mutants can be treated *in vitro* with D-lactic dehydrogenase, and they then become capable of amino acid and carbohydrate transport (15).

Other Systems

It appears that the transport of glucose in *Azotobacter vinelandii*, that of sugars and amino acids in *Arthrobacter*, and that of gluconate in *Pseudomonas* specifically require malate. Other transport systems in other organisms have been shown to specifically require alpha-glycerolphosphate, NADH, ethanol, or L-lactate. Thus, based upon vesicle studies, other transport coupling mechanisms exist. But what is interesting is the high degree of specificity of the substance required and in many cases the unusual nature of the required substance. Surely the mechanism of activation must be unique.

THEORIES OF MEMBRANE TRANSPORT

As earlier mentioned there are three theories of permeation. The first is the permease theory upon which the previous discussion is based. It assumes that there is a highly specific carrier protein (the "permease") located in the cell membrane which is responsible for the penetration of its specific substrate. This theory accounts for the specificity of the materials penetrating and, while not specifically mentioned, the permease must be vectorially organized since it passes materials in one direction only (from out to in for an "in" permease). While supported by a great deal of evidence, its principle difficulty is that it does not explain how energy is coupled to the process of active transport. While it was earlier assumed that energy-rich phosphate was involved, this is not a necessary part of the permease hypothesis.

The second theory, called by some the redox model, assumes that the carrier protein can be oxidized and reduced, presumably by a second highly specific protein which might be capable of oxidizing and reducing several related carrier proteins (as D-lactic dehydrogenase is involved in the transport of several amino acids). The carrier protein, in its oxidized state, for example, might have a high affinity for the penetrating substrate (presumably on the external surface of the membrane) but when reduced (at the interior surface) its affinity for its substrate is less and the material is released internally. Energy is coupled to transport by alterations in the configuration (and thus affinity) of the permease during transport. Again, the redox permease is organized vectorially in the membrane. The principle difficulty with the theory is that it does not explain the inhibition of transport by agents which uncouple oxidative phosphorylation.

The third theory, called chemoosmotic, assumes that during the oxidation of any of a variety of materials, hydrogen ion (*protons*) are removed from the cell and thus a gradient of ions is formed across the membrane by the vectorially oriented respiratory system. This gradient is used to drive active transport. For example, in the transport of lactose, the transport carrier is thought to be bifunctional and to have binding sites for both its transport substrate (lactose) and protons. When such a carrier exists in a membrane across which there are gradients of chemical and electrical potential, the flow of protons down this gradient into the cell is coupled to and thus serves to drive the active transport. It is thought that the redox and hydrolytic energies of the cell's metabolism can be transduced to form a transmembrane gradient of chemical, osmotic, or electrical potential and it is this potential which drives transport. Uncouplers short circuit this potential and thus inhibit transport. There is a great deal of evidence that during respiration and transport there are changes in the proton concentration internally and externally, but the principle difficulty is that the theory does not account for the high specificity of the substances to be transported nor the high specificity of the energy source required. One would suppose that if a respiratory hydrogen ion gradient occurred (which it does) and that if the various transport proteins were driven by a proton gradient, almost any energy source would do, yet perhaps the most striking thing about the transport mechanism is the high specificity of the energy source required.

While the chemoosmotic theory seems to be the most popular at the moment (at least one finds more papers and reviews devoted to it), it seems possible that the various aspects of permeation emphasized by each theory may have a common origin. One may suppose that in the membrane there exist specific proteins, controlled by specific genes. The function of these proteins is the penetration of those substances capable of actively penetrating the cell. All other substances either diffuse in and out freely (except if held by adsorption to protein or Donnan effects) or are excluded completely. These specific proteins we choose to call permeases to emphasize the high degree of specificity with respect to their substrate (the material transported). These permeases are oriented in the membrane and they are vectorial proteins, that is, they pass their substrate in only one direction. While the detailed mechanism of energy coupling is not known, a few things are known about it. First, the material which generates the energy is highly specific and other materials, fully capable of generating energy to the cell, will not suffice for transport. This implies that there is a close connection between the oxidation of D-lactate, for example, and the transport permease for proline. One can conceive of the oxidation enzyme as forming a high energy intermediate located at and not removable from the enzyme surface itself, from which energy can be transferred to a limited series of permeases without involving the mediation of energy-rich phosphate bonds. One may think of this energy-rich enzyme-bound intermediate as convertable to phosphate bond energy (and thus provide for oxidative phosphorylation) or convertable to osmotic or electrical potential gradients, but when this happens, those particular molecules are not available to energize transport. Whether it is the energy-rich intermediate which itself creates the transport, or whether it first creates an ionic gradient which then causes transport is not clear, but in all cases it is the vectorial positioning of the transport chain which is critical to the passage of material into (or out of) cells. Uncoupling agents react with the energy-rich intermediate, or damage the ultrastructure of the membrane which structure is essential before any of the above reactions can occur.

BINDING PROTEINS

In discussing permeation, we made two assumptions: (1) the bacterial membrane is not permeable to anything but water, and (2) if anything did get in it was because of a specific permease for that purpose. Under such assumptions, a great deal about the peculiarities of cellular penetration are explainable. So much so, indeed, that we were going to regard this hypothesis as pretty close to the truth. Using this hypothesis, many people have studied the permeation of a variety of substances, but with some rather peculiar results. In the penetration of sulfates, for example, much was in accord with the permease hypothesis until the cells were heat-shocked, and then, all of a sudden, all the material that was supposed to be inside the cell appeared in the external "shockate." This type of result led to the discovery of *binding proteins*, which bind various substances and hold them against diffusion. Binding proteins occur in the periplasmic space and some of them at least play a role in permeation. In the case of the galactose in *E. coli*, which results from a permease in the cell membrane as we have previously discussed, when it is osmotically shocked, galactose (at its usual concentrations) does not enter; but if the dialized shock fluid is added back, galactose penetrates normally (16). That is, there appears to be a galactose-binding protein that assists the galactose permease.

In the cases of permeases, several proteins may be involved in the permeation process. One can imagine a binding protein that reacts with a transport protein, which reacts with an energy-coupling protein, which might even react with a discharging protein on the inside of the cell. That is, the permease system could be more complex than just one protein; but not more than three or four proteins can be involved because of the nature of the genetics of permease.

Yet the binding proteins are of some interest. One isolated from *E. coli* K12 (17) binds leucine, isoleucine, and valine against concentration gradients. Many years ago (18) it was found that *Micrococcus lysodeikticus* could take up rather large quantities of lysine against a concentration gradient without concomitant energy release, and that this lysine was not inside the cell but was all released if one removed the cell wall. It was also known (19) that the citrate space (the volume into which citrate would penetrate) in *E. coli* is about 25 percent of the cell volume, yet *E. coli* will not metabolize citrate even though it has the internal enzymes capable of doing so; that is, the citrate was not penetrating the cell membrane. Clearly, these phenomena can be explained by the existence of binding proteins in the periplasmic space, which take up the substances even though the material does not penetrate.

There are other aspects of permeability that we ought to touch upon. As you know, bacteria can attack insoluble substances by "secreting" an enzyme that will hydrolyze them to soluble materials. But how does the enzyme get out? How, in fact, is it synthesized and then pushed through the cell membrane? Is there an "enzyme" permease? Penicillinase has been intensively studied. In some of the aerobic spore formers it actually appears to be formed in vesicles outside the membrane proper. In staphylococcus the penicillinase seems to be synthesized in or on the surface of the membrane. In fact, there is some evidence, which we shall take up later, that all proteins of the cell, not only those to be excreted, are made at or in the membrane. Some extracellular enzymes may be excreted in the absence of protein synthesis (20).

Mesosomes

If membranes are so important, why doesn't a cell have more of them? Certain cells do. Photosynthetic bacteria and autotrophic bacteria seem to be a veritable sack of membranes. Internal membranes have been seen in *E. coli* (21) and more clearly in *Azotobacter* (22). And many kinds of cells, at certain stages of the single-cell growth cycle, produce membranes internally. One of these, which seems to be associated with cell division, is the *mesosome*. It is found more frequently in gram-positive organisms, but has also been seen in gram-negative bacteria. It appears to consist of membranous material and to be attached to the osmotic membrane. Thin sections indicate that the mesosome is also attached to nuclear material and that in a cell about to undergo division there appears to be a mesosome at each end of the nuclear substance, which is stretched out into a rod-shaped form. At the middle of the long axis of the nuclear rod, the cell mass starts to invaginate; subsequently, new mesosomes develop close to the points of invagination. Upon division these attach to the new nuclear fragments so that each new cell has an old and a new mesosome connecting the nuclear material with the membrane. The mesosome appears to be attached to the inner layer of the osmotic membrane and not to the entire triple-layered membrane.

A penicillinase secreting staphylococcus contains a

mesosome, evidently bounded by an infolding osmotic membrane. After incubation in a glucose—phosphate medium in which about 30 percent of the previously bound penicillinase is released in a soluble form, these cocci show an almost complete conversion of mesosomes to empty vesicles. Presumably these vesicles originally contained mesosomes. It has been supposed that penicillinase is formed in the mesosome and inserted into invaginations of the plasma membrane that surround it. The enzyme is released into the medium when the membrane comes in contact with the outer environment.

Exposure of mesosome-containing bacteria to hypertonic solutions induces the extrusion of mesosome contents from the cytoplasm into the space between the membrane and the cell wall. These extrusions are in the form of tubules or vesicles, usually at the poles of the cell, which suggests the attachment of the mesosome to the cell wall; such attachment can be seen in some preparations. When protoplasts are made of dividing cells, there is a release of mesosome tubules into the medium. There is also evidence that some mesosomes, or mesosome-like structures, are sites of oxidative metabolism. The question of mesosome function may be clarified when techniques for their separation from cytoplasmic membranes are better refined. The lipid composition of the membranes comprising the mesosome and the osmotic membrane in staphylococci seems to be very similar (23).

PLASMOLYSIS

We should also say a few words about the plasmolysis of bacterial cells. The turbidity of a bacterial suspension depends upon the light scattered by the particles (and not upon either the light absorbed or the size of the particle). When bacteria are placed in a solution of higher osmotic pressure, water is withdrawn and the membrane is pulled away from its usual position, which is tight up against the cell wall. This creates another surface for light reflection and the apparent turbidity increases (24). It happens that *E. coli* has a glucose-6-phosphate dehydrogenase that is bound to the membrane and "reachable" from the outer surface. When the cell is plasmolyzed, the membrane appears to be stretched, evidently pulling the dehydrogenase away from its carrier system. Thus, putting *E. coli* in 0.5 *M* NaCl inhibits the dehydrogenase completely as long as the membrane is intact, but has no effect at all upon the isolated enzyme (25). These kinds of results suggest that in the membrane various enzymes occupy a specific and particular position, and that they are organized and physically located one to another so that reaction sequences can take place with less dependence on random diffusion.

Osmophilic yeasts have always been a physiological problem. They require sugars and are not aided by high salts. It occurred to us that they might have membranes impermeable to water. If so, osmotic pressure would not affect them; since no water could be pulled out. They could live on the water they produced in their own metabolism — metabolic water. Other creatures do, the moth, for example. We therefore gave such yeasts glucose tagged with tritium, which would be oxidized to water. If the membranes were not permeable to water, the tritium would remain inside the cell. Unfortunately for theory, about 25 percent of the tritium was outside as water, so we did not really learn much about the reasons for osmophilic yeasts.

REFERENCES

Membranes

Guidotti, G. 1972. Membrane proteins. *Ann. Rev. Biochem. 41*:731—752.

Machtiger, N. A., and C. F. Fox. 1973. Biochemistry of bacterial membranes. *Ann. Rev. Biochem. 42*:575—600.

Salton, M. R. J. 1967. Structure and function of bacterial cell membranes. *Ann. Rev. Microbiol. 21*: 417—422.

Salton, M. R. J. 1974. Membrane associated enzymes in bacteria. *Advan. Microbial Physiol. 11*:213—284.

Singer, S. J. 1974. Molecular organization of membranes. *Ann. Rev. Biochem. 43*:805—833.

Permeation Transport

Boos, W. 1974. Bacterial transport. *Ann. Rev. Biochem. 43*:123—146.

Kaback, H. R. 1970. Transport. *Ann. Rev. Biochem. 39*: 561—598.

Kaback, H. R. 1974. Transport studies in bacterial membrane vesicles. *Science 186*:882—891.

Oxender, D. L. 1972. Membrane transport. *Ann. Rev. Biochem. 41*:777—814.

Reeves, J. P. 1973. *Microbial Permeability.* Dowden, Hutchinson & Ross, Inc., Stroudsburg, Pa.

Sussman, A. J., and C. Gilvarg. 1971. Peptide transport and metabolism in bacteria. *Ann. Rev. Biochem. 40*:397—408.

Others

Goldfine, H. 1972. Comparative aspects of bacterial lipids. *Advan. Microbial Physiol. 8*:1—58.

Greenawalt, J. W., and T. L. Whiteside. 1975. Mesosomes: membranous bacterial organelles. *Bacteriol. Reviews 39*:405—463.

Ikawa, M. 1967. Bacterial phosphatides and natural relationships. *Bacteriol. Reviews 31*:54—64.

Ryter, A. 1968. Association of the nucleus and the membrane of bacteria: a morphological study. *Bacteriol. Reviews 32*:39—54.

Papers

1. *J. Bacteriol. 120*:994 (1974).
2. *Ann. Rev. Microbiol. 21*:417 (1967).
3. *J. Gen. Microbiol. 29*:39 (1962).
4. *J. Biol. Chem. 244*:2450 (1969).
5. *Bacteriol. Reviews 26*:42 (1962).
6. *J. Bacteriol 95*:2054; *100*:1342; *120*:99 (1974).
7. *Bacteriol. Reviews 31*:54 (1967).
8. *Bacteriol. Reviews 21*:169 (1957).
9. *J. Bacteriol 120*:1085 (1974).
10. *Proc. Natl. Acad. Sci. 54*:891; *58*:225; *58*:274 (1967).
11. *J. Biol. Chem. 244*:5981 (1969).
12. *J. Bacteriol. 122*:332 (1975).
13. *J. Biol. Chem. 246*:7023 (1971).
14. *Science 186*:882 (1974).
15. *Proc. Natl. Acad. Sci. 70*:1917 (1973).
16. *J. Biol. Chem. 242*:793 (1967).
17. *J. Biol. Chem. 241*:5732 (1966).
18. *J. Biol. Chem. 76*:281, 288 (1958).
19. *Can. J. Microbiol. 4*:109 (1958).
20. *J. Bacteriol. 122*:34 (1975).
21. *J. Bacteriol. 92*:780 (1966).
22. *J. Bacteriol. 114*:346 (1973).
23. *J. Bacteriol. 121*:137 (1975).
24. *J. Bacteriol. 87*:1266 (1964).
25. *J. Bacteriol. 87*:1274 (1964).

5
Other Cellular Structures

FLAGELLA

With the electron microscope one may see flagella on the outer surface of many bacteria. Bacterial flagella are composed of protein and are not covered by extensions of the membrane structure, as are flagella from eucaryotic cells. Bacterial flagella are antigenic (H antigens) and differ serologically from cell wall antigens (such as O antigens, lipopolysaccharides). They consist of a single protein composed of a subunit monomer of about 30,000—40,000 molecular weight. The composition of this protein, called *flagellin*, may differ from one organism to another. In *Proteus* and *Bacillus* the flagellin contains no histidine, tryptophan, proline, or cysteine, but some strains contain hydroxyproline, which is found elsewhere in nature only in collagen, actinomycins, and the cell wall of some fungi. Flagellin from *Salmonella typhimurium* contains ε-*N*-methyl lysine, and this is the only place that this substance has so far been detected in nature.

Flagellin can be isolated in a pure form. If the isolated protein is depolymerized (e.g., at 10 mg of protein per ml in 3 *mM* phosphate buffer at pH 7, heated to 50—65°C), one can obtain a solution of low viscosity. If short fragments of normal flagella are added, the solution will polymerize into long flagellar filaments indistinguishable from normal flagella in the electron microscope. It appears that in this case the structure of the flagella has been built into the molecular structure of the flagellin, and when it polymerizes it does so in the physical form of the final organelle. In synthesizing organelles it seems evident that, to make the three-dimensional structure of the organelle, one must either have this structure (and its assembly) already built into the molecular structure of the components or, in addition to the materials that enter the organelle, the cell must build enzymes to assemble the structure.

In the intact cell, flagella appear to arise from a small granule seemingly attached to an internal membrane from a basal structure sometimes appearing as two disks. A small thread penetrates the external membrane and the rigid cell wall, and then appears to enlarge into the flagellum proper. Most flagella seem to be spirals, but one can find organisms with straight, uncurved flagella. These are nonmotile. In a large *Spirillum*, true flagella can be seen in the dark field, and each flagellar bundle (dipolar flagella, multitrichous) rotates slowly in a direction opposite to that of the cell. Both flagellar bundles rotate in the same direction. When phenol is added, rotation of the flagella may be uncoordinated and the cell is motionless. Flagella of some bacteria may be broken off by treatment in a malted milk blender (for a few minutes) without killing the deflagellated cells. Some of the cells rotate slowly owing to the action of the small stubs of flagella remaining. In log phase cells, about half the flagella broken off can be regenerated, and they seem to grow gradually and not by sudden extrusion. It is alledged that flagella grow (elongate) from the *end* away from the *cell* (1). Protoplasts still possess flagella, but to the best of our knowledge protoplasts are not motile. This suggests that the cell wall serves as a base against which the flagellum exerts its force. As a matter of fact, just how a flagellum can cause motion in a bacterium is not exactly clear. Some have supposed that the fine thread which penetrates the cell wall and enlarges into the flagella actually rotates. Although this is generally accepted (2), if the flagella does rotate, it would be unique. Nature did not invent the wheel or the rotating shaft — that is man's contribution. So it would be highly unlikely that the flagella would rotate. Instead, flagella may be considered to operate something like an oar. They are anchored and pivoted at the rigid cell wall and activated by displacement of the underlying cytoplasmic membrane to which they are also firmly attached (3).

Flagella can be synthesized by cell-free protein synthesizing systems, and flagellin can be detected in the cell membrane before it has been assembled into flagella (4). It has been claimed that flagellin can be synthesized even if RNA synthesis is inhibited, and it has been suggested that mRNA for flagellin is stable. Flagellar synthesis can occur in the absence of cell growth; for example, tryptophan is necessary for cell growth but does not occur in flagella. Using mutants that require tryptophan, if tryptophan is withheld, the cell cannot grow, but flagella can readily be synthesized.

OTHER METHODS OF BACTERIAL TRANSLOCATION[1]

Bacterial surface translocation means surface spreading and motility on a solid surface. A spreading zone is a film, broad or narrow, of one or at most a few layers of cells extending from the edge of a colony or an area of confluent growth, and therefore frequently so thin that it is barely visible to the naked eye. This spreading zone has been shown to be caused by six different kinds of bacterial translocations, as follows:

1. *Swarming* is a surface translocation produced through the action of flagella but is different from swimming. The micromorphological pattern is highly organized in whirls and bands. The movement is continuous and regularly follows the long axis of the cells, which are predominately aggregated in bundles during the movement. It is dependent on excessive development of flagella and partly on cell-to-cell interaction. It is found in *Proteus mirabilis*, *P. vulgaris*, *Bacillus alvei*, *B. rotans*, *Clostridium tetani*, and *Cl. novyi*; they all possess peritrichous flagella and are able to swim in fluid media.

Surface-active substances in certain concentrations and EDTA inhibit swarming of proteus by inhibiting the synthesis of flagella and thus preventing the formation of the long cells. Flagellated nonswarming bacteria can be made to swarm by treatment with phage lysates of other phenotypically similar strains, which indicates that not only flagella but an excessive number of flagella are necessary for the organism to swarm. Cells move in the form of large rafts or bullet-shaped colonies. The size of the motile microcolonies depends on the amount of available moisture; the dryer the agar, the bigger the microcolonies. The spreading rate of a swarming strain depends on the humidity of the plate, growth rate of the organism, and incubation temperature. The long swarmed cell and short nonswarming cell both swim at 10—15 μm/s. On insufficiently dried plates the isolated swarmer cells are also seen to move, but not as fast as rafts of cells, and the rate of spreading may be accelerated so that the whole plate is covered in one wave. On very dry plates incubated at 22°C (conditions that inhibit or slow down swarming), wandering and rotating colonies are seen because the motile power of groups of cells is greater than that of single cells and probably also depends on the number of cells per group. Motility is confined to cells in contact with other cells but loosely bound (i.e., rafts); they move at 1 μm/s. The dryer the agar, the bigger the microcolonies, and under the microscope they can be seen to travel in a counterclockwise continuous curve.

2. *Swimming* is a surface translocation produced through

[1]Henrichsen, J. 1972. Bacterial surface translocation: a survey and a classification. *Bacteriol. Reviews 36*:478-503. (Reviewed by Patricia E. Kasenchar.)

the action of flagella, but is different from swarming, and only takes places when the film of the surface fluid is sufficiently thick. The micromorphological pattern is unorganized. The cells move independently and at random in the same manner as flagellated bacteria in wet mounts. With a concentration of 0.2 percent agar, the bacteria swim on top of the agar, but if oxidative metabolic pathways are not preferred, they swim throughout the depth of the agar or at the bottom. The most peripheral part of the spreading zone of *B. cereus* produced by swimming motility consists of closely set finger-like projections, which can be seen macroscopically. Spreading is very sensitive to dryness of medium and is therefore seen only on freshly poured agar plates. Different organisms have different numbers of peritrichous flagella.

3. *Gliding* is surface translocation produced by an unknown mechanism occurring in nonflagellated bacteria. The micromorphological pattern is highly organized in whirls and bands. The movement is continuous and regularly follows the long axis of the cells, which are predominantly aggregated in bundles during the movement. It is seen in several groups of microorganisms and is a characteristic associated with filamentous unicellular blue-green algae. There are three main groups of bacteria that show gliding: (1) filamentous, gliding bacteria (*Beggiatoa*), (2) fruiting myxobacteria (*Sporangium*), and (3) *Cytophaga*.

Gliding is usually confined to solid surfaces such as glass and agar gel but sometimes can also be observed on the surface film of a fluid or the air—water interface. Gliding on agar depends on the humidity of the plate even more so than swarming. Colony morphology is influenced by the amount of moisture present and the concentration of nutrients. Under optimal conditions, colonies will be seen as completely flat, rapidly spreading, almost invisible swarms or as a spreading, rhizoid growth with a honeycomb appearance. The direction of motility is a matter of chance, and the rate of translocation varies with the organisms and conditions. The movement may be due to fibrils associated with the inner layer of the outer membrane. The primary difference between swarming and gliding is that swarming is performed by bacteria having flagella and gliding by those without.

4. *Twitching* is produced by an unknown mechanism occurring in both flagellated and nonflagellated bacteria, but is not due to the action of flagella. The micromorphological pattern is varying but not as organized as in swarming or gliding. The cells move predominantly singularly, although smaller moving aggregates occur. The movement appears as intermittent and jerky and does not regularly follow the long axis of the cell. The direction of movement of individual cells and the number of jerks per time unit is quite haphazard. Twitching, like gliding, motility is dependent on the availability of moisture, with the greatest activity observed on media poor in nutrients with a relatively low agar concentration, freshly poured, dried for a very short period, and incubated in

a humid atmosphere.

5. *Sliding* is produced by the expansive forces in a growing culture in combination with special surface properties of the cells that result in reduced friction between cell and substrate. The micromorphological pattern is that of a uniform sheet of closely packed cells in a single layer. The sheet moves slowly as a unit. This movement is dependent on growth. The zones produced are macroscopically indistinguishable from spreading zones of twitching strains except that the spreading zone of a sliding organism consists of a single layer of densely packed cells and movement of single cells or groups of cells relative to other cells cannot be observed. Velocity values can vary from 2 to 25 μm/min. Sliding can be observed in *Streptococcus*, *Corynebacterium* sp., and *Moraxella phenylpyruvica*. It is not restricted to certain media and may be regarded as a special case of the normal process of colony formation of a particular strain on the surface of a particular solid medium.

6. *Darting* is produced by expansive forces developed in an aggregate of cells inside a common capsule that result in the ejection of cells from the aggregate. The micromorphological pattern is that of cells and aggregates of cells distributed at random with empty areas of agar in between. Neither cell pairs nor aggregates move except during ejection, which is observed as a flickering in the microscope. It is dependent on growth in capsulated aggregates. Darting can be observed in *Staphylococcus aureus* and *S. albus*. Optimal conditions for growth are a high atmospheric humidity. Microscopically the cells appear to be in clusters of cocci.

FIMBRIAE OR PILI

On the outer surface of some bacteria are projections called *fimbriae* or *pili*. Some are involved in chromosomal transfer, and the presence of pili is characteristic of the F^+ (male) strains. They seem to serve as a tube through which DNA is transferred to the F^- (recipient) cell.

NUCLEAR STRUCTURES

Within the cell itself are nuclear bodies, but for many years it was not clear whether or not bacteria had "nuclei," as such. Today it is clear that bacteria have nuclei, although not enclosed in a membrane. The evidence for a bacterial nucleus consists of the following:

1. The universal presence of DNA in living cells.
2. Areas in the cell where DNA is evidently localized observed both through staining reactions shown with the light microscope and as areas of altered electron density seen in the electron microscope.
3. The separation or division of such DNA material

before or simultaneously with cell divisions.

4. An array of genetic evidence which indicates that the bacterial cell acts as if it had a nucleus.

Thus there is morphological, chemical, and genetic evidence of a nucleus. Such an organ, however, does not appear to be surrounded by a membrane as is the case in higher cells; rather, the DNA is localized at least part of the time in a given area of the cytoplasm without being physically separated from it. As mentioned in the discussion on mesosomes, and as will be discussed elsewhere, DNA may also be found attached to membranes and sometimes enclosed within them; but not all of the DNA of the cell is so enclosed, and not all the time.

With the phase-contrast microscope one can see nuclear bodies that appear to be the same as those seen in stained preparations. These are thought to be nuclei for the following reasons:

1. The bodies seen in phase contrast divide regularly with each cell division.

2. In fixed preparations, the bodies stain characteristically as DNA. In stained preparations they are called *chromatin bodies*. They may be treated with ribonuclease (to remove RNA) and still stain, but are lost on treatment of the preparation with deoxyribonuclease. All, or nearly all the DNA of the cell is contained in them. Radioactive (tritiated) thymidine, which is incorporated only into DNA, concentrates in the same location as the chromatin bodies.

3. The morphology may change considerably as seen in phase contrast, depending on the environmental conditions; similarly, in stained preparations the morphology is variable. There may be more than one such body per cell, probably because cell division follows somewhat irregularly after nuclear division; the DNA per cell varies, whereas the DNA per chromatin body is constant. Spores have only one chromatin body, and the amount of DNA per spore is constant.

4. It is possible to separate a particle from *B. subtilis* when the protoplasts (free from cell wall) are treated with lipase. These particles contain DNA, RNA, and protein, and virtually all the DNA in the cell. The DNA is evidently on the surface of the particle, since it streams off during the preparation (5).

MITOCHONDRIA

In the animal and plant cell the respiratory enzymes are contained in separate organelles called *mitochondria*. The mitochondria are about the size of bacteria. It is possible to separate particles from bacteria that are proportionately smaller, which contain the respiratory enzymes, and it was originally thought that perhaps bacteria had mitochondria in the same fashion as the eucaryotic cells.

However, if bacteria are ruptured, by ultrasonic

vibration, for example, and the unbroken cells removed, the oxidative activity of the cell-free portion can be centrifuged down at relatively low speeds, indicating that it is associated with a relatively large particle. If further sonication is applied to the cell-free extract, the oxidative activity becomes more and more difficult to centrifuge down; that is, it is associated with smaller and smaller particles. Since membranes when prepared by osmotic shock from protoplasts contain essentially all the oxidative activity, it seems probable that the membrane has been broken into smaller and smaller pieces by repeated sonication. Although these fragments resemble mitochondria in function, they actually were originally part of the membrane structure and not comparable to animal mitochondria.

Indeed, it appears that mitochondria of eucaryotic cells may once have been bacteria (6,7). The mitochondria contain their own DNA, which differs from the nuclear DNA. The DNA from the mitochondria is circular. The mitochondria also contain their own ribosomes; these are the 70S ribosomes characteristic of bacteria, rather than the 80S characteristic of eucaryotes. Protein synthesis in mitochondria is inhibited by chloramphenicol (characteristic of bacterial protein synthesis), which substance does not inhibit protein synthesis in the eucaryotic cells. In short, the mitochondria appear to have many of the characters of procaryotic cells, and perhaps they are the residuum of some important invasion, eons ago. When we consider the process of respiration later, and particularly the toxicity of oxygen, such an invasion of a eucaryote by a procaryote, which has solved the oxygen toxicity problem, would seem to be a remarkable evolutionary advance. However, this implies that there were procaryotes and eucaryotes evolving independently, and not one from the other, a philosophical problem that we shall have to face later. One objection to the theory of the bacterial origin of the mitochondria is that the cytochromes of higher cells differ from those of bacteria. However, there is at least one bacterium, *Paracoccus denitrificans*, whose cytochromes and other features so closely resemble those of mitochondria that it might indeed have served as the ancestor to the mitochondira (8).

The remaining cellular structures (for a review, see *Ann. Rev. Microbiol. 28*:167, 1974) may be divided into organized and unorganized. Organized structures might first be represented by spores, but we shall defer consideration of them until Chapter 15.

ORGANIZED MATERIALS

Bacteriocins (those for *Escherichia coli* are called *colicins*) are proteins produced by various species of bacteria that act only upon strains of some or closely related species. Many strains possess them; in *E. coli* perhaps 20 percent of the strains produce such materials.

They are controlled by a transmissible plasmid, termed the C factor. They seem to adsorb to a specific site on the susceptible strains and invade the cell. The bacteriocins differ from phages in that they do not multiply in the cell after infection but only kill it. Strains producing bacteriocins have been found to possess structures within them some of which appear to be incomplete phages, but others seem to be nonsedimentable proteins of relatively lower molecular weight and are not necessarily visible within the cells.

Rhapidosomes seem to be headless bacteriophage tails and sediment as though they were a bacteriophage fragment. So far they have not been associated with biological activity. Similar tubules are occasionally found in other microbial (*Azotobacter* and some *Streptococcus*) cells, which may have a similar origin.

Cysts occur in *Azotobacter*. Certain of the cells, especially those containing relatively high amounts of poly-β-hydroxibutyrate, form a thick cell coat to produce a cyst that is more resistant to drying and to heat than the nonencysted cell. There are both normal-sized and giant cysts.

Ribosomes are organized particles, high in RNA, involved in protein synthesis. From the cytological point of view they are also of interest. Bacterial ribosomes are relatively stable particles (100—200 Å) composed mostly of RNA (60 percent) and containing 85—90 percent of the total RNA of the cells. They can be dissociated into two primary particles and separated and purified by centrifugation. One particle has a sedimentation constant of 30 Svedberg units (called 30S; molecular weight of approximately 1.3×10^6); the other 50S. These may be associated in various ways. A 30S and 50S particle associate to form a 70S particle, which is the entity presumably active in protein synthesis. A variety of other particles (100S, 150S, 400S, etc.) resulting from various types of association are known. There are also units of much larger size, called polysomes, which consist of several 70S units held together along a strand of mRNA. The ribosome particles of eucaryotic cells are 40S and 60S, which combine to form an 80S unit. From the viewpoint of cytology, it is evident that, although some ribosomes appear to be free in the cytoplasm, there tends to be a degree of organization in their location. They seem to be associated with the nuclear apparatus, near the mesosomes, near the external membranes, and in regular strands within the cytoplasm as though they might be attached to an endoplasmic reticulum not easily visible in electron micrographs. Electron micrographs show them reasonably uniformly distributed throughout the cytoplasm, but when cells are fractionated, especially if the cells are broken by gentle means, the ribosomes seem to be attached to membranes. We will discuss this discrepancy later.

Ribosomes contain several proteins. The 30S portion contains some 20 proteins, and the 50S even more (perhaps 30). In fact, there are so many that they cannot all be fitted into the same 30S or 50S piece, which means that not all the ribosomes are the same. Some have one protein that is lacking in another. Some of the proteins of the ribosome are not as firmly bound as others and can be more readily dissociated. Three of these on the 30S ribosome, S_2, S_3, and S_{14}, are involved in binding transfer RNA (tRNA); S_1 is involved in messenger binding. These four proteins may well constitute the site on the 30S ribosome where message and charged amino acid (tRNA) interact. Another, S_{12} reacts with streptomycin, and cells resistant to streptomycin lack this protein. The actual number of ribosomes per cell varies with the growth rate (9).

The 50S ribosome always contains two proteins, L_7 and L_{12}, which comprise about three molecules per ribosome. Most other proteins are present as one molecule or less. These two proteins are involved in translocation, react with GTP, and perhaps surprisingly, are contractile proteins (10). It is as if each ribosome has a built-in muscle, whose function is to pull the messenger along one triplet. We shall have more to say about ribosomes later.

There are still other organized structures found in particular cells: chromatophores in photosynthetic bacteria, "phycobilisomes" (tetrapyrolles attached to protein) in blue-green algae, and there is a "purple membrane" resembling retinal rods in halophilic bacteria (11).

UNORGANIZED MATERIALS

Under particular circumstances some bacteria may possess granules or globules, or even crystalline deposits within their cytoplasm. One type was recognized very early and called *metachromatic granules* (because they were stained *red* by aged alkaline methylene blue) or *volutin*; it was originally thought to be nucleic acid because of its affinity to basic dyes. In fact, these granules are composed of polyphosphates of relatively low molecular weight, but some may be as long as 500 residues per molecule. They may comprise as much as 40—50 percent of the cell. They are formed by a polyphosphate kinase

$$[ATP + (HPO_3)_n \rightleftarrows ADP + (HPO_3)_{n+1}]$$

and in some cases can serve as energy storage for the cell.

For years a refractive "fat" globule, staining well with fat stains, had been seen in many bacterial cells; it was thought to be a neutral fat, that is, a triglyceride of fatty acids. However, upon study it was found that many organisms store a poly-β-hydroxibutyrate which stains as "fat" cytochemically, and which is extractable with fat solvents. Much of what cytologists have called "fat" appears to be β-hydroxibutyrate polymer. In bacteria, glycogen is seldom found and starch rarely found. Glycogen is usually detected cytochemically by formation of a brown granule with iodine, whereas starch forms a blue complex. Such blue complexes are seen in some of the *Clostridia* when the cells are exposed to iodine, but starch cannot be isolated from them and the blue color appears to be due to a substance other than starch. The majority of such materials found in bacterial cells appear to be, by our present knowledge, reserve foods held within the cell. Any other functions that they may have (and we suspect they do have some) are matters for further study.

Some microorganisms contain crystals. In *B. thuringensis*, an insect toxin appears in the cytoplasm as a crystal, which may be the result of overproduction of spore proteins. In a mutant of *B. subtilis*, a crystal of RNA polymerase can be found, and in some autotrophs (*Thiobacillus neapolitanus*, for example) crystals of ribulose-1-5-diphosphocarboxylase are found. In addition, some organisms have gas vacuoles (12).

REFERENCES

Flagella

Iino, T. 1969. Genetics and chemistry of bacterial flagella. *Bacteriol. Reviews 33*:454—475.

Jahn, T. L., and E. C. Bovee. 1965. Movement and locomotion of microorganisms. *Ann. Rev. Microbiol. 19*:21—58.

Kushner, D. J. 1969. Self-assembly of biological structures. *Bacteriol. Reviews 33*:302—345.

Rhodes, M. E. 1965. Flagellation as a criterion for the classification of bacteria. *Bacteriol. Reviews 29*:442—465.

Smith, R. W., and H. Koffler. 1971. Bacterial flagella. *Advan. Microbial Physiol. 6*:219—340.

Nuclear Associated

Curtiss, R., III. 1969. Bacterial conjugation. *Ann. Rev. Microbiol. 23*:69—136.

Fuhs, G. W. 1965. Fine structure and replication of bacterial nucleoids. Symposium on the Fine Structure and Replication of Bacteria and Their Parts. *Bacteriol. Reviews 29*:277—298.

Valentine, R. C., P. M. Silverman, K. A. Ippress, and H. Mobach. 1969. The F-pilus of *Escherichia coli*. *Advan. Microbial Physiol. 3*:2—52.

Mitochondria

Ashwell, M., and T. S. Work. 1970. The biogenesis of mitochondria. *Ann. Rev. Biochem. 39*:251—290.

Borst, P. 1972. Mitochondrial nucleic acids. *Ann. Rev. Biochem. 41*:333—376.

Linnane, A. W., J. M. Haslam, H. B. Lukins, and P. Nagley. 1972. The biogenesis of mitochondria in microorganisms. *Ann. Rev. Microbiol. 26*:163—198.

Organized Materials

Bradley, D. E. 1967. Ultrastructure of bacteriophages and bacteriocins. *Bacteriol. Reviews 31*:230–314.
Kurland, C. G. 1972. Structure and function of the bacterial ribosome. *Ann. Rev. Biochem. 41*:377–408.
Nomura, M. 1967. Colicins and related bacteriocins. *Ann. Rev. Microbiol. 21*:257–284.
Nomura, M. 1970. Bacterial ribosome. *Bacteriol. Reviews 34*:228–277.
Reeves, P. 1965. The bacteriocins. *Bacteriol. Reviews 29*:25–45.
Schlessinger, D., and D. Apirion. 1969. *Escherichia coli* ribosomes: recent developments. *Ann. Rev. Microbiol 23*:387–426.
Shively, J. M. 1974. Inclusion bodies of prokaryotes. *Ann. Rev. Microbiol. 28*:167–187.

Unorganized Materials

Dawes, E. A., and P. J. Senior. 1973. The role and regulation of energy reserve polymers. *Advan. Microbial Physiol. 10*:136–266.
Rogoff, M. H., and A. A. Yousten. 1969. *Bacillus thuringiensis*: microbiological considerations. *Ann. Rev. Microbiol. 23*:357–386.

Papers

1. *Science 169*:90 (1970).
2. *Nature 254*:389 (1975).
3. *J. Bacteriol 100*:512 (1969).
4. *Arch. Biochem. Biophys. 139*:97 (1970).
5. *J. Bacteriol 75*:102, 369 (1958).
6. *Science 169*:641 (1970).
7. *Amer. Scientist 58*:281 (1970); *59*:230 (1971).
8. *Nature 254*:495 (1975).
9. *J. Bacteriol. 122*:89 (1975).
10. *Nature 242*:86 (1973).
11. *Nature 233*:238 (1971).
12. *Bacteriol. Reviews 31*:1 (1972).

6
Bacterial Nutrition

To grow, living cells need the following:

1. A suitable energy source. Most organisms will use glucose. We know of a few that do not. One organism studied would grow only on methane or methanol and not glucose. This inadequacy proved to be a lack of hexokinase (glucose + ATP → glucose-6-phosphate + ADP), for it would grow on glucose-6-phosphate (1).
2. A suitable source of elements is needed. In quantity, these are nitrogen, phosphorus, and sulfur. Trace elements (Fe, Mg, Mn, Mo) are frequently obtained from $MgSO_4$. Nutrient broth is frequently magnesium deficient.
3. A suitable environment; as to temperature, pH, and the presence or absence of oxygen.
4. Water.
5. A suitable source of those substances that it needs but is unable to make (if any), such as vitamins and amino acids.

To study microorganisms one wishes to grow them under defined and reproducible conditions. Since there are microorganisms capable of growth under very simple conditions, one may start with these. Table 6-1 lists several useful media. The AC medium is designed to provide a rich medium with enough buffering capacity so that even if all the sugar present (0.1 percent) is converted to lactic acid the pH will not go below 7. The Davis medium is widely used for growing *Escherichia coli* on a "synthetic" medium. It uses citrate to keep iron in solution, and hence is not especially good for primary isolation, say of organisms that attack some other substrate, phenols, or flavins. One ends up isolating organisms that attack citrate. Medium B is completely inorganic and has relatively low phosphate, since the latter tends to inhibit pseudomonads, which might be involved in the oxidation of more exotic substrates. Medium E of Vogel and Bonner is self-sterilizing; after standing for several days, 2 ml can be added aseptically to 100 ml of sterile water and a sterile medium obtained that has not been autoclaved.

TABLE 6-1. *Useful Media*

	g/liter	%
Nutrient broth		
Tryptone	5.0	0.5
Beef extract	3.0	0.3
AC (active cell) medium		
Tryptone	5.0	0.5
Yeast extract	5.0	0.5
Glucose	1.0	0.1
K_2HPO_4	5.0	0.5
Davis medium		
K_2HPO_4	7.0	0.7
KH_2PO_3	3.0	0.3
Sodium citrate ($2H_2O$)	0.5	0.05
$MgSO_4 \cdot 7H_2O$	0.1	0.01
$(NH_4)_2SO_4$	1.0	0.1
Glucose	5.0	0.5
Medium B (base medium, low phosphate)		
K_2HPO	1.0	Nitrogen free
$MgSO_4 \cdot 7H_2O$	0.3	10 ml of (1), (2),
NaCl	0.2	and 10 mg of $FeSO_4$
$FeSO_4$	0.01	
$(NH_4)_2SO_4$	1.0	With nitrogen (BN) + 10 ml of (3)

(1) 10 ml/liter of 10% K_2HPO_4, 2% NaCl
(2) 10 ml/liter of 3% $MgSO_4 \cdot 7H_2O$, trace of Mo, Mn
(3) 10 ml/liter of 10% $(NH_4)_2SO_4$

To hold iron etc. in solution can add 10 ml of 5% EDTA or 10 ml of 5% Na citrate, both adjusted to pH 7.

Media is at pH 7.4 to 7.6; 6 ml N/1 H_2SO_4/liter will bring pH to 5.0 to 5.5.

Medium E (*J. Biol. Chem.* *218*:19, 1956)

Water: 600 ml $MgSO_4 \cdot 7H_2O$, 10 g Citric acid H_2O, 100 g K_2HPO_4, 500 g $NaNH_4HPO_4 \cdot 4H_2O$, 175 mg (final volume about 1 liter)	Dissolve in order listed. This gives a 50-fold concentrate that is self-sterilizing 1 ml + 49 ml distilled water is pH 7 salts medium. Glucose (sterilized separately) usually added at 0.5%. Many *E. coli* and other organisms grow well on it.

To look in more detail at medium B, the K_2HPO_4 is a buffer and serves as a source of potassium and phosphorus; $MgSO_4$ provides magnesium, sulfur, and often trace elements. Ammonium sulfate serves as the nitrogen source. Such a medium is frequently deficient in manganese and molybdenum, sometimes biotin addition is helpful. It is easily made up by having available 10 percent solutions of the ingredients and mixing them as shown in Table 6-1. Sometimes one adds yeast extract as a stimulant; 0.05 percent yeast extract will normally permit growth to an optical density (O.D.) of about 0.2, owing to the yeast extract itself. Growth beyond this level is then presumed to be due to growth on whatever other substrate has been added.

To progress to a more complex medium, Table 6-2 shows a convenient way of making small quantities of Snell's *Lactobacillus* medium in which many quite fastidious organisms can grow. Normally, one provides glucose (0.5—1 percent final concentration). One may set up this medium as follows: mineral salts, sodium acetate, iron, glucose in all (see Table 6-2). Then, with addition designated +:

Tube	AA	PP	Vit	
1	+	+	+	All three requirements
2	—	+	+	Two out of three
3	+	—	+	Two out of three
4	+	+	—	Two out of three
5	+	-	-	Single requirement
6	-	+	-	Single requirement
7	-	-	+	Single requirement
8	-	-	-	No requirement

Should there be no growth in tube 1, it may be due to four situations:

1. Some additional factor is required: blood, yeast, and so on.
2. The physical conditions: pH, temperature, or oxygen level may be unsuitable.
3. The organism may not use glucose, acetate, or even AA as energy sources.
4. The medium may be too complex; might expect growth in tube 8. In any case, look at the environment from which the organism was isolated; if it was a rich environment, probably a rich medium was suitable; if poor, then poor. Try to simulate the natural condition if growth is not obtained in any of the eight media.

Beyond this there is little to say that is not pretty obvious, but we should spend a little time on the functions of accessory growth factors. These were thought of as chemical substances, differentiated from essential amino acids or energy sources on the basis of a very much smaller quantitative requirement. They are materials that the organism must find preformed in its environment and

TABLE 6-2. *Complex Synthetic Medium (Snell's Lactobacillus Medium)*

Code	Composition	For: 5 ml	10 ml	1 liter
MS	Mineral salts			
	1.0 g K_2HPO_4 (10 ml 10%)			
	1.0 g KH_2PO_4 (10 ml 10%)			
	0.1 g $MgSO_4 \cdot 7H_2O$ (1 ml 10%)			
	0.01 g NaCl (0.1 ml 10%)			
	Made to 200 ml	1.0 ml	2.0 ml	200 ml
AC	60 g Sodium acetate to 200 ml	0.1 ml	0.2 ml	20 ml
AA	Amino acids			
	5.0 g Casamino acids			
	50 mg Tryptophan			
	200 mg Cysteinc			
	Made to 200 ml	1.0 ml	2.0 ml	200 ml
	May be replaced by a synthetic mixture			
PP	Purines, pyrimidines			
	10 mg each: Adenine			
	Cytosine			
	Guanine	1.0 ml	2.0 ml	200 ml
	Uracil			
	Thymine			
	Boil with 100 ml H_2O and 10 ml concentrated HCl until dissolved.			
	Adjust to pH 7. Made to 200 ml.			
V	Vitamins			
	5.0 mg each: Thiamine			
	Riboflavin			
	Ca pantothenate			
	Niacin			
	Pyridoxamine			
	Thioctic acid			
	0.5 mg each (500 μg):			
	Folic acid			
	Biotin			
	Vitamin B_{12}			
	Made to 200 ml	0.1 ml	0.2 ml	20 ml
Fe	Made fresh			
	50 mg $FeSO_4 \cdot 7H_2O$ to 100 ml	0.1 ml	0.2 ml	20 ml

which play some essential role in its metabolism.

Studies began in 1901 when Wildiers, studying phosphorylation in yeast, showed that with small inocula there was no growth of yeast on continuous transfer in a mineral salts glucose medium, whereas with a greater inoculum there was good growth. Small inocula would grow if wort was added, so Wildiers postulated that wort contained an active substance, which he called "bios," necessary for the growth of yeast, and this substance was introduced with the inoculum when larger inocula were used. Pasteur had been able to grow yeast indefinitely in mineral salts media, but he had used a heavy inoculum. Up to the early 1930s there was a great deal of discussion as to whether "bios" existed, whether it was important, and whether Wildiers' experiments were properly conducted. Indeed, the early work was beset by difficulties. There was first a controversy over whether any such "bios" substances were "essential" or only stimulatory. Then the scale of the requirement was somewhat different from that which was common at the period. For example, even an amino acid, tryptophan, whose requirement was much higher than that of a "growth factor" would allow good growth at 4 ppm, yet 1/100 of this would permit visible growth. A further difficulty was that there were really no very good ways of purifying organic materials. Chromatography was not common, and most organic extracts were highly impure. Finally, there was an enormous "strain variation;" that is, there were marked differences in the nutrient requirements between strains that were indistinguishable in other ways. There was, of course, no concept of bacterial mutants, and indeed no way to approach their study.

Still, by the 1940s it was pretty clear that the accessory growth factors of bacteria were indeed the vitamins required in the diet of animals and man. Thiamine (vitamin B_1) was, for all practical purposes, the only vitamin that had been isolated on the basis of animal assay. From the early 1930s on, when an animal dietary factor became evident, a search was made for a microbial assay that could assist in its isolation and rapid quantitation. It would be of some value to look, at least briefly, at the functions of these vitamins.

Thiamine (vitamin B_1) exists in the coenzyme form thiamine pyrophosphate (Figure 6-1), which is normally attached to the enzyme. The coenzyme is capable of combining with keto acids and, together with the enzyme, leads to their decarboxylation. Thiamine pyrophosphate (also called cocarboxylase) is the coenzyme of organic acid decarboxylation, especially keto acid. Thiamine is heat labile, yet it may be autoclaved with media (it is not always completely destroyed), and most organisms will grow on it. Heating splits it into pyrimidine and thiazole, but many bacteria can use these to synthesize thiamine.

Riboflavin (vitamin B_2 or vitamin G) (Figure 6-2) is the ribitol derivative of isoalloxazine, and the coenzyme forms are FMN (flavin mononucleotide) but usually FAD

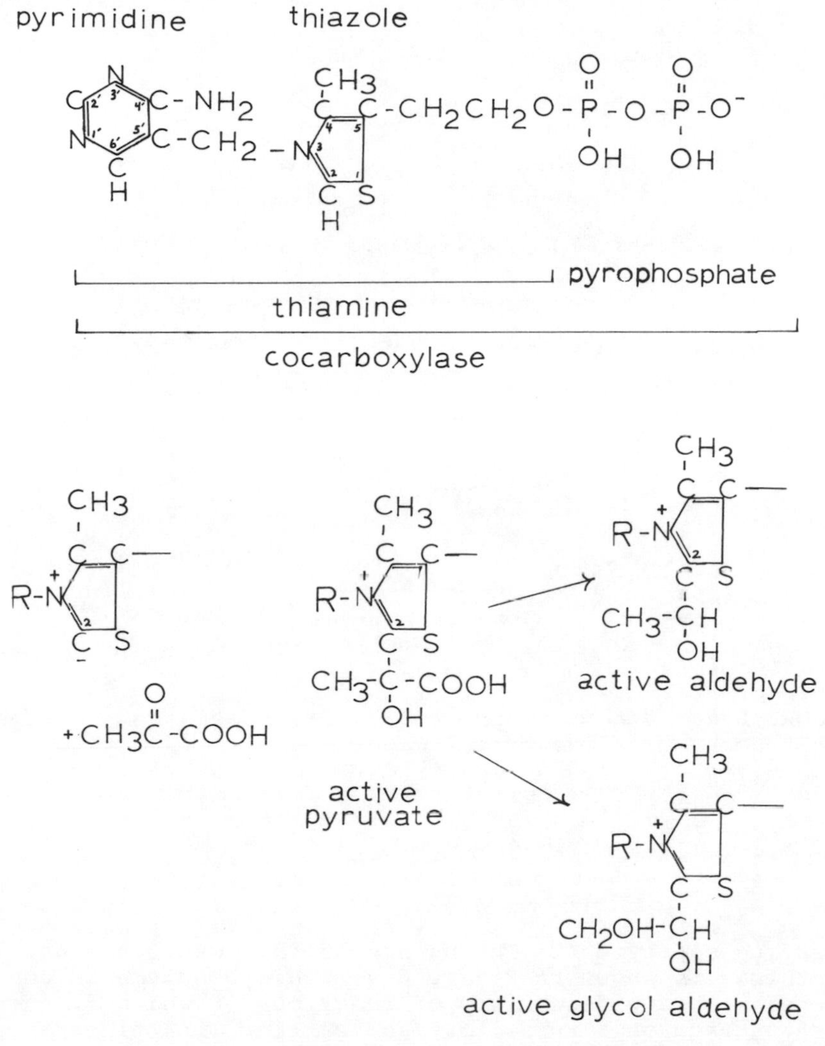

FIGURE 6-1. *Thiamine (vitamin B_1) and its coenzyme forms.*

(flavin adenine dinucleotide; i.e., flavin-phosphate-phosphate-ribose-adenine). It functions as a "transformer" since it is able to accept 2H but to release 1 electron at a time. The two hydrogens are absorbed at the positions marked and the nitrogen goes from a valence of 3 to 5. It can then give off an electron and a hydrogen ion, reverting back to a valence of 3, and similarly for the lower nitrogen of the center ring. Many of the flavins with this hydrogen-electron transport function are tightly bound to protein and are not usually readily dissociable coenzymes.

Pantothenic acid is an integral part of the acyl group

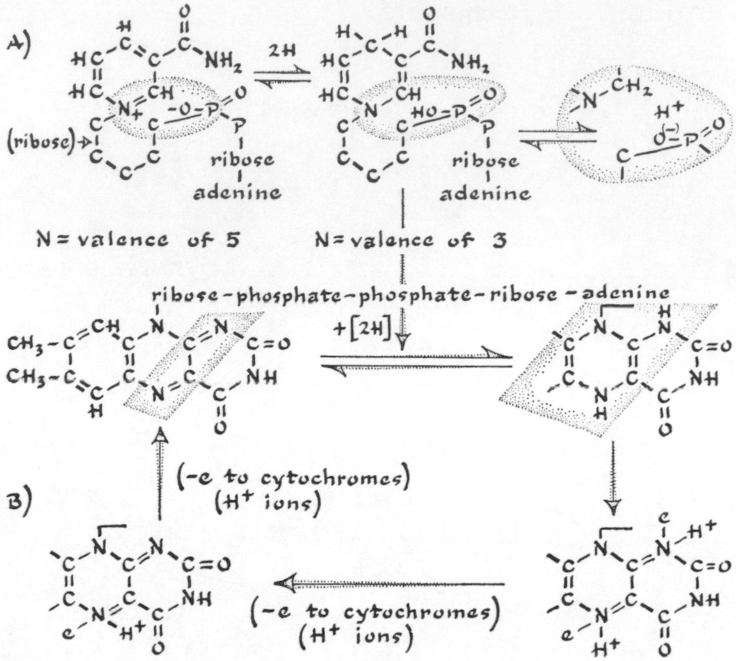

FIGURE 6-2. *Diagram representing the mechanism of hydrogen transport of the nicotinamide coenzyes (A) and the conversion of 2H to H^+ and electrons by the flavoproteins (B). From AN INTRODUCTION TO BACTERIAL PHYSIOLOGY, Second Edition, by Evelyn L. Oginsky and Wayne W. Umbreit. W. H. Freeman and Company. Copyright (c) 1959.*

carrying coenzyme A. The nature of this substance and its synthesis is shown in Figure 6-3. This coenzyme is concerned with a wide variety of reactions in which it serves the function of a repository and carrier of a wide variety of organic acids. Many of the components that go into its synthesis can serve as growth factors for microorganisms. For example, *Acetobacter suboxidans* requires pantoic acid, *Corynebacterium diphtheriae* requires β-alanine, *L. arabinosus* requires pantothenic acid, and *L. bulgaricus* requires pantotheine, each being able to carry out the reaction sequence from the point of entry onward.

Niacin (Figure 6-2) exists in the form of two hydrogen-carrying coenzymes, NAD and NADP. The nitrogen in the nicotinamide ring carries hydrogen, again because it can alternate between a valence of 3 or 5. If one chemically reduces NAD to NAD—D_2 (the D_2 standing for deuterium) and then allows it to be enzymatically oxidized (e.g., with alcohol dehydrogenase and acetaldehyde) only one half the deuterium is released. If the NAD is reduced by lactic dehydrogenase with deuterium-containing lactate, and then

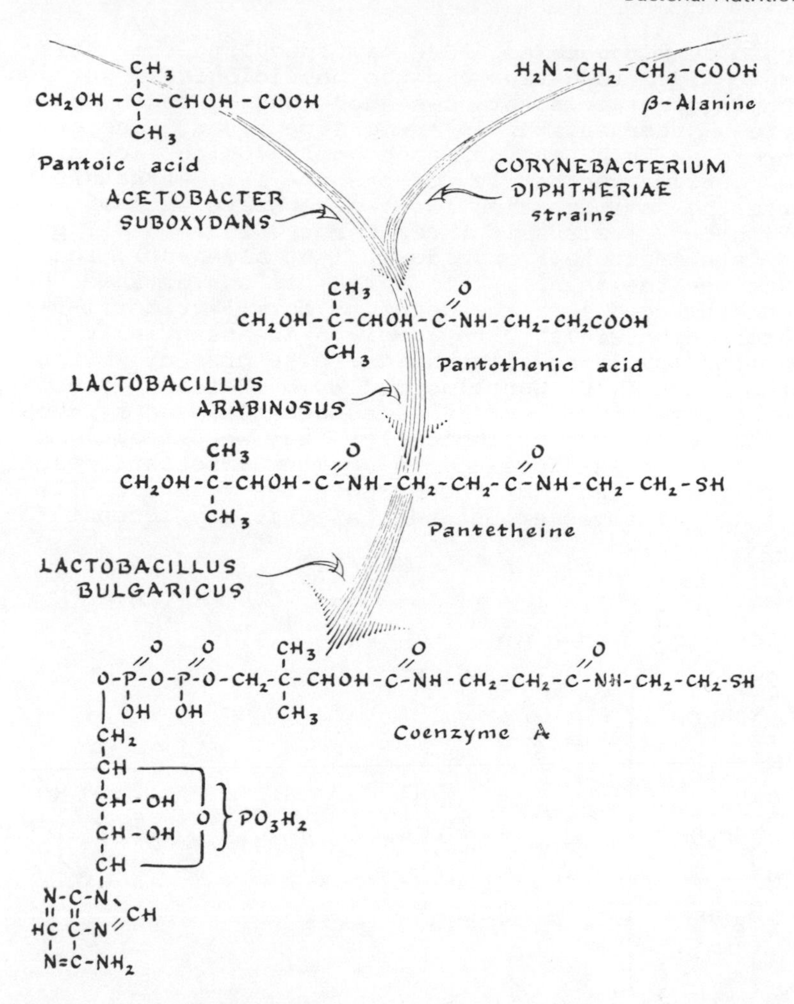

FIGURE 6-3. *Pantothenic acid and coenzyme A. Each organism can carry out the synthesis if provided the precursor noted. From AN INTRODUCTION TO BACTERIAL PHYSIOLOGY, Second Edition, by Evelyn L. Oginsky and Wayne W. Umbreit. W. H. Freeman and Company. Copyright (c) 1959.*

the NAD—D_2 is oxidized by alcohol dehydrogenase and acetaldehyde, all the deuterium is released. That is, there are two sides to the NAD molecule; the enzymes act on one side, but chemical reduction reduces both sides.

Certain enzymes show a similar stereospecificity with respect to NAD (and to NADP). If side A is the side reduced by alcohol dehydrogenase, this same side is reduced by D- or L-lactate, D-glycerate, and L-malate acted on by their respective dehydrogenases. The other side, the "side B," is reduced by α-glycerophosphate dehydrogenase, triose phosphate, D-glucose, L-glutamate, and dihydrolipoic dehydrogenases. One may speculate that this single-sided reduction has the physiological function of grouping certain substances about one hydrogen "pool," whereas other substances are grouped about, and exchange hydrogen within, another such pool. Using each side of the coenzyme separately permits the same structure to serve the same function for two separate sets of compounds.

Vitamin B_6 exists in three forms (Figure 6-4): pyridoxine (the 4-alcohol), pyridoxal (the aldehyde), and pyridoxamine (the amine). All three are convertible into each other and into the coenzyme form, pyridoxal-5-phosphate, which seems to be involved in essentially any reaction involving amino acids. The present writer, together with I. C. Gunsalus and W. D. Bellamy, discovered the coenzyme form over 30 years ago (2), and synthesized it chemically and biologically. But we certainly had no idea that it was involved in so many important reactions. It is postulated that the amino acids form a Schiff base with the aldehyde group and that this may account for the activity (3).

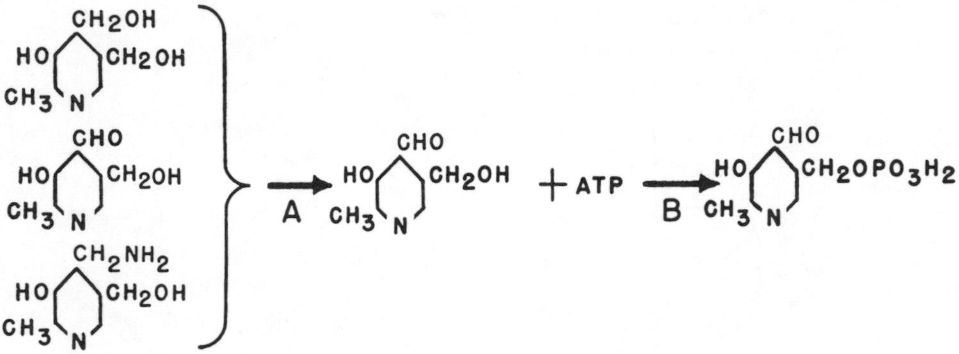

FIGURE 6-4. *The vitamin* B_6 *group. The three substances, pyridoxine, pyridoxal, and pyridoxamine, are converted (A) to pyridoxal which is phosphorylated (B) to the coenzyme form, pyridoxal-5-phosphate.*

Biotin (Figure 6-5) is so widely distributed in nature that a biotin-deficient diet, even a deficient microbiological medium, is very difficult to prepare. It is involved in fixation reactions of carbon dioxide, which is presumably added to the molecule as indicated in Figure 6-5. One might take biotin as an example of the different kind of approach taken by the biochemist and the physiologist. Early studies on biotin were confused by the fact that biotin has some effect on permeability. Indeed, it could be shown that in biotin-deficient cells the fatty acids were practically depleted from the lipopolysaccharide layer (4), and the composition of the lipid material was altered significantly, not only in location, but also in function. These kinds of results are not explained by the known biochemical reactions of biotin, and we still need to search for a physiological explanation.

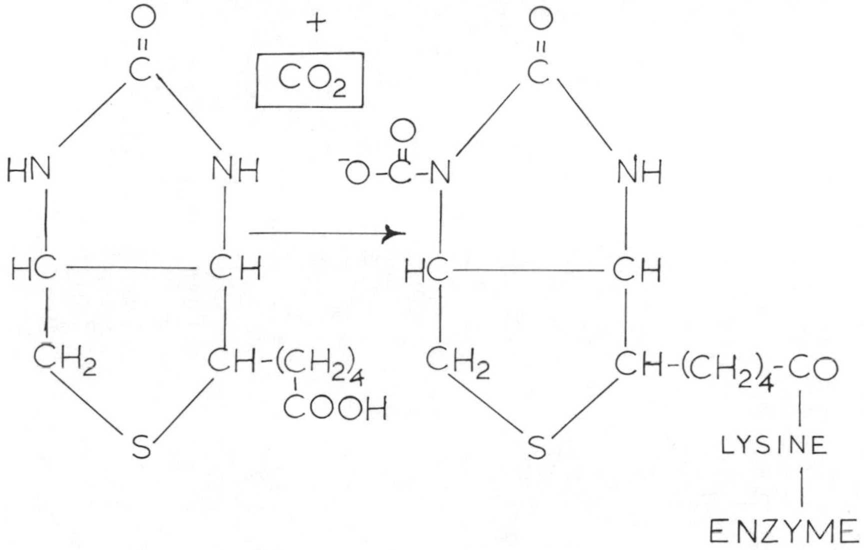

FIGURE 6-5. *Biotin. This substance appears to be attached to its enzyme proteins by way of a lysine bridge and under these circumstances is capable of adding carbon dioxide.*

Folic acid (Figure 6-6) is composed of pteridine, para-aminobenzoate, and glutamic; the coenzyme form is usually the tetrahydrofolate with four glutamics attached. As may be seen in Figure 6-6, it serves to oxidize or reduce one-carbon groups, and these attach to the N^5 or N^{10} or both. Of course, PABA (para-aminobenzoic acid) is usable as a growth factor for some microorganisms that require it to synthesize folic acid. One may also recall that it is in this path of synthesis of folate from PABA that sulfonamides act.

Lipoic acid (or *thioctic acid*) picks up the active aldehyde from the thiamine associated decarboxylating enzyme to oxidize it and transfer the acyl group to CoA; the hydrogen is removed to FAD, as illustrated for pyruvate in Figure 6-7. It thus serves as an acyl-group carrier and a hydrogen carrier. It is attached to the enzyme proteins via lysine and thus forms a very firm attachment to the enzymes.

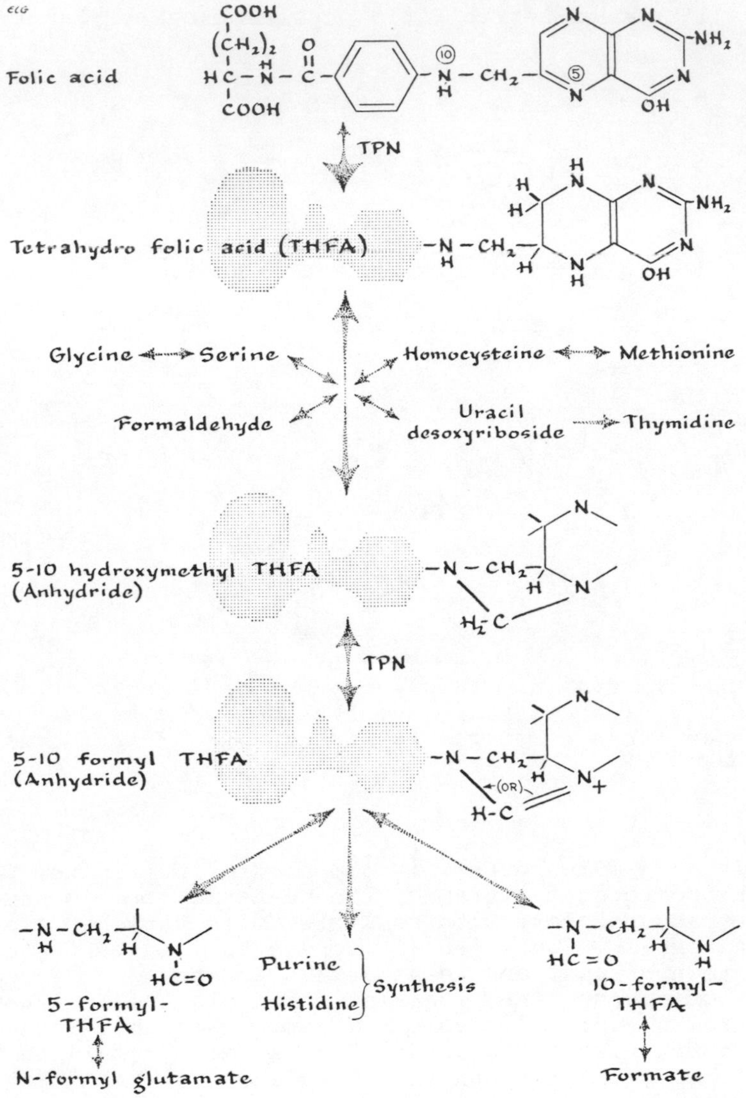

Folic acid

TPN

Tetrahydro folic acid (THFA)

Glycine ⟷ Serine

Homocysteine ⟷ Methionine

Formaldehyde

Uracil desoxyriboside → Thymidine

5-10 hydroxymethyl THFA (Anhydride)

TPN

5-10 formyl THFA (Anhydride)

5-formyl-THFA

N-formyl glutamate

Purine, Histidine — Synthesis

10-formyl-THFA

Formate

FIGURE 6-6. *The folic acid family. The coenzyme form is tetrahydrofolic acid with one or more glutamics attached. It plays a role in the oxido-reduction of one-carbon intermediates and in their exchange. From AN INTRODUCTION TO BACTERIAL PHYSIOLOGY, Second Edition, by Evelyn L. Oginsky and Wayne W. Umbreit. W. H. Freeman and Company. Copyright (c) 1959.*

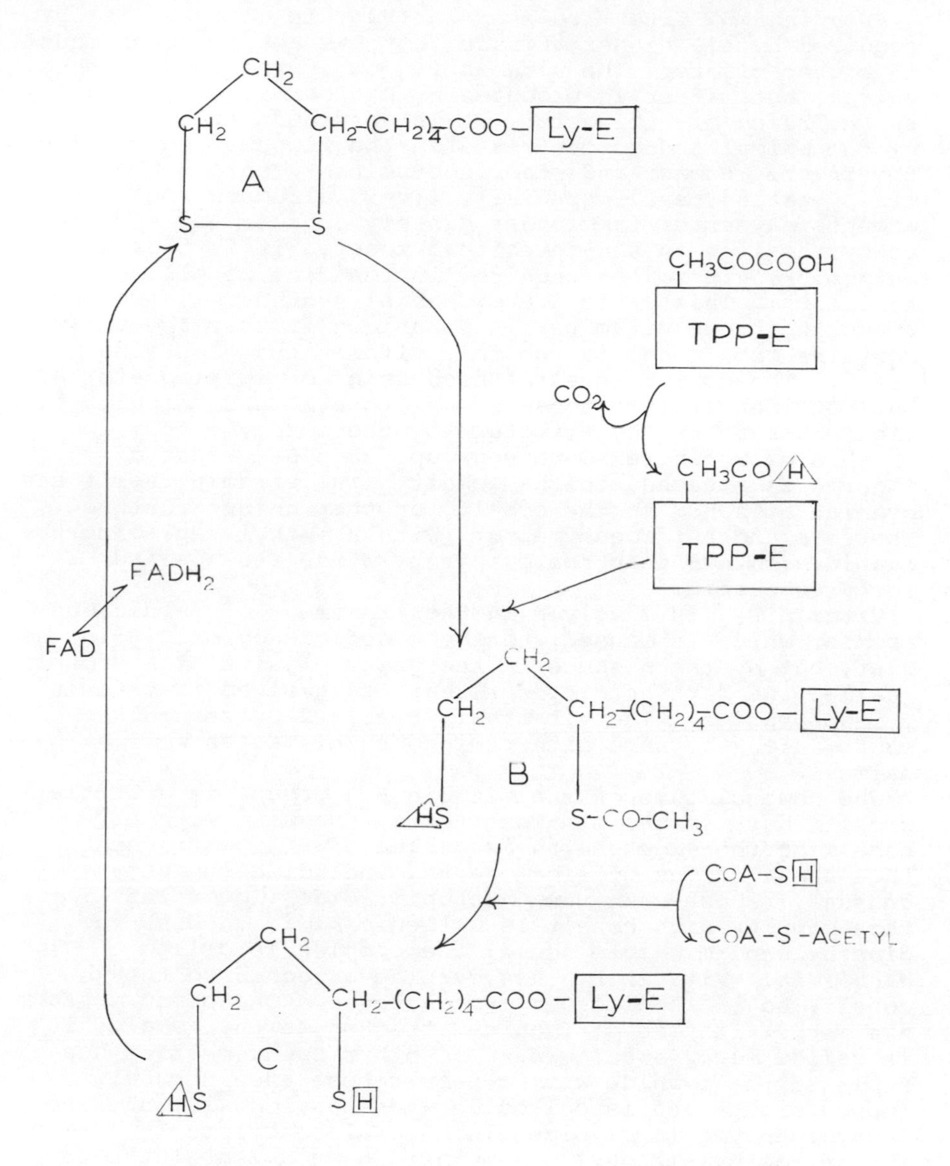

FIGURE 6-7. *Lipoic acid and its activity. Substance A is lipoic acid (also called thioctic acid, 6,8 dithio octanoic acid) which is attached it its enzyme protein by way of lysine (Ly-E). Pyruvate reacts with an enzyme containing thiamine pyrophosphate (TPP-E), loses carbon dioxide, and the "acetaldehyde" (enzyme bound) produced by this reaction is transferred to the lipoic to form substance B (6 acetyl dihydro lipoic), which transfers the acetyl group from the 6 position to coenzyme A, forming acetyl-CoA and substance C (dihydro lipoic) which is then oxidized, via FAD to form lipoic.*

Vitamin B_{12}, first isolated in 1948, is unique. It is required by all higher animals, but has not yet been found in higher plants. The vitamin is not synthesized by the animal, but rather is produced by microorganisms, usually in the rumen of ruminants. It is absorbed from the rumen by the animal and concentrated in the tissues, especially the liver. Humans and other nonruminants obtain the vitamin by eating meat, especially liver, although small amounts may enter from other dietary sources and some may be synthesized in the intestinal tract. It is thus an *animal protein factor* required in the diet of all nonruminants. This exclusive microbial synthesis is unique among the known vitamins. Also unique is that the vitamin contains cobalt and is the only vitamin that contains a metal. The cobalt is surrounded by a corrin ring similar but not identical to those of chlorophyll and hemoglobin. Figure 6-8 shows the structure of the coenzyme form in which a deoxyribose adenine group (at upper right of figure) is attached to the cobalt. The vitamin itself has cyanide attached to the cobalt, another unique feature. There is another coenzyme form with a methyl replacing the cyanide (rather than the deoxyribose adenine) called methylcobalamin.

Vitamin B_{12} is involved in the treatment of pernicious anemia, which is caused, not by a deficiency of B_{12} in the diet, but by the absence in the gastric juice of a protein (called the *intrinsic factor*) that is involved in vitamin B_{12} absorption. The vitamin is required by several bacteria and by some higher microorganisms, as well as by animals.

The nomenclature of the vitamin B_{12} group is a little complex but, in the simplest terms, the ring without cobalt or the side chains is called *corrin*. Attached to this are a series of side chains, including the very unusual, ribose—phosphoryl—propionamide (lower left); this complex with cobalt is called *cobamide*. With the dimethylbenzimidazole added, the complex is called *cobalamin*. Vitamin B_{12} has cyanide attached to the cobalt, so it is called *cyanocobalamin*. One coenzyme form has methyl instead of cyanide attached to the cobalt; it is called *methylcobalamin*. The other coenzyme form has replaced the cyanide with a deoxyribose adenine chain (upper right) and is called deoxyadenosylcobalamin. The deoxy coenzyme is involved in several reactions (glutamic mutase, methyl-malonyl-CoA mutase, α-methyleneglutarate mutase, diol-dehydrase glycerol dehydrase, ethanolamine deaminase, amino mutases, and ribonucleotide reductase), only two of which we shall examine. One of these, glutamic mutase (Figure 6-9) is so unusual as to require comment. Note that, in the conversion of glutamic acid to β-methyl aspartic acid, carbon 3, right in the center of the glutamic chain, is now located in the side chain, and there is a new link between carbons 2 and 4. Part B shows how this might be accomplished. Many of the reactions listed above involve similar *group transfers*. The other reaction (not shown), ribonucleotide reductase,

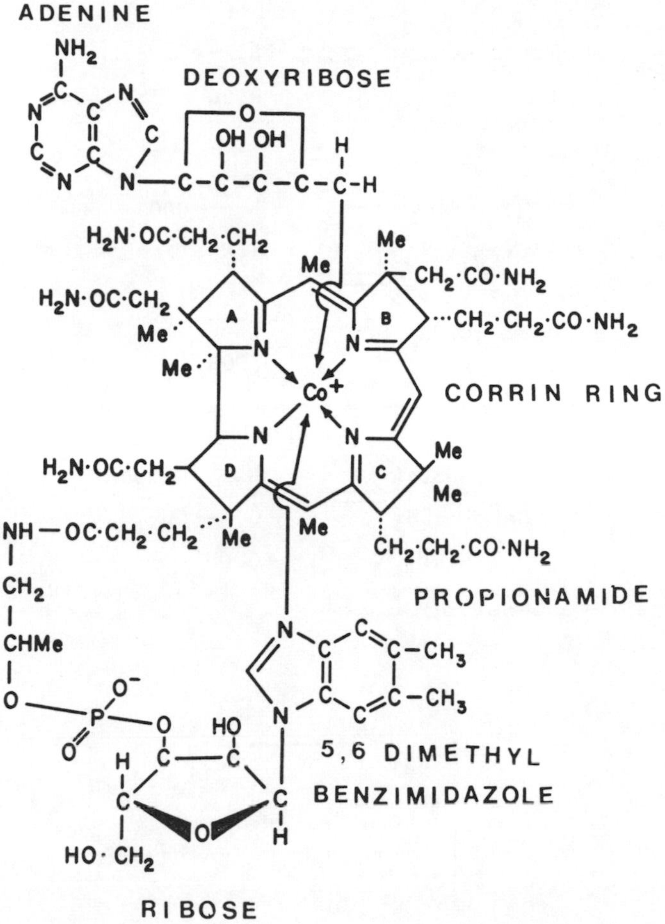

FIGURE 6-8. *The coenzyme form of vitamin B_{12}.*

involves the conversion of ribose nucleotide triphosphates (ATP, GTP, etc.) into the corresponding deoxyribose triphosphates (dATP, dGTP, etc.). The natural reductant is reduced thioredoxin, but other thiols will act, such as dihydrolipoic. This reaction serves as the primary source of the deoxynucleotides and thus of DNA.

The other coenzyme form, methylcobalamin, is involved in the methylation of homocysteine to form methionine, in methane formation (originates from carbon dioxide), and in acetate formation from carbon dioxide.

One may ask, are there any more? It is difficult to predict the future, but considering the amount of work that has been done on microbial nutrition it seems

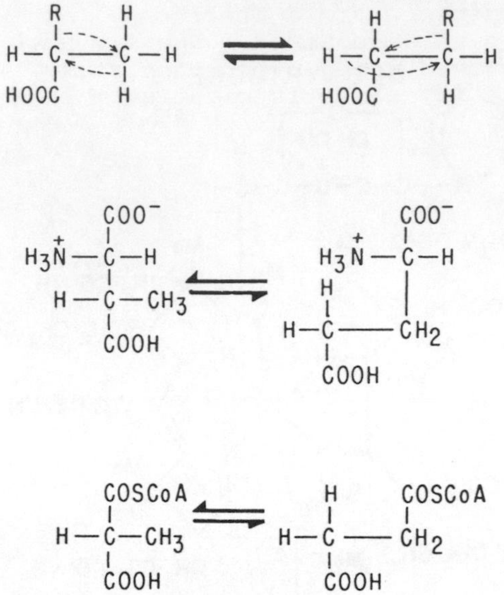

R H
H—C—C—H ⇌ H—C—C—H
HOOC H
H R
HOOC H

COO⁻
$H_3\overset{+}{N}$—C—H
H—C—CH_3 ⇌
COOH

COO⁻
$H_3\overset{+}{N}$—C—H
H
H—C—CH_2
COOH

COSCoA
H—C—CH_3 ⇌
COOH

H COSCoA
H—C—CH_2
COOH

Schematic representation (*top*) of the glutamate mutase and methylmalonyl CoA mutase reactions (*shown below*).

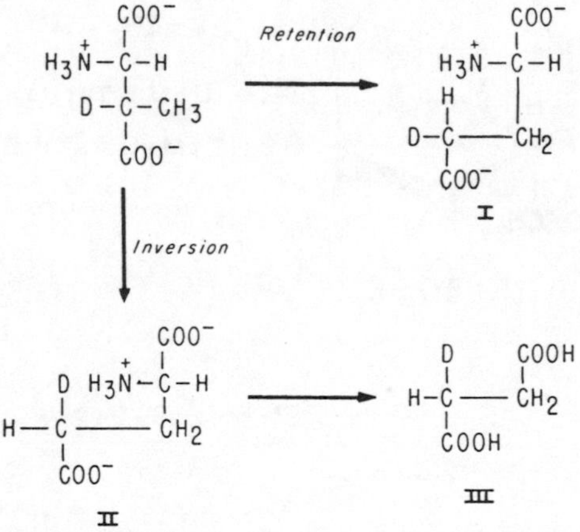

COO⁻
$H_3\overset{+}{N}$—C—H
D—C—CH_3
COO⁻

Retention

COO⁻
$H_3\overset{+}{N}$—C—H
H
D—C—CH_2
COO⁻
I

Inversion

COO⁻
D $H_3\overset{+}{N}$—C—H
H—C—CH_2
COO⁻
II

D COOH
H—C—CH_2
COOH
III

Possible stereochemical course of the glutamate mutase rearrangement.

FIGURE 6-9. *Reactions catalyzed by vitamin B_{12} containing enzymes. From Sprecher, M., and D. B. Sprinson, Ann. N. Y. Acad. Sci. 112:655–660 (1964). Used with permission.*

unlikely that there are many "unknown" factors. However, occasional ones are still found, usually for organisms with some specialized growth condition. For example, *Methanobacterium ruminantium*, a methane-producing bacterium, was recently shown to require 2-mercaptoethane sulfonic acid (5). While unusual and unstudied organisms may still turn up with other requirements, this seems unlikely for the better-known organisms; the vitamin requirements are known, and there are a few if any unknown nutritional factors yet to be discovered. Usually such a statement is immediately proved wrong by further study, which is a good reason to make it, since it might stimulate further discovery in this field.

REFERENCES

Reviews

Barker, H. A. 1972. Corrinoid-dependent enzymic reactions. *Ann. Rev. Biochem.* *41*:55—90.

Brown, C. S., D. S. Macdonald-Brown, and J. L. Meers. 1974. Physioloigcal aspects of microbial inorganic nitrogen metabolism. *Advan. Microbial. Physiol.* *11*: 1—52.

Demain, A. L. 1972. Riboflavin oversynthesis. *Ann. Rev. Microbiol* *26*:369—388.

Forrest, H. S., and C. Van Baalen. 1970. Microbiology of unconjugated pteridines. *Ann. Rev. Microbiol.* *24*: 91—108.

Friedmann, H. C., and L. M. Cagen. 1970. Microbial biosynthesis of B_{12}-like compounds. *Ann. Rev. Microbiol.* *24*:159—208.

Gibson, F., and J. Pittard. 1968. Pathways of biosynthesis of aromatic amino acids and vitamins and their control in microorganisms. *Bacteriol. Reviews* *32*:465—492.

Hutner, S. H. 1973. Inorganic nutrition. *Ann. Rev. Microbiol.* *26*:313—346.

Jensen, S. L. 1965. Biosynthesis and function of carotenoid pigments in microorganisms. *Ann. Rev. Microbiol.* *19*:163—182.

Knappe, K. 1970. Mechanism of biotin action. *Ann. Rev. Biochem.* *39*:757—776.

Koser, S. A. 1968. *Vitamin Requirements of Bacteria and Yeasts.* Charles C Thomas, Springfield, Ill.

Plaut, G. W. E., C. M. Smith, and W. L. Alworth. 1974. Biosynthesis of water-soluble vitamins. *Ann. Rev. Biochem.* *43*:899—922.

Snow, G. A. 1970. Mycobactins: iron-chelating growth factors from Mycobacteria. *Bacteriol. Reviews* *34*:99—125.

Papers

1. *Can. J. Microbiol. 18:*1907 (1973).
2. *J. Biol. Chem. 155:*685 (1944).
3. *Nature 235:*201 (1972).
4. *J. Bacteriol. 89:*437 (1965).
5. *J. Bacteriol. 120:*974 (1974).

7 Growth—The Recurring Miracle

When one studies bacterial growth, at least three aspects must be considered:

1. The growth and division of an individual cell.
2. The problem of population growth in the usual cultural conditions, that is, where concentrations of nutrient and end products are continually changing as nutrients are used up and waste products accumulate.
3. Population growth under conditions of continual replenishment of nutrient and continual removal of wastes (chemostat or turbidostat conditions).

These are really quite different aspects of the problem of growth; they have different problems and require different approaches; and they yield somewhat different information. It is not that one is a better approach than another; rather, each asks a somewhat different question and provides a somewhat different answer. They should be and they will be treated separately.

GROWTH AND DIVISION OF THE INDIVIDUAL CELL

Consider first the problem faced by the bacterial cell. To grow it must make new material, but it is enclosed in a rigid and unyielding cell wall. If this "corset" weakens, the cell may swell and burst. The cell must find a way to expand the cell wall while still retaining its rigidity and mechanical strength. It must also make new membrane for the new cell to be formed, new DNA, new ribosomes, and new proteins; in short, all the constituents for the new cell.

These cell constituents must be prepared and properly located somewhat synchronously. If a cell divides before it has enough cell wall, it may be unable to survive. If it weakens its muramic layer, it may emerge as a protoplast and burst. Indeed, early microbiological media had salt added to them, in amounts from 0.5 to 1.5 percent; the notion was prevalent that this permitted better growth from low inoculum. We still use about 0.5 percent NaCl in phage work. Then it was found that the presence of the salt tended to increase the number of peculiar forms found early in the growth; that is, during the lag phase it was thought that salt was toxic, and it was gradually reduced or eliminated. Actually, what was probably happening was that in the lag phase some of the cells were dividing before they had formed enough cell wall. The salt tended to assist the protoplast and thus would promote their growth from small inocula. Thus the aberrant forms were not really the result of toxicity but protoplasts and spheroplasts that were surviving.

We consider first how a bacterial cell is able to expand its cell wall. The newly synthesized acetyl—muramic—peptide must first be incorporated into the cell wall. This is done, as shown in Figure 7-1, by a transglycosylation to increase the length of the muramic—glucosamine chain (Part A, Figure 7-1), and then, by way of transpeptidation, to cross link the peptide (Part B, Figure 7-1). Both reactions occur by attachment to preexisting peptidoglycans.

There is also a possibility of insertion of newly synthesized material into a preexisting cross linked muramic complex. One may conceive of the murein layer as consisting of several three-layered units. One layer is a chain of repeating acetyl glucosamine—acetyl muramic acid chains (Figure 7-2). From the muramic acid there extends downward (or upward) a characteristic peptide, and these peptides are cross linked. Thus there is a second layer of cross linked peptides. Then is the third layer composed of chains of acetyl glucosamine—acetyl muramic acid repeating units attached to the now cross linked peptides. What can happen is the following. First, a link in the first murein chain can break leaving layers 2 and 3 still intact and thus capable of holding the cell. This break is accomplished by lysozyme-like enzymes located at the membrane, presumably held in some inactive form until cell wall expansion is initiated. A new acetyl glucosamine—acetyl muramic acid unit is then inserted and cemented in. The two new units are then cross linked and one has produced a new cell wall one murein unit longer; but at all times at least two of the three layers were intact and mechanically able to maintain the cell from bursting by osmotic forces.

Cell division itself begins by invagination of the membrane and the cell wall, frequently at the position of a mesosome (Figure 7-3). These penetrations are gradually built up from the outer wall and meet in the center (1). The cell is now separated into two cells, but they do not necessarily break apart, and chains of cells are common. New cell wall synthesis preliminary to division can be followed by immunofluorescent techniques. Two types of cell wall growth are found. In one, new cell wall synthesis occurs at one end of the cell only and elongation occurs from that end, a process somewhat analogous to budding. This type of growth accounts for instances in

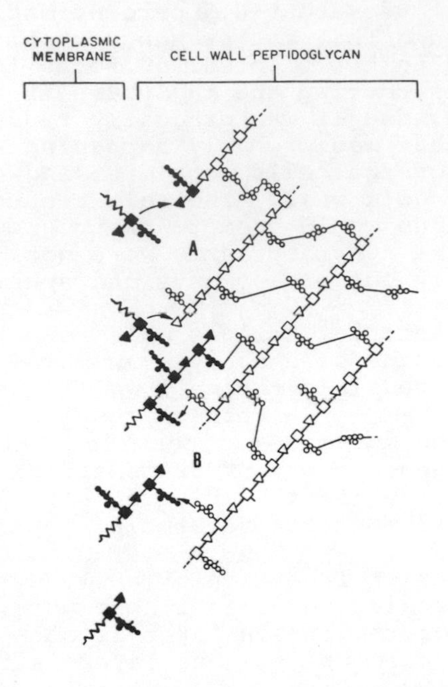

FIGURE 7-1. *Addition of muramic peptides to the cell wall. Part A illustrates transglycosylation; Part B, transpeptidation.*

The symbols are:
- Δ *Acetyl glucosamine,*
- *o individual amino acids,*
- □ *acetyl muramic.*

Newly synthesized material is black.

From Mirelman, D., R. Bracha, and N. Sharon, Ann. N. Y. Acad. Sci. 235:326—347 (1974). Used with permission.

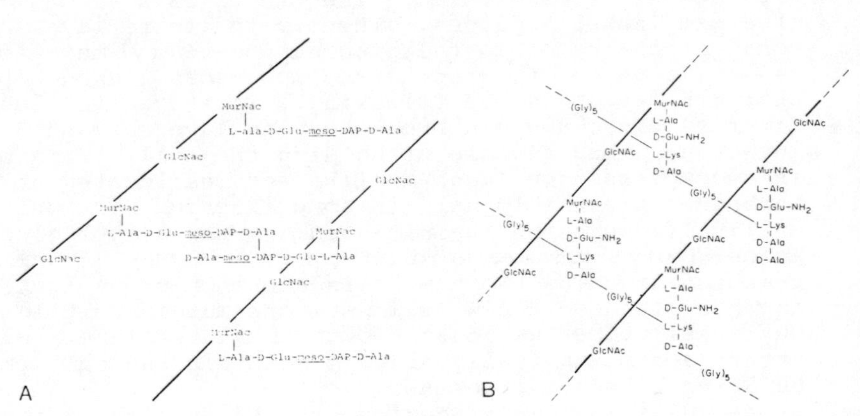

FIGURE 7-2. *Structure of peptidoglycan showing the linear acetyl—muramic—acetyl glucosamine strands together with peptide cross links. A is the peptidoglycan of Escherichia coli; B that of Staphylococcus aureus. From Blumberg, P. M., and J. L. Strominger, Bacterial Reviews 38:291—335 (1974). Used with permission.*

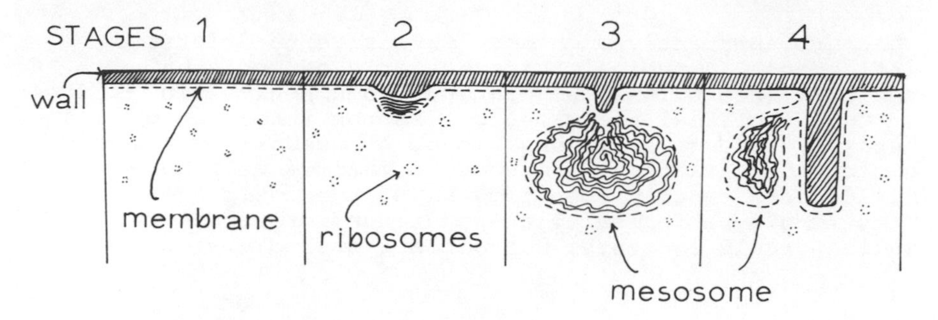

FIGURE 7-3. *Invagination of cell wall and cell membrane showing development of mesosomes. Modified from Rogers, H. J., Bacteriol. Reviews 34:194–214 (1970). Used with permission.*

which one of a pair of recently divided cells possesses flagella, whereas the other newly formed portion has not yet had time to develop them. In this case one cell has a cell wall composed primarily of older material, while the other cell contains the newly synthesized cell wall. The other growth mechanism involves elongation in both directions by the insertion of a new wall along the length of the cell, frequently starting in the middle, but also at intervals along the length. As such, each of the two daughter cells has roughly one half of its cell wall as older material and the other half as new material. Each mechanism seems to be characteristic of certain strains, and both types can occur within a single species. For example, *Escherichia coli* B/r (a radiation-resistant mutant of *E. coli* B) elongates at one end only, whereas its parent *E. coli* B elongates at both ends (as does *E. coli* K, including radiation-resistant mutants derived from it). Under rare circumstances in *E. coli*, but with somewhat greater frequency in several gram-positive organisms, the organism may continue to grow even if the cell wall is not entirely formed, and structures resembling "germination tubes" (i.e., club-like, deformed cells) may be seen. If there is overt cell wall damage, as with penicillin treatment, these may be increased considerably. The process of cell growth and division may be separated in time, as has been done in Figure 7-4.

Of course, initiation of invagination is not always followed by completion, and also completion of invagination is not always followed by cell separation. It is well known that these intricate processes of cell division are more sensitive to toxic influences than protein or DNA synthesis, or even membrane synthesis itself.

It is the impetus for and the process of cell division

that are apparently much more sensitive to deleterious influences. A typical characteristic of sublethal toxicity (too high a temperature, some penicillin, lack of Mg^2, sublethal phenol, etc.) is the formation of "snakes," enormously elongated cells, usually with "multinucleate" character. These may sometimes be induced to divide very rapidly upon removal of the toxic agent. These could be formed by the continued growth of the cell wall and cell contents, but without cell division.

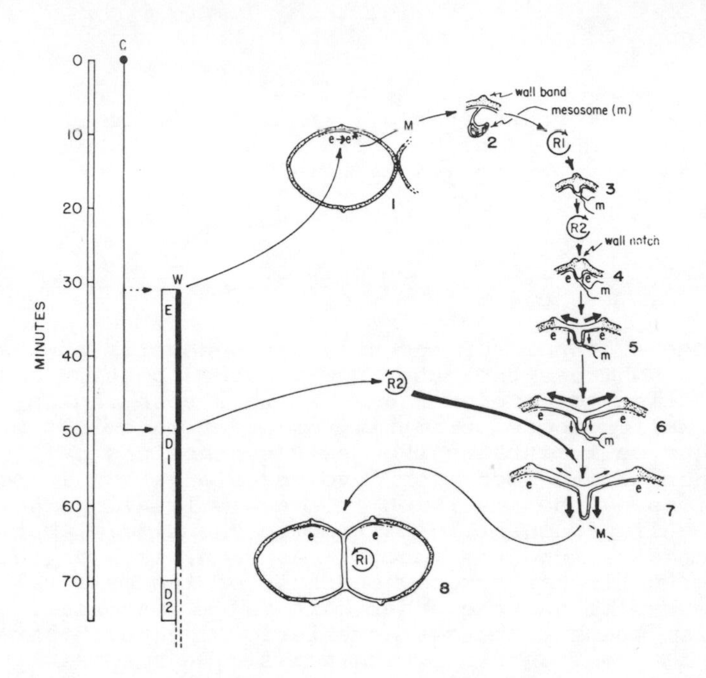

FIGURE 7-4. *Hypothesis for cell wall growth in Streptococcus faecalis. The entire cycle from division to division is about 75 min. During the first 31 min the chromosome is actively replicating but there is no cell wall enlargement. Enlargement (W) requires 19 min and is shown in steps 1 through 5. When the chromosome has finished replication it signals the beginning of the division cycle.* D_1 *(18-20 min) (steps 6,7), requires protein synthesis, but* D_2 *(about 5 min) (step 8) occurs in the absence of protein synthesis. From Shockman, G.D., L. Daneo-Moore, and M. L. Higgins. Ann. N. Y. Acad. Sci. 235:161—196 (1974). Used with permission of G. D. Shockman, L. Daneo-Moore, and M. L. Higgins.*

Among the gram-positives, nonspore forming mutants of *Bacillus cereus* (2) form septa, but there is no separation of cells. The septa occur at the usual cell length (3), but the zone between them never develops. Such long chains of cells can be disintegrated into individual cells by treatment with lysozyme, which suggests that even cell separation is enzymatically controlled. In *Clostridium acidiurici* (4) growth at 43°C results in long filaments that have nuclear material at evenly spaced intervals, are gram-negative, and contain muramic acid that has only one half its usual content of D-alanine. The cells are septate, but it appears to be only via membrane partitions; the muramic layer does not seem to be laid down between the membrane partitions. In lactobacilli grown deficient in vitamin B_{12}, filamentous forms are produced, but addition of this vitamin to the filamentous cells permits normal division after an interval (5). This is not to suggest that vitamin B_{12} is specifically concerned with cell division, but rather than when some function of the cell is slowed down, division cannot occur.

Certain mutants of *E. coli* K (6,7), after exposure to ultraviolet light, produce long nonseptate filaments that grow from both ends. For several hours protein, DNA, and RNA synthesis appear to continue normally, and division of the nuclear material occurs repeatedly during growth; that is, the filaments have many "nuclei" more or less uniformly throughout the length. The filaments reach 50 to 200 normal cell lengths before growth stops and lysis occurs. If such growing filaments are exposed to 42°C, septum formation is initiated at one or both ends. These short cells can grow and produce normal colonies of individual cells, which suggests that the location of the invagination points and the location of the DNA must be prefixed by some means, and that division does not occur unless DNA is on both sides of the division point. When *E. coli* is kept just above its minimum growth temperature, one gets long filaments with incomplete septum formation, yet the optical density continues to increase, and DNA, RNA, protein, and mucopeptide are still synthesized (8). These observations are cited not because they give a clear picture of what is happening, but because they give the impression that protein, DNA, RNA, mucopeptide, and membrane synthesis can continue to proceed independently; yet the coordination between them that results in invagination and the laying down of new membrane and new cell wall can be disrupted, leading to unusual forms.

There have been two general hypotheses regarding the division impetus for the individual cell. In one, it is supposed that during the log phase of growth the synthesis of all cellular components is continuous and exponential. When a cell reaches a given size it divides, presumably because the increased diffusion difficulty (within the cell itself) inherent in a larger size tends to inhibit certain internal reactions and favor certain external reactions (such as cell wall synthesis); thus more cell wall is formed and "pinches off" the cell so that there are now two cells. By the terms of this hypothesis, in a given medium the cell size at division should be constant; but the cell size could well vary from medium to medium, since the concentration of an intermediate at any point within the cell would depend not only on the location in the cell, its transport into the cell, and the rate at which it was used, but also upon the concentration of the intermediate or its precursor in the medium, since the substrate would need to diffuse to the permease. Presumably, in a richer medium one would expect to have a shorter generation time and a larger volume of cell at division, which is indeed found experimentally. For example, in the log phase of growth, *B. megaterium*, growing on a glycerol—mineral salts medium, had a generation time of 44 min and an average cell volume of 6.89 μm^3 (measured in a Coulter counter); in Penassay broth the generation time was 23 min and the cell volume, 11.09 μm^3.

The second hypothesis supposes that there must be a sequence of molecular events before division can occur, and that this sequence must be ordered. This sequence hypothesis is based upon studies on variation in generation time. In a given medium, if the process of A to B to C were dependent upon diffusion inside the cell, one would expect little variation in the diffusion rates, the cell size at division should be constant, and the time it took from one division to another should be constant. But experimentally these are not constant. In a chemostate, where nutrient is limiting, the variation of size at division may be relatively small, but in a turbidostat or in a normal culture in the log phase, the actual size of the cell at division, and the generation time for each cell, may vary widely. It has been supposed that the variation of the generation time is related to the number of synthetic steps that a cell has to go through to build its cell substance, and one of the earliest microbial physiologists, Otto Rahn, devised a formula for relating the coefficient of variation in generation time to the number of synthetic steps involved. It turns out that the number of synthetic steps, calculated from the variation in generation time, is about equal to the number of operative genes in a cell, about which, of course, no information at all was available in Rahn's day. The variation in generation time is greatest in rich media and less variation is observed in simpler media, which indicates that there are more synthetic steps required in the simple medium.

However, there are also other hypotheses that may be related to the two described above or that may be applicable in certain cases only. One is that a particular substance is necessary before cell division can occur and that conditions may be found where this substance may be absent or limiting. Knowledge of these materials arose during studies attempting to explain the lag phase as dependent upon inoculum size; that is, the lower the inoculum, the longer the lag. If sterile media from cultures on which the organism was in the log phase was added, lag was shortened, which suggests an external substance necessary for rapid cell division. Such a substance was named *schizokinin*, and proved to be iron-chelating hydroxamic acid (9,10). Other organisms seem to respond to other materials (11,12). Another hypothesis is based upon observations of a "white body" close to the invagination process, in which the cell wall, sometimes preceded by a membrane, cuts across a cell. The use of a higher resolution electron microscopy has shown that this "white body" is membranous; in fact, we have earlier called it a *mesosome*. Still another hypothesis is that division is dependent on the action of cyclic AMP (13).

POPULATION GROWTH IN THE USUAL CULTURAL CONDITIONS

It is clear that we know little about the details of growth of single cells, but we mostly deal with populations of cells, that is, with cell cultures. We normally inoculate a medium and incubate the culture, allowing the organism to grow. In this kind of a self-limited situation we get a "normal" growth curve typically like that shown in Figure 7-5, in which we may distinguish four phases.

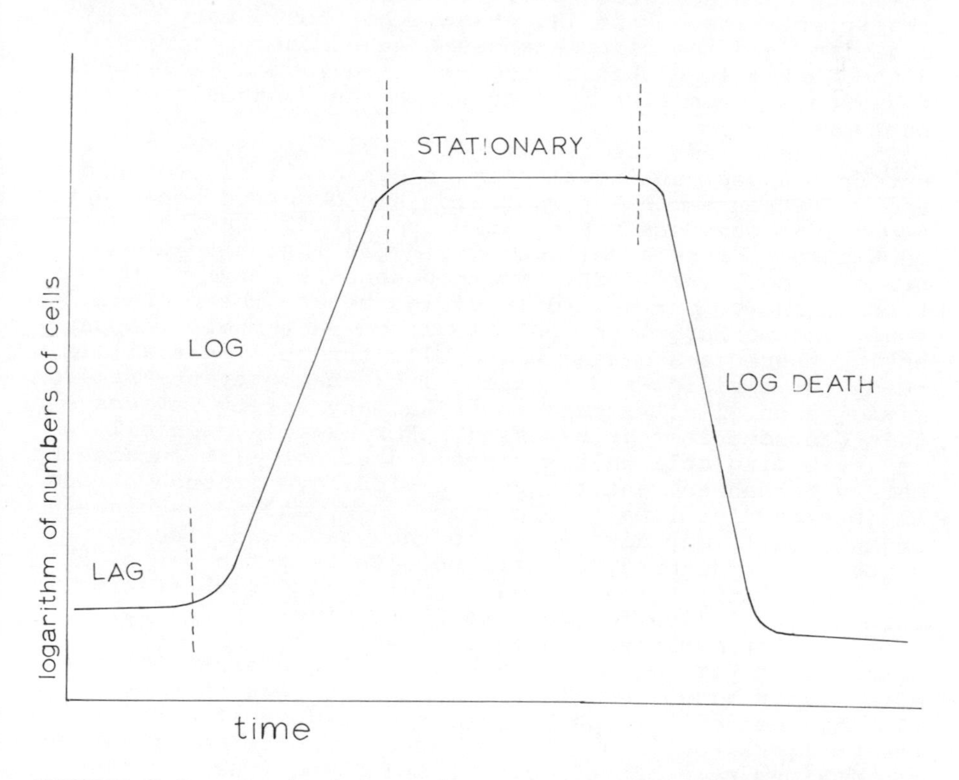

FIGURE 7-5. *Typical growth curve of a bacterial culture in a constant volume of medium. Cell numbers are per milliliter or other constant volume.*

The Lag Phase

If a reasonably low inoculum of cells from the stationary phase before it has reached the death phase are put into the same medium in which they were growing, the lack of multiplication during the early part of the lag phase need not be paralleled by lack of cell growth; that is, in the synthesis of new protoplasm. Under these circumstances, the lag is lag in cell division, not a lag in the synthesis of new protoplasm; the cells grow larger, but do not divide. They may become multinuclear, may contain more than one spheroplast, and normally show increased metabolic activity per cell without an increased metabolic activity per unit protoplast (i.e., measured by cellular DNA, or protein).

However, if cells from the stationary phase are placed into a different medium, or at a different temperature, or if they are taken after longer periods of the stationary phase (from "older cultures"), there may be a lag in the synthesis of protoplasm as well as a lag in division (reflected in decreased cell numbers). The nature and characteristics of the lag phase, therefore, vary with organisms and the circumstances of the observation; and although the lag, when it occurs, is always a lag in cell division, it can also be a delay in the synthesis of new protoplasm.

The reasons for the lag in cell division may be different from those causing the lag in synthesis of protoplasm, and a certain amount of confusion has occurred because the two situations were not clearly separated. But, as mentioned, before a cell can divide it must have a membrane, a cell wall, DNA, RNA (ribosomes), enzymes (proteins), and so on; should it divide before these are all ready and workable, the cell would be in trouble. Apparently, there is a signal that tells the cell that all is ready for division and a safety lock that prevents it from dividing until all is ready. Like many safety devices, this can sometimes be bypassed. For example, penicillin in inhibiting cell wall synthesis does not also engage the safety mechanism, and the cell divides even though it has an imperfect cell wall. In some cases, called *unbalanced growth*, cells may divide before they have made the necessary structures, usually because they run out of a given constituent in the medium required for this structure (e.g., lysine for cell wall, thymine for DNA, or inositol for membranes in yeast).

Earlier physiologists regarded the lag phase to be due essentially to three factors. The first was thought to be the time needed for the diffusion of new substrates into the cell, the build up of the necessary intermediates for the maximum rate of metabolism, and the diffusion of products from the old culture out of the cell. The second was thought to be the time required to form new ribosomes or to replace those which may have been damaged or inactivated in the stationary phase. The older term was "recovery from injury," but we have phrased the same concept in modern terms without implying injury in the sense of gross damage, merely the clogging up of the machinery of synthesis, as might well occur when growth is stopped in the stationary phase. The third factor was considered to be the time involved in adaptation to the new environment, which today we might consider as largely the time required for messenger RNA to be formed and for the enzymes whose formation it directs, to become effective contributors to the cell's metabolic machinery. Changing the medium would certainly involve a modification of the machinery. "Step-down" conditions (i.e., inoculation into a poorer medium) require first the synthesis of messenger RNA and then the formation of the enzymes necessary to make the substances formerly obtained from the richer medium. Even using an inoculum from the log phase, which normally shows no lag when transferred to the same medium, may show a lag in step-down conditions. "Step-up" conditions, while shutting off messenger formation, require the synthesis of more ribosomes before growth can occur at the rate characteristic of the new medium. Many of the contradictory viewpoints on the lag phase are, in fact, correct, but applicable to only particular conditions, and distinctly different "causes" may result in the delay in cell division characteristic of the lag phase.

To summarize, one may obtain a lag when the following occur:

1. The inoculum is taken from the stationary phase or beyond and transferred to the same medium.

2. The inoculum is taken from any phase and transferred to a richer medium (a step-up experiment, Figure 7-6). Here the "lag" is really a matter of semantics. The rate continues for a time at the rate characteristic of the old medium and then increases to the rate characteristic of the new medium. The rate of growth of cells is dependent on their rate of protein synthesis, and this in turn depends on the number of ribosomes. It takes time t before the ribosome content of each cell can be increased.

3. The inoculum is taken from any phase and transferred to a poorer medium. Growth stops (Figure 7-7) until the enzymes can be made that are capable of synthesis of the materials present in the rich medium that were now lacking in the poor medium. When these have been synthesized, growth occurs at the rate characteristic of the poor medium, not that of the rich medium.

The Log Phase

The log phase is a consequence of binary fission: 1—2—4—8—16—32, and so on; this is $2^0 = 1$, $2^1 = 2$, $2^2 = 4$, $2^3 = 8$, and so on. If we plot the exponent (the logarithm), we obtain a straight line. For any generation the number of bacteria, starting with *one* organism, will be 2^n, where n, is the *number* of generations. If we start with a cells and each reproduces in a constant

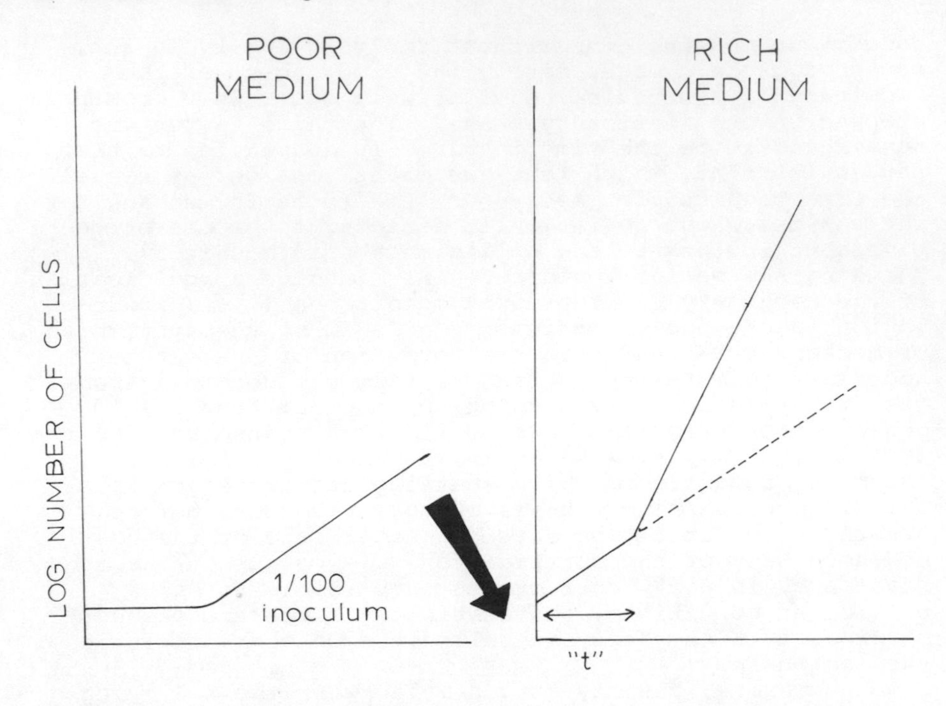

FIGURE 7-6. *"Lag" phase in "step-up" conditions. Note that growth continues at the "poor" medium rate and the lag is only the time it takes to grow faster.*

fashion, as they do in the log phase, the organisms at any generation will be $a \times 2^n$.

The generation time (or doubling time) is

$$\frac{\text{time elapsed}}{\text{number of generation}} = \frac{t}{n}$$

If the number of bacteria at n generations is B_n and the number of bacteria at the start is B_0

$$B_n = B_0(2^n)$$

$$\log B_n = \log B_0(2^n) = \log B_0 + n \log 2$$

$$n = \frac{\log B_n - \log B_0}{\log 2}$$

Hence

$$G \text{ (generation time)} = \frac{t \log 2}{\log B_n - \log B_0}$$

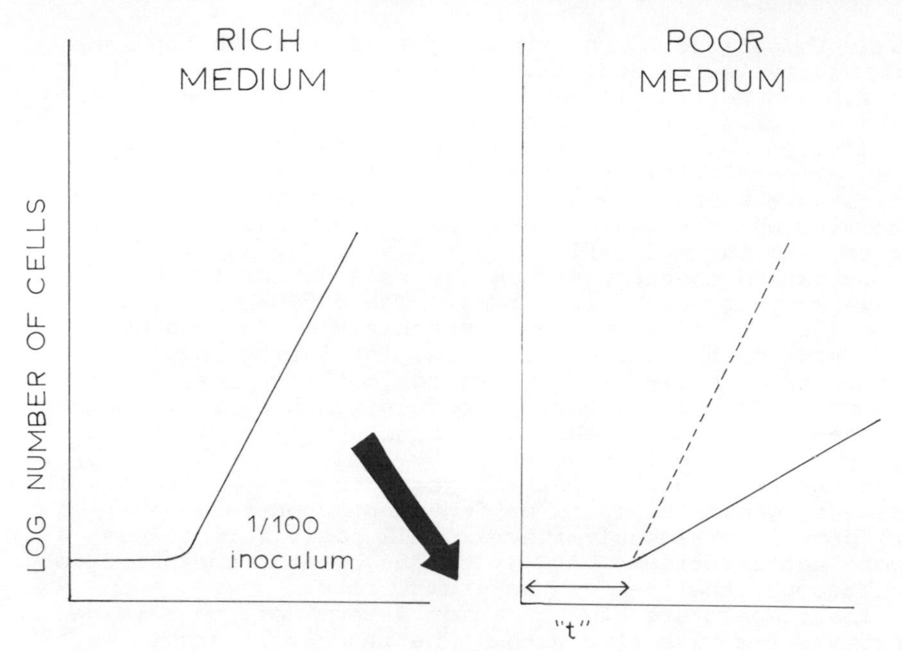

FIGURE 7-7. *A "step-down" condition. Note that growth stops entirely and then proceeds at a slower rate characteristic of the poorer medium.*

The following factors control the generation time:

1. The nature of the organism. Nobody, for example, can make the organism causing tuberculosis (*Mycobacterium tuberculosis*) grow as fast as optimum growth of the organism causing diphtheria (*Corynebacterium diphtheriae*), in the laboratory, at least.
2. The temperature. Within the possible limits, for each 10°C rise in temperature one can expect to double the rate of growth.
3. The rate of penetration of medium constituents. This is influenced by the concentration of materials in the medium and by stirring.

Since an increase in one log represents a ten-fold increase in cells (i.e., when one goes from a population of log 6 to log 7, one is going from 10^6 organisms to 10^7, which is a ten-fold increase), a doubling time is represented by the time taken to double (i.e., the log of 2, which is 0.301). Thus the time it takes to increase by 0.3 log is the doubling time or the generation time. This value may be determined very simply. One way, for example, is to have organisms growing in the log phase, and when they reach a given turbidity, say an optical density of 0.8 (assuming that the stationary phase would reach an optical density of 1 or more), the medium is diluted with an equal volume of the same medium at the same temperature. The time required to reach the same turbidity is the generation time.

The generation time during the log phase remains constant and is determined by the interplay between the inherited characteristics of the organism and the environmental conditions. *Escherichia coli*, under the most favorable conditions, exhibits a generation time of about 20 min, but the mycobacteria may require several hours, even with the most effective media yet devised. It is interesting to speculate on the reasons for the wide diversity in generation times. The differences are probably due to the rate of synthesis of new protoplasm, and not to the rate of division, for during the log phase increase in cellular nitrogen content is paralleled by increase in cell numbers. Speculation has arisen on the mechanisms by which the rate of growth is controlled. Do only a few reactions control the rate of synthesis, or is the entire metabolic pattern geared to a lower rate? The rate of growth seems to be dependent upon the number of ribosomes, presumably they are all active or at least a constant fraction is active. The greater the number of ribosomes, the greater the growth rate.

The temperature of incubation determines, to a large extent, the rate of growth. The process of growth is a chemical one, and the rate of chemical reactions in the biological range is doubled by a 10°C rise in temperature. Similarly, the rate of mesophilic bacterial growth between 20 and 40°C has a temperature coefficient of about 2. Within these limits there is little destruction or denaturation of proteins; but if the temperature is raised much above 40°C, the thermal inactivation of enzyme protein becomes important in limiting the rate of growth. The optimum temperature, at which the generation time is shortest, is therefore a compromise between the stimulating effect of increasing temperature on chemical reactions and its destructive effect on the enzyme proteins; for the majority of bacteria, termed *mesophilic*, the optimum lies between 30 and 42°C.

The generation time in the log phase is controlled not only by the inherent character of the bacteria and by the temperature, but also by the constituents of the medium. It is precisely because variation of the source of nitrogen, carbon, or vitamins in the medium is reflected in the rate of bacterial growth that the methodology of bacterial nutrition exists. A compound not synthesized at all by the cell must be added to the medium and is therefore termed, in ordinary usage, an *essential metabolite*. Such substances should rather be called *essential nutrients*, for actually all the compounds required for cell protoplasm synthesis are essential metabolites; some are synthesized rapidly from the medium constituents, others slowly, and others not at all. If, for instance, the synthesis of some amino acids from $(NH_4)_2SO_4$ in the medium proceeds slowly, growth will be slow and the generation time long; if the amino acids are added to the medium, this metabolic hurdle is bypassed, growth is more rapid, and the generation time is shortened. Comparison of the rate of growth of the two cultures, at suitable times during incubation, indicates that supplementation with these amino acids, although not necessary, stimulates growth. Similarly, the nature of the source of carbon, which furnishes not only building units for protoplasm but also the energy for building, will affect the rate of multiplication. If the cell possesses mechanisms for using molecular oxygen in respiration to obtain energy, aeration of the culture leads to increased energy production per carbon atom utilized and to faster growth. The concentration of the nutrient limits, primarily, the total amount of growth, but at low concentrations it also affects the rate of growth.

Thus, by altering the medium, or by decreasing the concentration of nutrient, one can lengthen the generation time. Analysis of the RNA content and of the protein nitrogen of such bacterial cultures shows that *the ratio of RNA to protein is inversely proportional to the generation time*; that is, the faster the cell can form RNA (ribosomes), the more rapid its growth.

Variations in cell properties during the log phase. In the usual type of culture, in which the composition of the environment is continually changing, one would expect that the bacteria would also have different properties when taken from different portions of the log phase, even though their rate of division were constant. After all, they are growing in different media (which they themselves have modified from the original medium) and certainly this would influence their chemical and enzymatic composition. Yet specific data on this point are hard to come by. There are data, of course, from continuous cultures which show that different media and different growth rates do produce cells of different properties, but this is not exactly a comparable situation. There are a few studies (e.g., 14), but usually such studies are incidental to some other work. The age of the culture influences the activity of the ribosomes. For example, during the early phases of log growth in *E. coli*, cell-free extracts capable of protein synthesis respond readily to the addition of the artifical messenger, poly-U, but about one third of the way up the log curve they fail to respond appreciably (Figure 7-8a) (15). The indications are that in the early log phase there are more ribosomes than message; thus ribosomes can respond to the *in vitro* addition of external message. This is also shown by the decreasing sensitivity of the cell-free system to streptomycin as one proceeds up the growth curve, and it is known that the 30S ribosome is protected from the action of streptomycin by the presence of messenger RNA.

In *Streptococcus faecalis* (Table 7-1), as growth proceeds, more of the ribosomes become associated with the membrane and their ability to synthesize protein *in vitro* surprisingly increases. Even in the stationary phase, when cellular *in vivo* synthesis of protein has stopped, the ribosomes maintain a very high level of protein-synthesizing activity, a property that we shall discuss later.

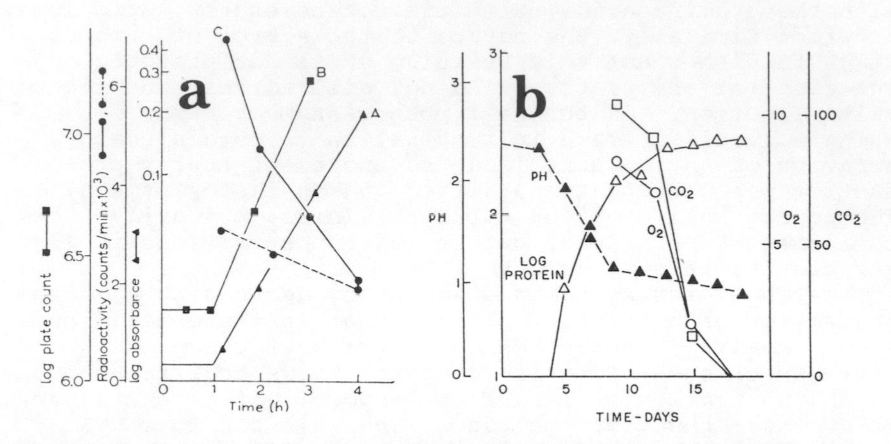

FIGURE 7-8. *Different properties of cells at different growth phases. Part a shows the differing response of protein synthesis by cell-free extracts of Escherichia coli to the addition of poly-U (and streptomycin -- dotted line). A is the log of the optical density, B is the log of the plate count, and C is the incorporation of radioactive phenylalanine into protein. Part b shows the respiratory activity and CO_2 fixation in cultures of Thiobacillus thiooxidans. Part a is from Li, L., and W. W. Umbreit, Biochim. Biophys. Acta 119:392–399 (1966). Part b is reproduced by permission of the National Research Council of Canada from the Canadian Journal of Microbiology, Volume 20, 1974, pp. 1709–1712.*

TABLE 7-1. *Growth of Streptococcus faecalis in Tryptone, Yeast Extract, and Glucose Phosphate Medium*

Time (min)	Phase	Growth (O.D.)	Protein Synthesis[a]	Ribosomes (%) Free	Ribosomes (%) Membrane	30S[b] (%)
75	Lag	0.08	45,000	38	44	18
150	Log	0.30	67,000	10	63	18
210	Late log	0.77	78,000	12	71	17
11 h	Stationary	0.74	126,000	9	77	14

[a]Cell-free protein synthesis in counts/min/mg of ribosomal RNA.

[b]There is reason to believe that these are damaged and inactive ribosomes.

Source: Data from *Science 154*:1350 (1966).

If one wishes to obtain large numbers of cells with high metabolic activity, one normally takes cells from the late log phase (60—70 percent of the final stationary phase turbidity), because cells in the stationary phase are much less active metabolically. We recently ran across a dramatic example of this phenomenon in *Thiobacillus thiooxidans* in which respiratory activity practically ceased at late log (Figure 7-8b).

The Stationary Phase

The chemostat and turbidostat both show that a culture can continue growing in the log phase "forever," providing the medium is continually replenished. The cells do not require a "rest period." It is true that some cultures "degenerate" under continuous growth conditions, but this appears to be due to the selection of a mutant rather than the degeneration of all the cells in a culture.

If growth can continue forever, what stops it? Obviously, the medium, since if we continue to replenish the medium, growth does not stop. The level of the population at which it stops depends on the medium. In general, the richer the medium, the higher the population it will support. How high can that population go? Is it crowding that stops growth? The usual nutrient broth culture of *E. coli* contains, in the stationary phase, about 10^9 organisms/ml. A much richer medium (i.e., 1 percent each of yeast extract, tryptone, glucose and 0.5 percent of K_2HPO_4 with 0.01 percent $MgSO_4 \cdot 7H_2O$) with good aeration may approach 10^{10}, a ten-fold increase in cells. But if into the same amount of medium, a dialyzing sack is placed and the smaller volume of medium *inside* the dialyzing sack is inoculated, the population may be brought to 10^{11} or 10^{12}, a thousand-fold increase over that of the usual culture. Thus, if 100 ml of nutrient broth supported a population of 10^9 cells/ml, there were 10^{11} cells in the entire 100 ml. In comparison, if there were 10 ml in the dialyzing sack, its population might be of the order of 10^{10} cells/ml, yielding 10^{11} cells for the sack and also for the entire 100 ml of media used. Thus, a given amount of material in a given volume of fluid supported the same number of cells whether they were allowed to grow throughout it or were confined to a small portion of the medium but had access to the nutrients of the remainder. It is not cell crowding that limits the population but the concentration of nutrient and the concentration of waste products.

Clearly, the reason why log growth stops is due to the medium. Just as clearly there are either of two factors involved. The first is the exhaustion of nutrient. This could be as simple as utilization of all the glucose, or it could be the exhaustion of some vitamin or mineral. After all, the use of microorganisms for bioassay of amino acids or vitamins is based upon providing a medium in which one nutrient is limiting. The amount of growth, and sometimes its rate, is proportional to the amount of the limiting nutrient. But what about the region of excess nutrient? Either some other nutrient becomes limiting or the second factor limiting microbial cultures is controlling. This second factor is the accumulation of toxic products. These might be acid (and thus an unfavorable pH), but more often they are more subtle. Frequently, keto acids accumulate that seem to inhibit growth. Sometimes greatly increased growth can be obtained by dialysis culture, whereby such toxic products are removed (17). Under most laboratory conditions we supply excess nutrient so that most laboratory cultures stop log growth because of inhibitory factors.

GROWTH YIELD

Suppose that we have a situation in which all (or most) of the substrate produces energy and this energy is used for the cell's growth. Then the *yield* of cells is a measure of the energy produced. One may express this as

$$\underset{\text{(yield of cells)}}{Y} = \frac{\mu\text{g dry weight of cells}}{\mu\text{M of substrate used}} = \frac{\text{grams}}{\text{mole}}$$

As shown in Table 7-2, the yield of cells of *Str. faecalis* is about 20 g/mole of glucose, and about 10 on arginine. This organism takes glucose to two lactic acid, and this process is known to form two ATPs. The yield per ATP is $Y_{ATP} = 10$. Arginine goes to citrulline and this to ornithine, the latter step generating one ATP, ($Y_{ATP} = 10$). For many organisms, the Y_{ATP} seems to come out about 10. This value is useful in considering metabolic pathways. For example, *Leuconostoc mesenteroides* has a growth yield of 14.1; that is, it must form 1.4 ATPs/mole of glucose. However, it produces one lactic, one CO_2, and one ethanol per mole of glucose, which should yield only one ATP. But it yields 1.4; therefore, something else is happening. There must be in this pathway a source of more energy than the known development of only one ATP. In contrast, *E. coli* under aerobic conditions has a *Y* of 43.2 (glucose), 40.4 (fructose), and 31.5 (xylose). Glucose to pyruvate + 2(2H) would yield two ATPs. There remain 4(2H) to be disposed of by respiration. If *E. coli* thus forms four ATPs (growth is a yield of 40), following the pathways described from glucose to acetic acid, the oxidation of its 4(2H) must yield but little energy. Preseumably, *E. coli* does not follow the pathway described, or respiration is not generating energy. On the other hand, *Str. faecalis* when grown aerobically yields more than it ought to, which is taken as indicative of oxidative phosphorylation (18). Such calculations are applicable only when most of the substrate is used for energy, which is converted efficiently into cells. If there are large amounts of polysaccharide formation or other substances, the calculations have much less meaning. (For a review, see 19,20; or for the oxidation of other substrates, see 21,22).

TABLE 7-2. *Growth Yields at the Stationary Phase*

Organism	Substrate	Y	Y_{ATP}	
Streptococcus faecalis	Glucose	22.0	11.0	Synthetic medium
	Glucose	23.0	11.5	Complex medium
	Arginine	10.5	10.5	Synthetic medium
	Arginine	10.0	10.0	Complex medium
Saccharomyces cerevisiae	Glucose	21.0	10.5	Synthetic medium
Pseudomonas lindneri	Glucose	8.3	0.3	Complex medium

A factor called *energy charge*, which is the ATP + 1/2 ADP divided by the sum of ATP + ADP + AMP, has been proposed (23). During growth the energy charge is about 0.8, but upon reaching the stationary phase it drops to roughly 0.5 and then may fall rapidly. Cells die at values much below 0.5.

A further factor has been proposed. This is used especially for aerobic transformations on substrates whose metabolic pathway may not be known (24). This value is γ (av-e), which is the yield in grams of dry cells per molar equivalent of available electrons (av-e) utilized. For a variety of organisms under a wide variety of conditions and substrates, the value of γ (av-e) has been shown to be 3.14 ± 0.11. An average of data from a variety of laboratories give 3.0 ± 0.28. Bacterial cells grown on many different substrates have a heat of combustion of 5,270 cal/g. The passage of an electron mole at the oxidation state of NADH to oxygen yields 17,412 cal/electron (34,825 cal/2 electrons based on the oxidation-reduction potentials of oxygen E_o, pH 7 + 0.816, NADH, E_o pH 7 - 0.32). If there are 3 g of cells formed per molar electron, these cells contain 15,810 cal. Furthermore, by making some assumptions regarding the P/O ratio on electron passage, the γ (av-e) may be converted to γ ATP; when this is done, using several hundred estimates employing many different organisms and media, the Y ATP calculates as 10.7

One may also look at this factor in another way. We know that one ATP yields 10 g of cells. If an electron mole utilized forms 3 g of cells, each electron must be capable of generating 0.3 ATP. To oxidize 2H to water requires two electrons and one O (1/2 O_2), and thus should be capable of generating 0.6 ATP. As such, the P/O ratio for oxidative phosphorylation should be 0.6, which compares favorably with the actual ratios found.

MOLECULAR BIOLOGY OF THE STATIONARY PHASE

The transition phase from log to stationary represents a somewhat drastic alteration in the metabolic interactions, and by the time stationary phase is reached, the cells have fewer ribosomes (possibly less than half), and the production of RNA, DNA, and protein virtually ceases. How long the culture remains fully viable in the stationary phase depends upon how toxic the medium is and the nature of the toxicity. That is, in some cases there may be mere inhibition of growth (or, of course, a lack of some necessary precursor) with little overt destruction of the cell. Indeed, some cultures slowly increase in numbers during the stationary phase, usually in a linear rather than a logarithmetic fashion. It was thought by some that, since the actual mass sometimes increases, there was continual growth and continual death in the stationary phase and the two balanced each other, the increase in mass being attributed to the accumulation of dead cells. While undoubtedly some cells grow and some die, this turnover appears to be only a small part of the population, and the increase in mass is usually the accumulation of other materials, such as polysaccharides, associated with each cell.

One process that stops dramatically is protein synthesis. Furthermore, cell-free systems for protein synthesis prepared from cells in the stationary phase are usually inactive, which is somewhat surprising since whatever might be the limiting factors inside of the cell, such factors could presumably be supplied to the cell-free preparations. When the ribosomes themselves are removed from the cell extract and reconstituted using log phase supernates, they are fully active (Table 7-1), which suggests that there is an "inhibitor" in the stationary phase supernates. In most cases this inhibitor is removable by dialysis, and proves to be due to an increased amino acid pool, which results in an artifact of measurement and not inhibition at all. The amino acid pool increases in the stationary phase because amino acids are still transported but are not utilized for protein synthesis. In a few cases, reasonably specific inhibitors of protein synthesis have been found in extracts of the stationary phase cells (25,26), but in most cases, when properly handled, the ribosomes taken from the stationary phase are fully functional and the necessary cofactors are fully active. Protein synthesis appears to stop not because the protein-synthesizing machinery has been damaged, but rather because there is no messenger RNA (27). Because messenger RNA is low, there are relatively few polysomes (28).

Evidently, at the stationary phase the transcription process is shut off. It is not that the enzymes are destroyed, since, if an inducer is supplied, induced enzymes can be synthesized, which shows that the transcription and translation mechanisms are at least partially functional (the amount of induced enzyme synthesized is much smaller than in the log phase but it is definite). Just what stops transcription is not known.

Certain strains of *E. coli* in which RNA synthesis ceases when the amino acids are exhausted (*stringent* strains) produce one (sometimes two) ^{32}P-labeled compounds in response to amino acid limitation, that is, when they reach the stationary phase owing to exhaustion of one or more amino acids. Initially, these substances were labeled Magic Spot (MS) 1 and MS 2; they proved to be the tetraphosphates of guanosine (ppGpp) (29,30). At first there was the suspicion that the magic spot might be a substance actually inhibiting growth, since it only appeared when growth was stopped; but further study showed that it came from GTP and seemed to be due to the ribosome "idling on," using up its GTP and converting it to ppGpp, because a particular amino acid required for protein synthesis was missing, and translation of the messenger RNA was held up at that point (31,32). The ppGpp inhibits ribosomal RNA synthesis (33).

PRESERVATION OF CULTURES

To extend the length of the stationary phase before it enters the death phase is a matter of some practical importance, especially in preserving cultures. Inasmuch as in most laboratory conditions it is the accumulation of toxic products that limits growth, one should attempt to minimize the toxicity. Acid, for example, can be held to a minimum by utilizing an adequate buffer (as is done in the AC medium described earlier), or one may add calcium carbonate to the medium. One should also provide more medium per cell, that is, a *thin* streak on a slant or a stab.

Two physical factors influence the length of the stationary phase. The lower the temperature, the slower the growth and the slower the deterioration, and thus the longer the stationary phase. In fact, over the biological temperature range, the time over which the pattern of a growth curve occurs is dependent on temperature, although the growth curve looks much the same. Lower temperature is a way of extending time — longer lag, longer log, longer stationary, slower death, longer survival.

The second factor is air supply. In many (but not all) cases, this acts in a somewhat peculiar fashion. One might suppose that if an organism used air (or really the oxygen contained in it) then air was really a nutrient like any other nutrient; if this were cut down, the organism would "smother" or starve, decreasing the length of the stationary phase. Actually, however, if the oxygen is cut down a little, although not entirely eliminated, the stationary phase is frequently extended. Apparently, in many cases the lack of all but a little oxygen slows down overall metabolism and thus extends the stationary phase in time. The situation is really a little more complicated: what lack of oxygen appears to do is to slow down the death phase and thus permit extension of the time that a larger proportion of the cells in a culture will stay alive, even if not preserving it at its maximum population.

The medium employed to "carry" cultures is not necessarily that which provides the fastest growth. In the case of stock cultures, we wish to have a medium on which the organism survives longest. For example, *E. coli* will grow well on *nutrient agar*, a beef extract with peptone base. If we add 1 percent glucose to this medium, the *E. coli* will grow faster and produce much more growth, but the medium becomes acid, and this acid causes the organism to die off more rapidly. If a 0.1 percent glucose is added, with enough buffer to take care of the acid (roughly 0.4 percent K_2HPO_4), more growth and longer survival are possible than on nutrient agar alone.

When cultures are removed from stock, they are generally given two to three transfers without refrigeration to "pep them up" before use. Although there may be a certain element of habit in this process — and we know of no ready explanation for the result obtained — it does seem to be a useful practice.

In *slant cultures* we simply make a narrow streak on a slant of suitable medium, allow the organism to grow, and after a suitable interval transfer it to another slant. Of course, some organisms (anaerobes, many streptococci, etc.) do not grow well on slants, and for these organisms deep stabs are used. In general, we choose a medium that is good for the organism involved, prepare a slant, refrigerate the slant after initial growth, and retransfer it after an interval chosen by experience. Some organisms do not survive long, and transfers must be made every two to three days; for others, transfers once a month are adequate.

Deep stabs are not only suitable for organisms to which air is somewhat harmful, but also are useful to organisms that thrive on air, partly because of the greater amount of media per organism. Such cultures, having less surface exposed to air, do not dry out as readily as slants.

It has been found that, if a layer of sterile *mineral oil* is put over a slant or a stab culture, the organisms beneath will survive (at room temperature) for a very long time. By this process we can increase enormously the effective life of a stock culture. A culture that might die out in a slant at room temperature within two weeks might be preserved under oil for a year or more. These cultures have the advantage not only of longer survival but they are also accessible for any limited number of transfer during this period. They are not as widely used as they might be, partly because they are a little messy to prepare and use.

Stock cultures take much space in a refrigerator when there are many of them. Furthermore, cotton and plastic plugs tend to become damp, and contamination, especially molds, may then enter. One may use smaller tubes or screw-cap tubes; but the size of tube and thus the amount of media cannot be cut down too far. We want the organism to survive, not to starve, and we must provide adequate medium per cell. A solution to the problem is *lyophilization*, a process of removing water from a bacterial or other preparation when it is in the frozen state. Under these circumstances a considerable number of the bacteria present survive. The dried powder is stable and can be kept at room temperature for years, and the organism can be recovered by placing the powder in a medium suitable for growth.

LOG DEATH PHASE

Eventually, even under the best of circumstances, the cells of a bacterial population begin to die. If the logarithms of the numbers are plotted, a straight line results; a log death phase. It is easy to understand a log growth phase: 2 organisms to 4, to 8, 16, 32, 64, and so on. But why a log death phase? Of 64 organisms, 32 die the first interval, 16 the next, 8 the next, 4 the next, and 2 the next. This seems most peculiar. Many multicellular organisms, seeds of plants, for example, or even many single-celled populations composed of different kinds of cells die off quite differently.

In Figure 7-9 we have plotted the usual death curves for multicellular organisms compared to bacteria; the insert bars give the number of organisms killed during the respective interval. Why is there this difference?

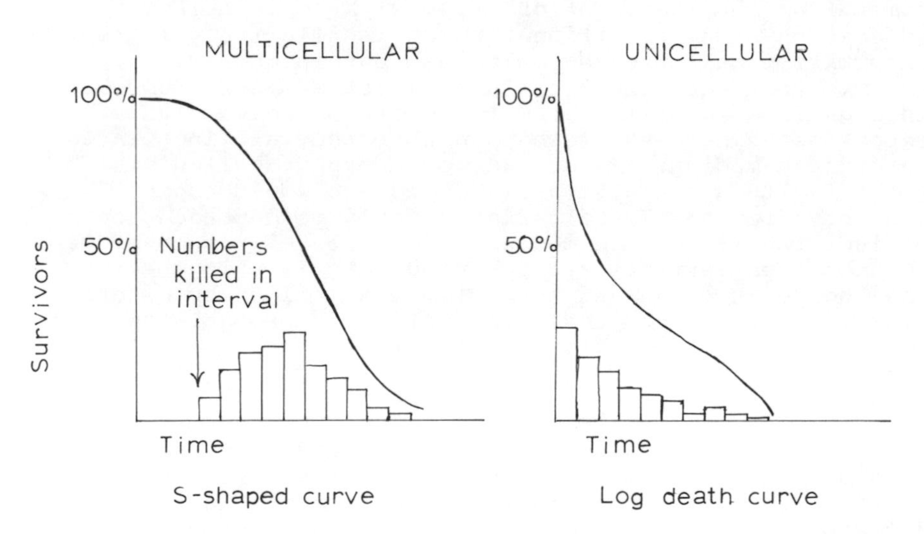

FIGURE 7-9. *Comparison of death rates of unicellular and multicellular organisms. The bars represent the number of organisms killed in the time interval.*

First, one might suggest that, in a multicellular organism if one of the cells is damaged, or even killed, others can replace it, and the total organism can survive. The die-away curve of a unicellular organism is $-dn/dt = kn$ or $n_1/n_2 = e^{-kt}$; that is, the loss of numbers of bacteria per unit time is directly proportional to their number. This rate is very much like that of a *unimolecular reaction*, a rate dependent solely on one molecule, as with ^{14}C (radioactive carbon). This corresponds to a *first-order reaction rate*, similar to the curve of denaturation (or partial destruction) of proteins in pure solutions, and has led some to assume that the death of bacteria results because of the loss of a particular protein in the organism. Of course, it would not have to be one protein, nor would it have to be a single molecule.

This does not necessarily mean that it is always the *same* protein, since the loss of any one of a number of proteins acting in series one after the other will stop the whole chain. This kind of concept will fit at least some of the data, but, although possible, it is hardly believable. That is, it is almost impossible to suppose that, of the billions of protein *molecules* in a cell, a sufficiently large number of one kind should be destroyed in the same instant. In short, there are several instances in which log death actually does occur, and in most cases the death rates are close to logarithmic, if not exactly so, but there is no really satisfactory explanation of this fact.

Actually, most of the information on log death curves comes from the treatment of cells with toxic materials or adverse physical conditions; there is relatively little information on untreated cultures, but it is tacitly assumed that the situation is at least similar. However, in organisms taken from the log death phase we might expect certain things to happen when they are inoculated into fresh medium. First, we would expect cell division to be faulty and irregular. Second, we would expect that recovery from the lethal effects would be dependent on the medium: the richer the medium, the more cells recovered. Both of these effects are observed both in cultures in the log death phase and in cells being killed by external toxicity; hence, the tacit assumption is probably correct.

PHYSIOLOGICAL YOUTH

In bacteria the problem of "age" of the individual cell is somewhat different than in other organisms. Presumably, the age of a cell is reckoned from its last division, but since it divided by binary fission, presumably half of it is "old" and half is newly synthesized. In an effort to apply criteria other than time since cell division to the problem of age, Sherman and Albus in the early 1920s, reasoning that the young of any species were usually more sensitive to harmful influences than the more mature members, subjected cultures at the end of the lag phase and in the stationary phase to a variety of sublethal toxic chemicals and environments and found, indeed, that the cultures from the lag phase were more sensitive. They therefore concluded that each culture went through a period of *physiological youth*. Further study has shown that the increased sensitivity to sublethal toxicity is not confined to the lag phase, but beginning at the late lag phase it extends throughout the log phase, that is, just during the period of most rapid division (Figure 7-10). The phenomenon, which is readily reproducible, seems to be a reflection of the known fact that the processes of cell division are more sensitive to harmful effects than are the processes of growth. Since, during the period of rapid division, the cells are younger, in the sense that the time since the last division was short, they are "youthful," and physiological youth is another way to measure this condition.

CONTINUOUS GROWTH IN CHEMOSTATS AND TURBIDOSTATS

In these cases one continually replaces the medium and draws away the end products. In chemostats, the rate of growth is determined by a limiting factor (e.g., a limiting amino acid), and thus growth rate is dependent upon the flow rate of the fresh medium. In turbidostats a complete medium is supplied and the culture maintained at a defined turbidity by rate of flow of the medium. These kinds of study show that the rate of growth depends upon medium (not an unexpected conclusion) and that growth can continue in the log phase essentially forever.

Although very elaborate chemostat systems have been devised, the situation can be simulated with simple means. For example, the data of Table 7-3 were obtained by allowing the organism (*E. coli*) to grow in 1 liter of the designated medium until it reached optical density 0.8 at which point 500 ml was withdrawn and used for analysis and 500 ml of fresh sterile medium replaced it, and so on for seven such transfers (seven generations). The data obtained (Table 7-3) show that the richer the medium, the faster the cells grow, the larger the cell, the more ATP per cell (but essentially constant ATP per gram of tissue), and the more ribosomes (i.e., RNA).

A chemostat system also permits one to do step-up and step-down experiments (for a review, see 34). Suppose that one has a culture growing at a constant log rate in a chemostat with an amino acid as the restricting nutrient, and suppose that at some instant we halve the amino acid concentration. Within 1 to 10 min (roughly one tenth of the generation time) the following happens:

1. A decreased rate of protein synthesis because less of the restricting amino acid is available and thus less of its specific charged tRNA.
2. A derepression of specific amino acid biosynthesizing enzymes, which results in an increase in the synthesis of the restricted amino acids and its precursors.

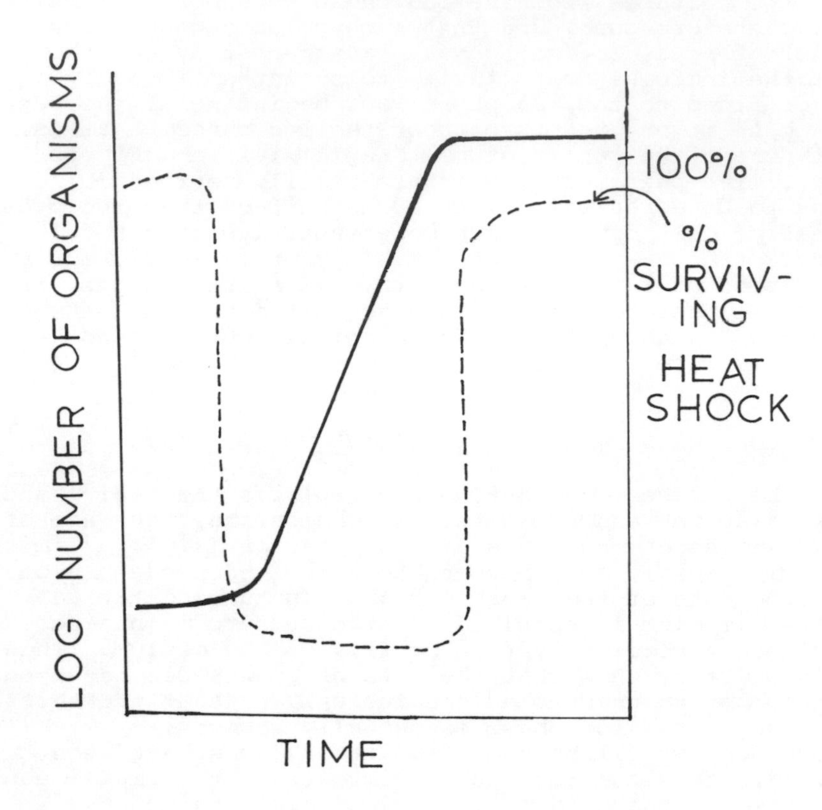

FIGURE 7-10. *Survival of bacteria taken from various phases of the growth cycle and subject to mildly toxic conditions, i.e., 55°C, 30 min.*

These effects are due to a decrease in the activity of the corepressor and a release of the allosteric inhibition of the synthetic enzymes already present.

3. A decreased rate of synthesis of DNA and RNA, both of which are dependent upon protein synthesis. DNA synthesis may continue on for a short interval, which seems to represent the completion of the DNA already started.

4. A decreased rate of growth; there are fewer ribosomes, less protein synthesis, and a diversion of synthesis to new enzymes.

In fact, it appears that the composition of the cells (the DNA, RNA, protein, mucopeptide, enzymes, etc.) is fixed by the circumstances — the rate of dilution, the composition of the medium, the degree of aeration, and the physical parameters; the cells' response to changes in these is rapid and dramatic. Chemostat studies lend themselves to mathematical descriptions, and a great deal of mathematical analysis has been done without, I believe, clarifying the situation during log growth. But what is perhaps surprising is that the faster the growth rate, the larger the cell. Look at cells growing on glucose (Table 7-3). Every 56 min they divide, and 10^{14} cells weigh 50 g. Add casamino acids and they divide every 28 min (twice as fast), and each 10^{14} cells weighs 92 g (twice as much) (evidently they were synthesizing cell materials four times as fast). One cannot reconcile these observations with the idea that a cell divides when it reaches a certain size, or that it takes a given time interval before division. One reasonable hypothesis is that both types of cells, no matter what their size, had to have the same complement of genes to pass on to the daughters. Possibly, then, the generation time is dependent on the time required to replicate the DNA. In the poorer medium it took 56 min; in the richer, half that. But in the richer medium the synthesis of substances other than DNA went even faster than the DNA synthesis, so the cell was larger at division.

TABLE 7-3. *Influence of the Medium Upon Generation Time, Cell Size, and ATP Content in Escherichia coli B*

Medium	Generation Time (min)	Wet Weight (g/10^{14} cells)	ATP (mg/10^{14} cells)	RNA per Cell
Lactate	82	39.6	135	Low
Glucose	56	50.7	173	
Glucose + 8 amino acids	43	58.7	247	
Glucose + casamino acids	28	92.2	415	
Glucose + casamino acids + malt extract	25	117.0	457	High

Source: Data recalculated from *J. Biol. Chem. 236*:515 (1961) and *Science 158*:658 (1967).

SYNCHRONOUS CULTURES

If one could start with a culture in which all the cells had just divided and were growing in the log phase, then at one generation time later, *all* the cells in a culture would divide and suddenly the population would double. Such cultures are called synchronous cultures (35). There are fundamentally three methods of experimentally obtaining *synchrony*.

1. Stop one factor essential for division; let all the cells reach this point and resupply the essential factor. For example, in bacteria requiring thymine (a unique component of DNA), when such strains are grown in thymine-deprived medium and the thymine runs out, the DNA synthesis already started continues, but since there is no thymine, the DNA is nonsense and the cell dies (thymineless death, see later). But about 5 percent of the cells do not die, that is, they are not synthesizing DNA (having just divided), and these will not initiate DNA synthesis until thymine is present. When thymine is added, these cells begin to grow; since they all start at the same time, the cells divide simultaneously. Or, one may make use of the fact that the temperature coefficient of DNA and protein synthesis are different. For example, *B. megaterium*, when chilled to 12°C, stops DNA synthesis, but protein synthesis continues. If left there for 30 min and then put at 34°C, after an interval of 40 min, the cell population suddenly doubles, and in 32 min doubles again (36). If *E. coli* growing at 37°C is suddenly shifted to 10°C, there is a long lag period of 4—5 h, and during this interval there is no synthesis of DNA, RNA, or protein; but if growth is started at 10°C, all can be synthesized (37).

2. One may select out cells at some stage of division. For example, one may filter a log phase culture through a 1.2 μ millipore filter. Only the smaller cells will go through; but inasmuch as these are the ones that have just divided, they are synchronous (38). An even simpler method is to take 1 liter of cells on synthetic medium (optical density = 0.15 = about 10^8 cells/ml), centrifuge, and put the thick suspension onto a filter paper stack containing 20 sheets of filter paper, and wash them with 70 ml of medium. The larger cells stay near the top, and the other cells are distributed down the stack roughly in proportion to their size. If a filter paper (say the sixth from the top) is used as inoculum, it has cells roughly the same size, that is, roughly at the same state with respect to division, and the culture resulting would be synchronous (39). Cells could be separated on a sucrose density gradient, and thus cells of a similar given size could be selected (40).

3. When cells enter the stationary phase, the growth tends to slow down; but, by the same token, the ones arriving there first wait for the others, so that cells taken from the late log phase provide synchronality (41).

Synchronous cells tend to run out of synchronous growth after two to four divisions since not all cells grow at exactly the same rate, and randomization of division times occurs. However, synchrony can be kept going much longer by continuous phased growth (42). Medium runs into a vessel (Figure 7-11) at such a rate that at one generation time it fills the dosing vessel completely, which then siphons over into the culture, thus diluting it by an equal volume, half of which goes immediately to harvest vessel to be analyzed. That is, a total food ration arrives at one time to a synchronized population, and the cells grow and divide at the same time over the next doubling period; at this point another ration arrives, and the cells again grow to division; at that point another ration arrives. If this arrival is repeated at similar intervals, an ordered procession should continue, and synchrony continues (partly because those cells that did not quite divide on time tend to get washed out of the system).

Such synchronous cultures permit one to approach a problem considered earlier: what happens to a cell between one division and the next? Here one has billions of cells all in the same state, and thus readily analyzed at any time period. In fact, between one division and another, synthesis DNA, protein, and RNA seem to be largely continuous (41), in spite of the fact that other types of data suggest that the protein synthesis must take place before DNA synthesis, and that from genetic considerations one would expect DNA synthesis to be delayed after division and to stop at some interval before the new division.

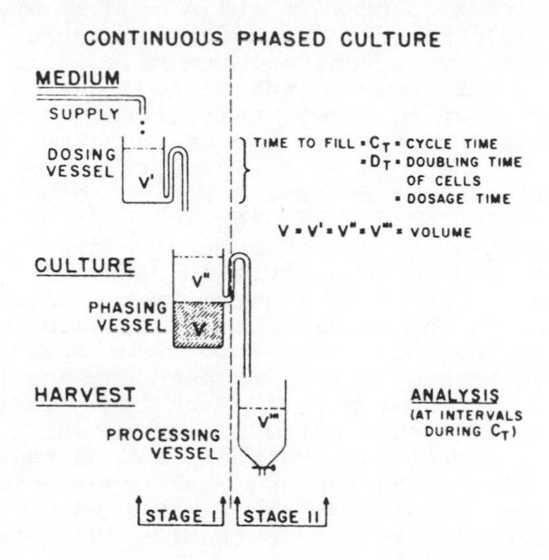

FIGURE 7-11. *Principles of operation of a continuous phased culture. From Dawson, P. S. Reproduced by permission of the National Research Council of Canada from the Canadian Journal of Microbiology, Volume 11, 1965, pp. 893–903.*

BACTERIAL GROWTH RATES IN NATURE[1]

The survival of a species ultimately depends on a growth rate sufficient to balance the death rate. In studying the ecology of a given species, one should not only study

[1]Brock, T. D. 1971. Microbial growth rates in nature. *Bacteriol. Reviews 35*:39–58. (Reviewed by Steven Mamber.)

the organism in the laboratory, but also in its natural habitat.

Two types of microbial populations can be easily recognized in terms of their growth kinetics. The first is the exponentially growing population, a closed population of which all members are growing continuously with no loss or gain of cells to or from the environment. This is not common in nature.

The second is the steady-state population, in which the cells may be growing continuously, but cells are lost from the population at the same rate that they are being added. In between are many gradations between exponential and steady-state populations.

In studying microbial growth rates it is necessary to assess population numbers. The bacteriologist cannot rely on direct counts primarily because of the small size of bacteria, and must rely instead on viable counts, which are not suited for ecological studies. Viable counts cannot reveal how rapidly organisms are growing in nature. For assessing bacterial numbers in a natural habitat, one can utilize direct counts, if cells studied are large and distinct, and fluorescent antibody procedures, which can be used to identify as well as enumerate cells. Here, listed briefly, are some of the methods used or being developed for studying growth rates in mixed populations:

1. Artificial substrates: the use of glass slides or microcapillary tubes placed directly in the environment and removed at various intervals for microscopic study.
2. Microscopic techniques: for studying growth on artificial substrates, or for immersion directly into the environment to study the activities of cells *in situ*.
3. Autoradiographic techniques: DNA labeling with radioactive thymidine is valuable in studying growth rates because DNA is the only cell macromolecule that does not turn over, and because synthesis does not occur in nongrowing cells (if eucaryotic). Such techniques have been used primarily with eucaryotic microbes, which exhibit a regular cell cycle of DNA synthesis and mitosis, but methods have been found for use with procaryotes, even though procaryotic cells do not exhibit the typical cell cycle and, moreover, do not synthesize DNA actively except at cell division.
4. Chemostats: used as excellent laboratory models of steady-state systems but have been applied to the study of natural systems; requires knowledge of flow rates and turnover time in the system.
5. Genetic techniques: by introducing a nonreplicating genetic marker into a bacterial colony before inoculation into a habitat. At each division the population containing the marker will be halved, and doubling time can subsequently be calculated.

A knowledge of growth rates in the environment may provide an important insight as to how microorganisms succeed in their natural habitats.

REFERENCES

Reviews

Dawson, P. S. S. 1974. *Microbial Growth.* Dowden, Hutchinson & Ross, Inc., Stroudsburg, Pa.
Halvorson, H. O., B. L. A. Carter, and P. Tauro. 1971. Synthesis of enzymes during the cell cycle. *Advan. Microbial Physiol.* *6*L47—106.
Harrison, A. P., JR. 1967. Survival of bacteria. *Ann. Rev. Microbiol.* *21*:143—156.
Hartwell, L. H. 1974. *Saccharomyces cerevisiae* cell cycle. *Bacteriol. Reviews* *38*:164—198.
Helmstetter, C. E. 1969. Sequence of bacterial reproduction. *Ann. Rev. Microbiol.* *23*:223—238.
Jannasch, H. W., and R. I. Mateles. 1974. Experimental bacterial ecology studied in continuous culture. *Advan. Microbial Physiol.* *11*:165—212.
Koch, A. L. 1971. The adaptive responses of *Escherichia coli* to a feast and famine existence. *Advan.Microbial Physiol.* *6*:147—218.
Mandelstram, J., and K. McQuillen. 1968. *Biochemistry of Bacterial Growth.* Wiley, New York.
Painter, P. R., and A. G. Marr. 1968. Mathematics of microbial populations. *Ann. Rev. Microbiol.* *22*:519—548.
Pato, M. L. 1972. Regulation of chromosome replication and the bacterial cell cycle. *Ann. Rev. Microbiol.* *26*:347—368.
Postgate, J. R. 1967. Viability measurements and the survival of microbes under minimum stress. *Advan. Microbial Physiol.* *1*:1—24.
Schlegel, H. G., and H. W. Jannasch. 1967. Enrichment cultures. *Ann. Rev. Microbiol.* *21*:49—70.
Schultz, J. S., and P. Gerhardt. 1969. Dialysis culture of microorganisms: design, theory, and results. *Bacteriol. Reviews* *33*:1—47.
Slater, M., and M. Schaechter. 1974. Control of cell division in bacteria. *Bacteriol. Reviews* *38*:199—221.
Strange, R. E. 1972. Rapid detection and assessment of sparse microbial populations. *Advan. Microbial Physiol.* *8*:105—142.
Sussman, M. 1965. Developmental phenomena in microorganisms and in higher forms of life. *Ann. Rev. Microbiol.* *19*:59—78.
Tempest, D. W. 1970. The place of continuous culture in microbiological research. *Advan. Microbial Physiol.* *4*:223—251.
van Uden, N. 1969. Kinetics of nutrient-limited growth. *Ann. Rev. Microbiol.* *23*:473—486.

Papers

1. *J. Bacteriol.* *104*:1052 (1970).
2. *J. Bacteriol.* *90*:1426 (1965).

3. *J. Gen. Microbiol. 49*:59 (1967).
4. *J. Bacteriol. 91*:1618 (1966).
5. *J. Gen. Appl. Microbiol. Japan 8*:246 (1962).
6. *J. Bacteriol. 90*:223 (1965).
7. *J. Bacteriol. 95*:123 (1968).
8. *J. Bacteriol. 95*:221 (1968).
9. *J. Bacteriol. 93*:286 (1967).
10. *Biochemistry 10*:4894 (1971).
11. *J. Gen. Microbiol. 49*:461; *50*:383 (1968).
12. *J. Biol. Chem. 245*:2922 (1970).
13. *Proc. Natl. Acad. Sci. 69*:3024 (1972).
14. *J. Bacteriol. 112*:715 (1972).
15. *Biochim. Biophys Acta 119*:392 (1966).
16. *Can. J. Microbiol. 20*:1709 (1974).
17. *Bacteriol. Reviews 33*:1 (1969).
18. *J. Bacteriol. 96*:1595 (1968).
19. *Bacteriol. Reviews 26*:95 (1962).
20. *Ann. Rev. Microbiol. 24*:17 (1970).
21. *Appl. Microbiol. 15*:1332 (1967).
22. *J. Gen. Microbiol. 55*:341; *63*:333 (1970).
23. *J. Bacteriol. 108*:1072 (1971).
24. *J. Bacteriol. 96*:1424 (1968).
25. *Can. J. Microbiol. 15*:159 (1969).
26. *Biochem. Biophys. Res. Comm. 24*:541 (1966).
27. *J. Biol. Chem. 247*:2289 (1972).
28. *Biochim. Biophys. Acta 161*:368; *166*:760 (1968).
29. *J. Biol. Chem. 246*:4381 (1971).
30. *Nature 238*:370 (1972).
31. *Nature 238*:38 (1972).
32. *J. Biol. Chem. 247*:6055 (1972).
33. *Nature 254*:530 (1975).
34. *Bacteriol. Reviews 30*:701 (1967).
35. *J. Bacteriol. 103*:789 (1970).
36. *J. Bacteriol. 71*:17 (1956).
37. *J. Bacteriol. 94*:157 (1967).
38. *Science 131*:1098 (1968).
39. *J. Bacteriol. 90*:84; *94*:1264 (1967).
40. *Proc. Natl. Acad. Sci. 57*:1611 (1967).
41. *J. Bacteriol. 91*:469 (1966).
42. *Can. J. Microbiol. 11*:893 (1965).

8
Molecular Biology and Physiology

DOGMA

We may review the dogma of molecular biology in terms of 10 propositions, which we think are likely to be true. We assume that you are familiar with this information and thus the propositions serve as a convenient summary.

Proposition 1. A living cell is made by and controlled by its enzymes. The enzymes process materials usable by the cell into energy and into cell substances.

Proposition 2. The cell has a "book" or "tape" that tells it what to do and how to do it. This tape is made of DNA, which is in the form of a double-stranded helix with base pairing. Thus for each adenine there is a thymine; for each cytosine, a guanine.

Proposition 3. It is the *order* of the bases in the DNA structure that contains the instructions.

Proposition 4. DNA has two functions: (1) it must replicate itself (proposition 5), and (2) it must give information to the cell (propositions 6—10).

Proposition 5. In bacterial cells, DNA is one long molecule that happens to be circular. Enzymes replicate it (discussed later), and the replication starts at a given point and goes all the way around the circle. The two strands of the DNA separate, and each is replicated, so that one finishes with two strands, each containing one strand of the old and one strand of the new DNA.

Proposition 6. DNA gives information to the cell by way of RNA. RNA is formed on the double-stranded DNA while the two strands are together. The *order* of the bases in the RNA is complementary to the order of *one* strand of DNA. This RNA, when synthesized, leaves the DNA and goes to the ribosomes, where it codes for the production of protein. It is called messenger RNA. The mRNA may be a long molecule, but it is much shorter than the DNA. One strand is long enough to code for a protein. A different mRNA strand is used for each different protein.

Proposition 7. In the cell exist protein manufacturing

sites called *ribosomes*. In procaryotic cells there are two types, one, with a sedimentation characteristic of 30S to which the mRNA attaches. The other is 50S, to which the other components attach. The two associate to form an "active" unit, called 70S.

Proposition 8. In the cell, and presumably manufactured on some part of the DNA, is a low-molecular-weight RNA (molecular weight of 24,000—30,000), transfer RNA. When "charged" with the amino acid, it attaches, under suitable conditions, to the 50S part of the ribosome and adds its amino acid to the growing peptide chain. The *order* in which the amino acids are introduced *via* the tRNA is dependent upon the order of the bases in the mRNA. There is a specific tRNA for each kind of amino acid. The tRNAs are "charged as follows:

1. Amino acid + ATP = adenylo-amino acid + pyrophosphate.
2. Adenylo-amino acid + specific tRNA = charged tRNA + adenylic acid.

The charged tRNA has the amino acid attached to it. There is a specific enzyme for each amino acid and for each of reactions (1) and (2).

Proposition 9. The ribosome is composed of protein and RNA, and the ribosomal RNA, which comprises 80—85 percent of the cell's RNA, is made on DNA. The ribosomes are not all alike, but they can accept any mRNA and the corresponding tRNAs and manufacture protein.

Proposition 10. The proteins are long peptides comprising chains of amino acids, linked via peptide bonds, such chains being over 100 amino acids long. This is the *primary structure*. The sequence of the amino acids comprises the *secondary structure*, and proteins differ from one another in the sequence of the amino acids that they contain. This sequence also determines their *tertiary structure*, that is, their configuration in space — their shape, surface contour, and so on. Most cellular proteins are enzymes, and with these enzymes the cell can synthesize whatever it needs (proposition 1).

In *Escherichia coli*, the DNA is 3 percent of the dry weight of the cell and thus weighs 7×10^{-15} g, which is equivalent to 1.6×10^7 nucleotides or 0.8×10^7 base pairs. This is also equivalent in length, since it is one string 2000 μm long of 5×10^6 triplet codes. Since the average gene (coding for one protein) is 2.5×10^3 triplets, the DNA in *E. coli* could code for 6,400 proteins.

In the following sections we shall take up only a very small part of molecular biology, that part having a specific interest for physiology. Table 8-1 lists some properties of bacterial nucleic acids that are frequently useful to have available.

TABLE 8-1. *Nucleic Acid Distribution in a Bacterium*

Volume of cell	10^{-12} ml
Dry matter per cell	2.5×10^{-13} g (1.5×10^{11} daltons)
DNA per cell: % of dry weight	2—4% (say 3.3%)
weight	5×10^{9} daltons
no. of nucleotides	1.6×10^{7}
length	2000 μm
Size and weight of average structural gene	2,500 nucleotides, 7.7×10^{5} daltons
Total no. of genes per cell	6,400
Genes for tRNA	50 (0.02% of total DNA)
Genes for 16S rRNA	10 (0.1% of total DNA)
Genes for 23S rRNA	10 (0.2% of total DNA)
Genes for mRNA	Thousands

	Sedimentation Coefficient	Molecular Weight (daltons $\times 10^{-3}$)	Nucleotides per Molecule	Total No. Molecules per Cell	Total RNA %	Total Nucleotides $\times 10^{-6}$	Total Weight (daltons $\times 10^{-8}$)
rRNA	16S	550	1,700	12,000	80	64	200
	23S	1,100	3,400	12,000			
tRNA	4—5S	25	70—80	100,000	10	8	25
mRNA	8—30S	100—1,500	600—5,000	500—1,000	1—2	0.8—1.6	2.5—5
				Total RNA		80	250

DNA

We shall assume that although you may never have had a formal course in molecular biology you still know a considerable amount about DNA, RNA, and protein synthesis and function. With respect to DNA we shall point out only three aspects that have a special importance in discussions of bacterial physiology.

In the bacterial cell, the DNA exists as one long circular strand consisting of about 1.6×10^7 bases; it may be as much as 2,000 μm long. There is no evidence that this long strand contains non-DNA sections, and presumably it requires 1.6×10^7 bases to provide the code for manufacture of a bacterial cell. Replication inside the cell of this extremely long, presumably intact DNA molecule must be an unusual mechanism, especially since the molecule must break into two separate chains, each of which is the template for the copy of its partner.

The single long strand of the DNA molecule seems to be organized in the form of a cable with perhaps 100 threads, but it is not known how the cable and threads serve as templates. Presumably, not all areas of the long strand are active at the same time. When DNA is isolated from phage or bacteria in the usual fashion, it has a molecular weight of about 10^6. The amount of DNA in phage T_2, for example, is 2×10^{-16} g. If this were all in one molecule, it would have a molecular weight of close to 10^8. Therefore, in phage T_2 there is either one molecule of DNA of a molecular weight of 10^8 or 100 molecules of molecular weight of 10^6. If phage labeled with tritium is carefully isolated and spread in a photographic emulsion, its length proves to be 52 μm, which corresponds to a molecular weight of 1.1×10^8; that is, there is one molecule, not 100. Furthermore, if DNA is carefully isolated, its molecular weight is 1.3×10^8, and if this DNA is stirred at high speed or subjected to shearing forces (even if pulled rapidly through a pipette or injection needle), it breaks into units of molecular weight of about 10^6. Evidently, then, at about every 0.5 μm (the length of a strand with molecular weight of 10^6), a weakness in the ribbon permits breakage; but within the cell presumably DNA exists as a long single molecule.

IN VITRO SYNTHESIS OF DNA

Especially through the studies of A. Kornberg, we know something of an enzyme capable of *in vitro* synthesis of specific DNA. This enzyme (obtained, e.g., from *E. coli*) will condense deoxyATP, deoxyGTP, deoxyCTP, and deoxyTTP (all must be present) into DNA, providing some DNA is present. This *priming* DNA has a given purine—pyrimidine ratio, and the DNA synthesized has the same ratio, which suggests that it is the replication of the DNA supplied which is occurring rather than a random assembly of nucleotides. The DNA used for priming must be single-stranded or broken. Native double-stranded DNA does not serve as primer. This suggests that the DNA must split apart into two single strands before this enzyme can act on it. The reaction releases inorganic pyrophosphate and continues until one of the triphosphates or the primer is exhausted. Since the deoxytriphosphates are all 5'-triphosphates, they must bond to the 3'-hydroxyl of the preceding deoxyribose in the chain, and the synthesis proceeds by extending the chain at the 3' end; it must proceed in the "downward" direction on the left side of the chain and in the "upward" direction on the right side, because these are the only deoxyriboses containing free 3'-hydroxyl groups. The point is that replications proceeds in *both* directions along the two strands of the replicated primer double helix.

However, it has become evident that the Kornberg enzyme (called pol-I for polymerase-1) is probably *not* the enzyme involved in DNA synthesis in the intact cell; it may have a function in *de novo* synthesis, but its main function is the repair of "nicked" or otherwise damaged DNA. Mutants of *E. coli* have been found that lack pol-I, yet grow normally. These contain two other DNA polymerizing enzymes (DNA replicases) called pol-II and pol-III. Additional mutants have been obtained that are deficient in pol-II, but which still grow normally. However, thermosensitive mutants, which lack pol-III only at their restrictive temperature (where they grow poorly), have been found, which suggests that it is pol-III that actually synthesizes DNA in the cell. It is thought by many (1) that DNA is synthesized in three stages. The first stage involves the synthesis by pol-III of pieces with a molecular weight of about 10^6 (and about 20S density), sometimes called Okazaki pieces. The evidence indicates that this particular replicase is associated with and bound to the membrane. At the second stage, 50—250 of the Okazaki pieces are joined together to give a unit of 50 to 250 x 10^6 molecular weight (density 70S—120S). It appears that the Kornberg enzyme (pol-I) may be involved here, but it is also backed up by pol-II and pol-III, which take over this function if pol-I is missing. Finally, even these longer strands are joined together to form a bacterial chromosome of 2500 x 10^6 molecular weight (or 2.5 x 10^9; density of 130S) found in the whole cell. Pol-III also appears to be involved in this process. Clearly, pol-III is the critical enzyme in DNA synthesis, and curiously it is associated with the membrane. There is evidence, based on very short times of pulse labeling, that DNA is synthesized in short pieces, Okazaki pieces, varying from 0.3 to 0.6 x 10^6 molecular weight, and that these are then attached to each other. These smaller pieces are identified in the following way. A pulse of radioactive thymine is given; a short interval thereafter the reaction is stopped and the DNA isolated. The DNA is then placed in dilute alkali, which causes the strands to separate. By labeling for up to 30 s, one obtains (on analysis by density gradient centrifugation) the original single strand of DNA containing no radioactivity (i.e., the old strand) and very small pieces of DNA that contain the radioactivity. Evidently, they are rather firmly attached to the DNA (as they would be if they are the complementary strand), but when split away from the old strand by alkali they appear as short-length Okazaki pieces. After 30—40 s, large labeled strands of DNA are also isolated. Thus the smaller pieces are evidently attached together to form the complementary strand of DNA.

RELATION OF DNA SYNTHESIS TO CELL DIVISION

Before a cell can divide it must replicate its DNA so that it may pass the genetic code to its progeny. Some knowledge of what must happen in the individual cell has been obtained by studying cell populations of a particular mutant of *E. coli*, 15K, which requires thymine, uracil, and arginine for its growth. Thus the ability to synthesize DNA, RNA, and protein can be controlled by the presence or absence of one or more of these constituents in the medium. If this mutant is placed in a medium containing arginine and uracil but no thymine, most of the cells die; but about 5 percent do not. The thymineless death results, presumably, because in the absence of thymine DNA synthesis goes on, but without thymine the DNA provides merely a nonsense code. The 5 percent that survived are presumably not making DNA. If, however, the same culture is placed in a medium with thymine but no uracil and arginine, and then subsequently transferred to a medium that contains none of these, the organism does not grow in any of the media; but it also does not die from lack of thymine. The longer it has been exposed to a lack of uracil and arginine, the fewer cells that die from lack of thymine. It is therefore supposed that when arginine and uracil are absent,but thymine is present, there is synthesis of DNA but none of proteins or RNA, and indeed a 50 percent increase in DNA is found. This is just the amount of increase one would expect in a nonsynchronized population if the DNA synthesis already initiated was allowed to run to completion, but that no new DNA could be initiated in the absence of protein synthesis. The 5 percent of survivors in the earlier experiment would be the number that had just completed their DNA cycle before the thymine was withdrawn and had not reinitiated it.

This information suggests that once initiated DNA synthesis runs to completion. If necessary bases are not present, the cell makes a nonsense DNA, which may kill it. But once it has completed its DNA cycle, it does not initiate a new one unless RNA and protein can be synthesized.

SPECULATION ON THE BACTERIAL NUCLEUS

It is possible to devise a model of how DNA operates in the "nucleus" of a bacterial cell; the model is capable of "explaining" two of the main problems that are as yet unsolved: (1) how such a long molecule of DNA is packed into a "nucleus" yet available for replication, but also available for mRNA production, and (2) how such a long molecule, a spiral, can unwind for replication fast enough to account for its duplication in a generation's time. The model will also "explain" a wide variety of observations and is thus useful. However, to devise the model, we have had to include three kinds of information; one is fact, a second is reasonable conjecture, and a third we just had to make up to make the rest fit. We may be a long way from the truth, and would not be surprised if some parts prove to be in error, but with this model a great deal of present information can be correlated.

It is possible to see a nuclear body in bacterial cells. By staining, the nuclear body appears to be somewhat variable in shape. In sections, in electron micrographs, one obtains the impression of a bundle of fibrils. For several reasons we suppose that the nuclear apparatus consists of a *cylinder* (a drum) with a diameter of about 0.5 μm and a length (in nature) of 0.5 μm. The drum is composed of protein and RNA. The DNA is placed on the outside of this cylinder or drum. Such particles have been isolated from *Bacillus megaterium* by lipase degradation of protoplasts (2). Allowing for some swelling due to treatment, the particles could be fitted within these dimensions. The particles isolated had one DNA to one RNA to three protein, and the DNA was on the outside and streamed off in the form of long threads. We suppose that the variously shaped "nucleus" seen in the microscope is due to viewing this drum from different angles and as it turns over and over in the cell (Figure 8-1A,B, and C).

As mentioned, one problem in the replication of DNA is how the long helical strand unwinds to permit separation of the strands for replication, and how indeed it can do it fast enough. We suppose that in the cell, when attached to the nuclear apparatus, the double strands are not twisted, but are anchored to "projections" or "stanchions" on the nuclear drum (Figure 8-1D and E). When removed from this projection, such as is done on isolation, the strands do curl up and form a twisted ribbon, the helix (Figure 8-1F), but inside the cell we postulate that the ribbon is flat, held by projections at about 0.5 μm intervals along the DNA. This gives one a strand of a molecular weight of about 1.5×10^6, the molecular weight of isolated DNA that has been subject to shear forces and broken. The "anchor points" are the points at which the DNA breaks. Such a 0.5 μm strand would have 1.6×10^3 base pairs, and if three base pairs are needed to code one amino acid into protein, this length would be sufficient to code for about 500 amino acids, plenty for a protein. We suppose then, that each length from anchor to

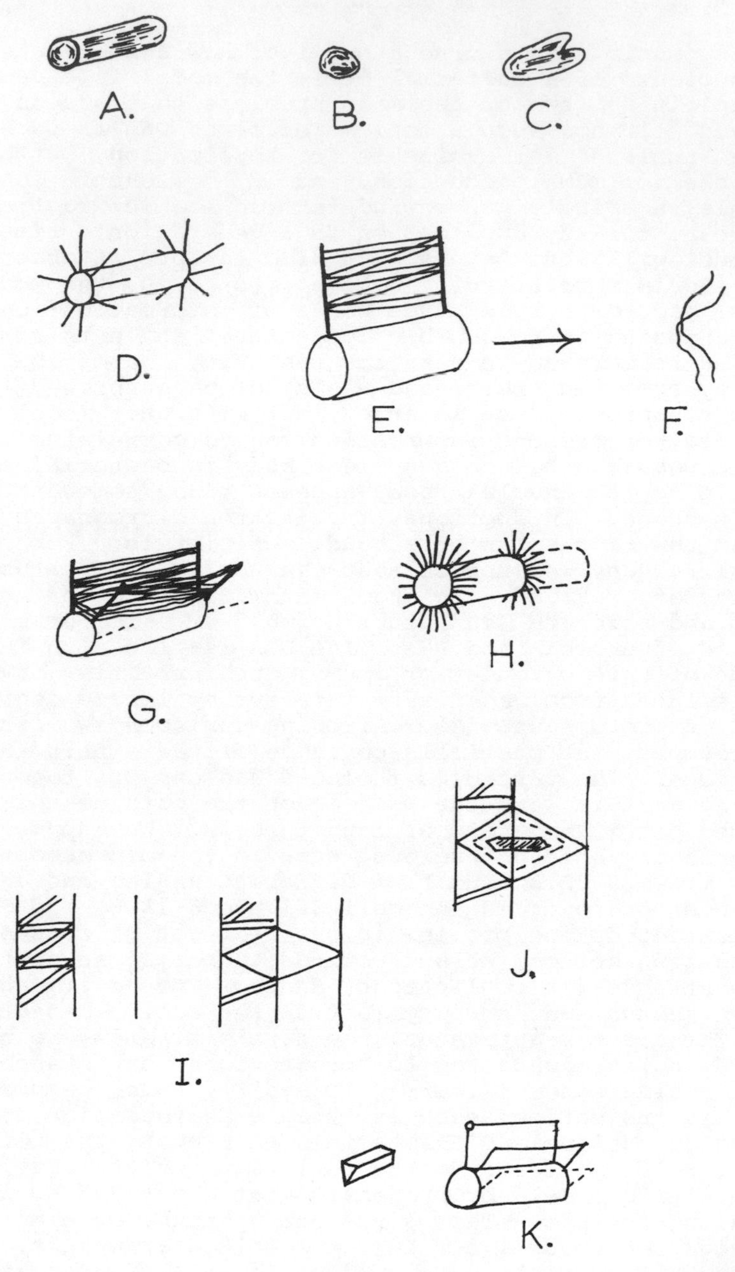

FIGURE 8-1. *Hypothetical model of how DNA is organized in a bacterial nucleus.*

anchor is the code for a single protein.

Since in a double-stranded helix of DNA each turn of the helix is 34 Å, a ribbon roughly 0.5 μm long would have to maintain itself against 140 turns. I do not know whether this is physically possible, although it seems unlikely. Nevertheless, on the theory that the bumblebee flies because it does not know aerodynamics, let us assume that it can be done — 140 twists can be held as a flat ribbon — and then see what one might deduce. Enzymes have been obtained after phage infection that are necessary for phage DNA synthesis but which do not by themselves synthesize DNA and are presumably involved in DNA unwinding (they are called DNA *unwindase*). However, in a phage, DNA is clearly packaged differently than in procaryotic cells; therefore, such unwindase systems may apply to phages but not necessarily to bacterial cells.

With a drum 0.5 μm in diameter (and 0.5 μm in length) there is not enough space on the circumference to accommodate all the DNA if it were there as a single layer. We therefore postulate that the DNA is strung across projections from the drum, as shown in Figure 8-1E and G, and we think that on the average there are about five DNA sections (0.5 μm long) on each projection, the amount that could code five proteins. With this postulate, assuming the DNA strand to be about 10 Å wide, the DNA would project upward from the drum about 50 Å. Particles isolated from *B. megaterium* possessed a layer of DNA about 500 Å, but we assume that since the DNA was streaming off of them, considerable swelling had taken place. At any rate, with an arrangement of this sort we can account for the enormous length of DNA packed into the "nucleus." Since the DNA is roughly 2,000 μm long and we are dividing it into 0.5-μm sections by binding it at its anchor point (4,000 sections), and are putting an average of five sections on each projection, we will need space on the circumference for 800 projections at each end of the drum. At the edge of the drum with a circle 0.5 μm in diameter (or 5,000 Å), the circumference of the circle would be 15,700 A. If we assume each projection to be 10 Å wide, we have plenty of room for the 800 required; so we postulate that there is a certain amount of room *between* each projection, presumably of the order of 10 Å.

Before replication of DNA can take place, protein synthesis and RNA synthesis must occur. This is because, we postulate, a new drum must be formed adjacent to the old (Figure 8-1H). One strand of the DNA is removed from its anchor point and pulled over the new drum (Figure 8-1I), a primitive type of mitosis. The DNA is then replicated by the enzyme building new DNA on the outstretched bands (they do not need to unwind) (Figure 8-1J). The length of the new DNA would be less than 0.5 μm and have roughly a molecular weight of 0.6×10^6. They do not extend the entire length of the ribbon. These are just the length required for the Okazaki pieces synthesized by the membrane-associated polymerase (pol-III), and they represent the DNA required for a single structural gene. These

Okazaki pieces are then joined end to end (or, in our model, around the corner made by two DNA strands bending at the anchor point) presumably by the Kornberg enzyme (pol-I). These joining sequences are relatively similar and are a little more "brittle" than the rest of the DNA strand, which accounts for the DNA breaking up into pieces of molecular weight of 10^6 when subjected to shearing forces. One also has semiconservative replication (one new strand attached to one old strand), which experiment shows to be correct. Indeed, with the drum arrangement, while DNA synthesis is going on in one part of the drum, mRNA synthetic can be going on at several other parts of the drum, a situation that almost certainly occurs.

We might further postulate that mRNA is made on the paddle wheel by the transciptase enzyme (Figure 8-1K), and which we have pictured as sort of a wedge that slides between the upright planes bearing the ribbons of DNA and thus makes mRNA. And to add another notion, the little point labeled "O" in Figure 8-1K represents the operator, and it may control the entry of the transcriptase into the appropriate slot. With this somewhat too mechanical system all the proteins controlled by a given operator (the operon) are represented on the DNA strands between two anchor points (stanchions). From the actual physical dimensions, one could postulate that there could easily be five to six ribbons between any two stanchions, so that we would not expect to find more than five to six proteins under the control of a single operator. We regard histidine, one of the exceptions, as strung on two related projection planes.

By the same kind of speculation, however, one could suggest that the Okazaki pieces are the replication of single genes, and this should provide a way of isolating the DNA of single genes. It will be interesting to see whether this prediction will, in fact, prove to be experimentally demonstrated.

This is probably as far as one should go with such speculation, and we, of course, present it tentatively, but it does nonetheless serve to correlate and coordinate a wide variety of facts. We assume that you realize that this is a picture of what might happen and not a claim that this is what does happen.

PROTEIN SYNTHESIS

We assume that you are familiar with the dogma that DNA is the template to prepare mRNA, which goes to the 30S portion of the ribosome and there codes for the order in which the amino acids are assembled into protein.

There are two kinds of enzymes involved in RNA synthesis. One, ribonucleic acid polymerase or transcriptase, is dependent upon DNA as a primer and seems to be the enzyme synthesizing the mRNA, that is, the RNA whose base sequence bears a precise relationship to the DNA of the cell. This requires the ribose triphosphates (ATP, UTP,

GTP, and CTP) and needs all four of them before it will operate. The product is RNA with the base sequence complementary to the DNA supplied and pyrophosphate. One should note that, in cells infected with an RNA virus, a new ribonucleic acid polymerase (RNA synthetase) appears, which is dependent on RNA as a primer.

The other enzyme capable of RNA synthesis is nucleotide phosphorylase or polynucleotide phosphorylase and uses as substrate the ribotide diphosphates. Any combination of ribotide diphosphates will work, and the RNA formed contains a random sequence of the bases used to form it. This enzyme yields the nucleotide polymer and inorganic phosphate. It does not require nor is it essentially influenced by any DNA or RNA primer. In most bacteria it is located in the membrane. It is a very important enzyme from the viewpoint of research, since with it one can make artificial RNA polymers with known base sequences, and these can act as messengers in protein synthesis. The availability of these polymers permitted an approach to knowledge of the base sequences required for each amino acid, and thus a knowledge of the genetic code. From the viewpoint of the cell, there is the strong suggestion that polynucleotide phosphorylase is involved in the *destruction* of mRNA.

Because of lack of time we must simply pass over a great deal of the information accumulated about RNA synthesis and breakdown in cells, except to say that 80—85 percent of the RNA in the cell is rRNA and some 10 percent is tRNA, both of which are relatively stable in contrast to mRNA, which is rapidly broken down (in bacteria, but not necessarily in eucaryotes) after use; that is, each mRNA serves to code the synthesis of 5 to 10 protein molecules and is then destroyed.

We also assume that the dogma of protein synthesis is familiar to you. In summary, the mRNA goes to the 30S ribosome where it serves as an attractant to charged tRNA, and, in the presence of various factors and GTP, adds its specific amino acid to the growing peptide chain (actually, the peptide is added to the amino acid); the entire tRNA with new peptide moves on one triplet codon, where the process is repeated. The charged tRNA is formed by a reaction between the amino acid and ATP, catalyzed by a different enzyme for each amino acid, and then the adenyloamino acid formed is transferred to a tRNA to make an aminoacyl tRNA, that is, a charged tRNA.

A mRNA may be engaged with more than one ribosome, and the long thread of mRNA may have several ribosomes "rolling down" it. These combinations of several ribosomes with mRNA are called *polysomes*.

BACTERIAL RIBOSOMES

Transfer of information from RNA to the amino acid sequences, and the formation of peptide bonds, takes place on the ribonucleoprotein particles, the ribosomes. Ribosomes are made of ribosomal proteins (one third) and rRNA (two thirds). Ribosomes dissociate into subunits, associate into dimers, trimers, and so on, and aggregate into larger polydisperse masses, or polysomes. These clusters of ribosomes, the polysomes, appear to be the active units in protein synthesis.

The ribosomes of microorganisms fall into two classes:

1. 70S: Most bacterial ribosomes are of the 70S type. The 70S ribosome is made up of one subunit of 30S and one of 50S. The 70S ribosome is usually considered to be the metabolically active particle. Dissociation of 70S occurs when the medium contains too little magnesium to maintain the integrity of the 70S particle.

2. 80S: Other kinds of microorganisms, such as *Azotobacter*, resemble the higher plants and animals in having ribosomes with sedimentation coefficient close to 80S, and this dissociates to give subunits of about 40S and 60S.

In a number of bacteria, rapid incorporation of amino acids has been found in coarse particulate fractions that sediment at low speed and are derived from cell membranes. These include *Alcaligenes faecalis*, *Azotobacter*, *B. megaterium*, and *E. coli*. In some cases ribosomes were obtained from these fractions. These results indicate that some of the ribosomes are associated with cell membranes; but whether these ribosomes are the most active *in vivo* or whether they are the most resistant to damage after cell disruption is not known. Perhaps polysome clusters, which may be the active units in protein synthesis, are preserved by attachment to membranes.

Ribosomes from *E. coli* contain RNAase and also a phosphodiesterase, which split RNA to 5'-mononucleotides. These enzymes are latent and inactive as long as the ribosomal structure remains intact. Once degradation has begun, however, some of the RNAase becomes active, and the result is an autocatalytic breakdown of the ribosomes.

Ribosomes isolated from *E. coli* contained a small amount of tRNA associated with the 50S. In the absence of protein synthesis, each 50S subunit in the 70S ribosomes reversibly binds one tRNA molecule.

[1]Nomura, M. 1970. The bacterial ribosome. *Bacteriol. Reviews 34*:228—277.

Kurland, C. G. 1972. Structure and function of the bacterial ribosome. *Ann. Rev. Biochem. 41*:377—408.

Haselkorn, R., and L. B. Rothman-Dengs. 1973. Protein synthesis. *Ann. Rev. Biochem. 42*:397—438.

(Reviewed by Vera Vasquez-Ortegon and Betsy S. Powell.)

Protein Synthesis

In *E. coli*, protein synthesis takes place in four steps: activation of amino acids, initiation, elongation (growing peptide chain), and termination (when protein is released).

The 30S portion provides a binding site for the mRNA, and 50S provides a binding site for peptidyl tRNA, the so-called P or D site. A second binding site for accepting aminoacyl tRNA prior to the formation of the peptide bond was also postulated, A site. The formation of the peptide bond is a cyclic process in which the state of ribosomes is identical at the beginning and end of each step of protein synthesis after the formation of the peptide bond. During the process there is the addition of an amino acid to the nascent protein, the presence of a different tRNA attached to the protein in P site, and the advance of the mRNA so that a new codon is available to direct the selection of a new aminoacyl tRNA for A site. Translocation occurs, that is, translocation of the peptidyl tRNA from A to P site as well as the movement of the mRNA by one codon length. The cycle is then completed. It is thought that the ribosomes remain immobile. mRNA moves and it must be coordinated with the formation of the peptide bond and translocation step.

Activation of Amino Acids

As discussed previously, the amino acid reacts with ATP to form an adenylo-amino acid, which reacts with a free tRNA specific for that amino acid, to become an amino acid carrying tRNA, which we call *charged* tRNA.

Initiation of Protein Synthesis

The mRNA attaches to the 30S ribosome. For protein synthesis to start, however, a specific triplet codon GUG is required. One will find AUG listed in even very modern monographs; if AUG is found at the beginning of the tRNA, it will unite with the UAC tRNA (complementary to GUG): if internal it will be recognized as AUG. GUG is the code for formyl methionine, and *every* protein starts with formyl methionine (see later). On the ribosome there are two sites for the absorption of tRNA. The A site, entry site, and the P site (sometimes called D site), the peptide or "donor" site. Charged tRNA can only adsorb to the 50S ribosome part at the A site, and its adsorption depends upon the specific mRNA triplet that occupies the mRNA on the 30S. These sites are thought to extend from the 50S to the 30S. But if one has a 30S ribosome with a GUG or AUG over the P site, the formyl-methionyl-tRNA will attach to the P (rather than the A) site on the 30S portion. This attachment then attracts a 50S ribosome, and a second charged tRNA (usually serine) now enters the

A site of the newly constituted 70S ribosome. After peptide bond formation occurs (the formyl methionyl being transferred to the seryl tRNA), the uncharged ("form-met") tRNA is released and protein synthesis continues by chain elongation. The dipeptide tRNA now moves to the P site, and the mRNA moves one triplet codon, using one GTP in the process. The initiation process requires three protein factors plus GTP. These factors are abbreviated IF for *initiation factors*.

IF_3 promotes proper binding of mRNA to the 30S subunit. IF_2 directs binding of initiator tRNA (form-met tRNA). It also controls the hydrolysis of GTP. IF_1 promotes the catalytic use of IF_2, which in turn controls hydrolysis of GTP. IF_1 also promotes dissociation of IF_2 and presumably GDP from the initiation complex.

GTP must be hydrolyzed. If it is not, IF_2 remains bound and peptide bond formation is blocked. The GTP may also be needed to move, bend, or twist the end of the tRNA closer to peptidyl transferase. This is called *accommodation*.

Chain Elongation

Chain elongation occurs in three steps: (1) codon-directed binding of the aminoacyl tRNA, (2) peptide bond formation, and (3) translocation.

There are two proposed sites on the 70S ribosome particle, P (donor) and A (entry). P is the site where the peptidyl tRNA, which will donate the peptide, is bound. A is the site where charged tRNA is bound at the time it accepts the peptide chain.

Codon-directed binding of the charged tRNA to the A site is catalyzed by the protein elongation factor T (EFT). This is composed of two parts: Tu (unstable) and Ts (stable). Tu binds with GTP, which complex binds charged tRNA. Next this combination attaches to the 70S form-met tRNA complex, releasing Tu GDP + Pi. Ts participates in the recycling of the Tu GDP to Tu GTP, so that the process can be repeated.

Peptide bond formation is catalyzed by a protein associated with the 50S ribosome called peptidyl transferase. In this transfer process the peptide chain is moved from the tRNA in the P site to the charged tRNA in the A site.

Translocation requires GTP and factor G. In translocation the peptidyl tRNA moves from the A site to the P site, replacing the deacylated tRNA. The entire complex moves down the mRNA one codon. The function of factor G is unknown. However, it must be released after binding for protein synthesis to continue.

Termination

When translocation places one of the nonsense codons UAA, UGA, or UAG in the A site, the ribosome binds a protein, R_1 or R_2, at an unknown site. This activates peptidyl transferase, which hydrolyzes the bond joining the polypeptide to tRNA in the P site. R_1 recognizes UAA or UAG; R_2 recognizes UAA or UGA.

It is not known how the deacylated tRNA is freed from the ribosomes, nor how the ribosome dissociates into subunits. It has been suggested that IF_3 may play a role. Some research indicates that IF_3 cannot initiate the dissociation but can only keep the subunits apart once they have been separated. It is interesting to note that the R (release) factors have not yet been tested for these activities.

A DIFFERENT VIEW OF BACTERIAL PROTEIN SYNTHESIS

It is generally assumed that these protein-synthesizing reactions take place in the cytoplasm, and electron micrographs of thin sections of bacterial cells show ribosomes distributed throughout the cytoplasm in an apparently random order. On the other hand, our own experience has led us to a different view as to the location and operation of the ribosomes. We shall outline this view, which we believe to be correct, although it is not yet generally accepted.

L. Daneo-Moore found (3) in *Streptococcus faecalis* that hexokinase and some of the enzymes of the galactose pathway were always associated with the membrane area; when released from the membrane by EDTA, they remained associated with an RNA component. Furthermore, other studies showed that in this organism rapidly labeled RNA, presumably mRNA, was firmly bound to the membrane fraction. It was found indeed that the membrane fraction had much more protein synthesis per unit RNA than did the ribosomal fraction, and that in this organism if one threw away the membrane fractions, as is customary when preparing ribosomes, one would throw away 90 percent of the poly-U-directed activity (i.e., a measure of protein synthesis). It was found that this organism could be broken by very gentle treatment. Cells near the end of the log growth phase, if treated with 20 mg of lysozyme/g of wet weight, were not converted into protoplasts, although they did lose some 70 percent of their hexosamine. They were stable enough to be washed in dilute saline and to be freed from lysozyme yet they were easily broken. Initially, we used a French press or grinding with alumina, but eventually found that passage three times through a large bore hypodermic syringe needle was sufficient to break the cells. We suspect that it is this gentle treatment during cell breakage which permits the ribosome to remain attached to the membranes and that more drastic treatments such as are normally applied to other bacteria, break ribosomes away from the membrane and thus they appear in the free state or as polysomes. We found that we could separate the membrane (particulate) fraction on a 10—70 percent sucrose gradient (rather than the usual 5—30 percent sucrose). A typical separation is shown in Figure 8-2. In such high level sucrose gradients, the 70S fraction comes to about fraction 25 (at fraction "D" in Figure 8-2). The solid line with solid circles represents the RNA content of gradient fractions and shows a remarkable peak at about 150S (point A, about fraction 20). This material does not incorporate protein and contains mostly 30S ribosomes. Electron micrographs show it to be ribosomes evidently attached to some kind of fibrous material, and we have evidence to indicate that this is composed of "polysomes" in which 30S ribosomes are attached to mRNA (roughly five to six ribosomes per message). Such a polysome, lacking 50S ribosomes, is incapable of protein synthesis. Protein synthesis itself is shown in the dashed line with the open circles. There seem to be three areas of endogenous protein synthesis (no mRNA added); area B at fraction 10 (about 400S), area C at fraction 15 (about 300S), and area D at fraction 25 (about 70S). The latter representing the synthesis presumably accomplished by the free ribosomes. However, the synthesis at fraction 10 (B) is remarkable, first because the RNA content is low, second because the peak is relatively sharp, that is, this does not seem to be a random group of particles but a group of definite size, roughly 400S. There is phospholipid present in the fraction and we think of it as a piece of the membrane which contains the protein synthesizing machinery. Since it seems to be of uniform size we think that it exists as a part of the mosaic structure of the cell membrane which differs in composition from the other membrane mosaic pieces about it (that are concerned with other functions) and thus breaks out on gentle treatment as a particle of relatively uniform size.

Because of these results (and others that have not been mentioned), we have evolved the following hypothesis as shown in Figure 8-3. We suppose that we have separated out a particle of reasonably uniform size which contains approximately eight 50S ribosomes rather firmly attached to the membrane but removable from it by mechanical or enzymatic means (especially by phospholipases). This particle also contains polynucleotide phosphorylase. The 150S A fraction contains mRNA plus 30S ribosomes and these may add to the particle to form 70S ribosomes all lined up in a row. The message moves down the ribosomal chain, the 30S ribosome remaining fixed to the 50S to which it originally attached. There are always some free 30S ribosomes which now are attracted to the mRNA (especially since its starting end seems to have a special affinity for 30S). As the message moves along, peptides are formed as in the conventional concept of protein synthesis, but as the message reaches the end of the membrane particle it is destroyed by an enzyme which we think to be polynucleotide phosphorylase since it is on this particle that most of it is concentrated. As the message pulls on through, it releases 30S particles that now dissociate (such dissociation occurs even in adequate magnesium

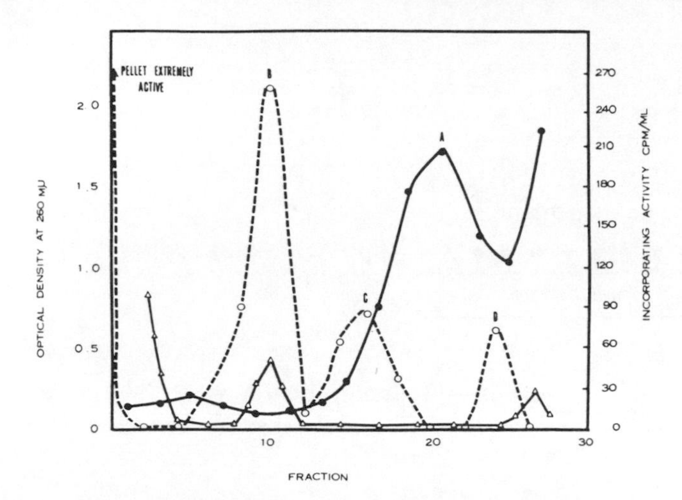

FIGURE 8-2. *Separation of the particulate fraction of Streptococcus faecalis on high level sucrose density gradients. Solid line with closed circles represents optical density at 260 nm as a measure of RNA. Dashed line represents incorporation of ^{14}C valine in the presence of nonradioactive amino acids as a measure of protein synthesizing ability. Solid line with triangles represents the incorporation of ^{14}C valine in the absence of other amino acids as a measure of "nonspecific" condensation or the completion of peptides already synthesized. From Herson, D. S., and C. L. Pon, J. Bacteriol. 100:1350–1354 (1969). Used with permission.*

because of the fixed position of the 50S portion), go to the DNA and there combine with mRNA, removing the mRNA from the DNA and protecting it from nucleases until it can reach another 400S section free from message and 30S material. It is the mRNA-30S complex which one sees as the 150S fraction (A). If so, since it has message and 30S it should be able to stimulate protein synthesis by the 400S particle. This it does indeed do: addition of the 150S portion to the 400S particle (B) increases amino acid incorporation two-fold and when added to fraction C (300S) the increase is four-fold (4). This suggests that fraction B is composed of particles having 30S and mRNA attached and the addition of mRNA-30S complex keeps the system going when the mRNA initially present has run out, whereas the 300S particle (C) represents the 400S particle minus its mRNA and 30S ribosomes, since protein synthesizing activity is so markedly increased by the mRNA-30S complex.

The large membrane fraction at the bottom (the pellet) of such gradients is also capable of protein synthesis but if it is sonicated or even syringed, and then recentrifuged through a gradient, a further 400S fraction can be isolated which is stimulated two to four-fold by the

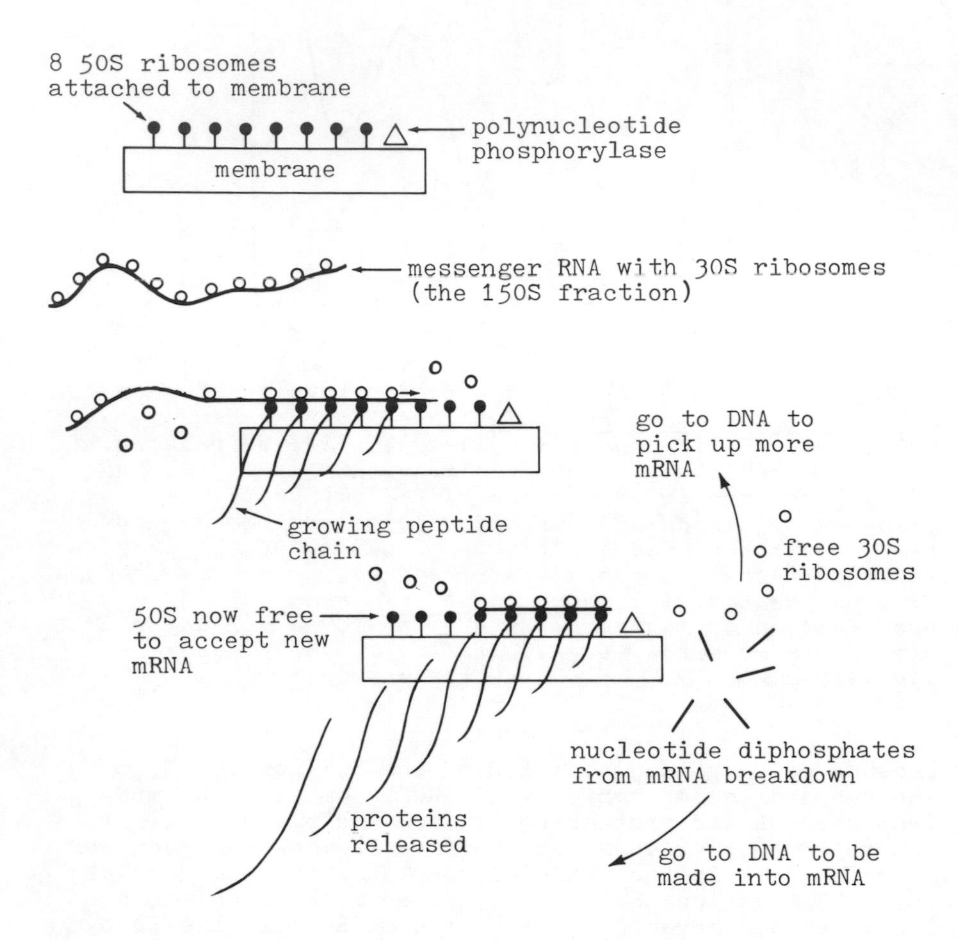

FIGURE 8-3. *A model of membrane associated protein synthesis.*

150S portion. Again it appears that one is breaking out a particle of uniform size which contains the protein synthesizing enzymes. These phenomena seem to apply to other organisms as well, not only to gram positives. If *E. coli* or *Pseudomonas* are broken with as gentle a system as possible (French press at 8,000 p.s.i.) we find a similar particle at about 400S and we find that between 50—80 percent of the protein synthesizing ability of the extract is associated with the membrane fractions rather than the ribosomal fractions. We, therefore, suppose that in the intact cell, in contrast to the rather drastically treated cell-free preparations, the 50S ribosomes are fixed in a definite order in special sections of the membrane, that mRNA-30S units come to these fixed ribosomes, and that most protein synthesis takes place at membrane surfaces. This orderly arrangement of the ribosomes represents a more efficient and more organized method of protein synthesis than that which we see in the more drastically treated cell-free preparations.

THE GENETIC CODE

We have mentioned that polynucleotide phosphorylase was able to synthesize "artificial" RNA, since it would condense any mixture of nucleotide diphosphates supplied to it. Thus in the presence of UDP it made a chain consisting solely of uridine residues, poly-U. If both ADP and UDP were supplied (in roughly equal proportions), the chain formed had UU, AU, UA, and AA sequences and lesser proportions of UUU, UUA, and so on, sequences. The probability of any given sequence could be calculated from the composition of the mixture of diphosphates used to make the polymer. It was then discovered that many of these polymers could serve as mRNA in the cell-free protein synthesis systems. It was found that if poly-U was supplied a peptide was formed (frequently large enough to be considered a protein) that contained only phenylalanine. A polynucleotide made from U:C at a 3 to 1 ratio would be expected to have many UUU sequences (and thus code for phenylalanine), but also UUC, UCU, or CUU sequences about one third as frequently. Thus, since we find that such a polymer incorporates serine at about one third the amount of phenylalanine, we presumed that serine had one (or all) of the above codes. The sequence (i.e., is it UUC or UCU or CUU?) one did not know.

It became possible by chemical synthesis to make triplet nucleotides in which the sequence was precisely known. It was found, for example, that $^{5}UCU^{3}$ (the first U being at the 5' end, the first position, and the second U at the 3' end) bound serine tRNA when it was supplied to a 70S ribosomal system, and it promoted the incorporation of serine when added to a system capable of synthesizing peptides. By this means the genetic code (i.e., which codon was associated with which amino acid) was worked out experimentally. However, more than one codon may stimulate the incorporation of a given amino acid (this is called *code degeneracy*). For example, UUU and UUC both code for phenylalanine, and UCU and UCC code for serine. The data of Table 8-2 suggest that, when the first two nucleotides are identical, the third may be either U or C, and that when the second codon is C the third may be any of the four. It further turned out that in several cases separate and distinct tRNAs were found to respond to the different codons. For example, there is a leucine tRNA that responds to UUG and a second leucine tRNA that responds to CUU. In fact, the term "degenerate" is rather an unfortunate one, since the code seems to be a bit more subtle than the three letter "words" might suggest; the fact that a given amino acid is coded for by more than one triplet means indeed that the code is more sophisticated.

TABLE 8-2. *The Genetic Code, or the Nucleotides in mRNA that Control the Incorporation of Specific Amino Acids*[a]

Second Letter	Third Letter	First Letter U	C	A	G
U	U	Phe	Leu	Isl	Val
	C	Phe	|	↓	↓
	A	Leu	|	Isl	Val
	G	Leu	↓	Met	Val, initial (f-met)
C	U	Ser	Pro	Thr	Ala
	C	|	|	|	|
	A	|	|	|	|
	G	↓	↓	↓	↓
A	U	Tyr	His	Asn	Asp
	C	Tyr	His	Asn	Asp
	A	Term(O)	Gln	Lys	Asp
	G	Term(A)	Gln	Lys	Glu
G	U	Cys	Arg	Ser	Gly
	C	Cys	|	Ser	|
	A	Cys, terminal	|	Arg	|
	G	tyr	↓	Arg	↓

[a]Initiator code: GUG, formyl methionine. Termination codes: UGA, usual; UAG, amber mutants; UAA, ochre mutants.

A case in point is that of methionine. At the beginning of every message there is a specific codon GUG (sometimes AUG) coupled with a tRNA for methionine. This methionine tRNA is capable of being formylated (from formyl tetrahydrofolic acid); thus formyl methionine seems to be the initiating amino acid in most bacterial proteins. Indeed, analysis of bacterial proteins show that over 40 percent have methionine as the *N*-terminal amino acid, about 30 percent serine as the terminal amino acid, and about 10 percent have alanine. Furthermore, several proteins have methionine—serine—alanine as the *N*-terminal sequence. This makes it appear that methionine is the first amino acid found in all proteins, the formyl group having been removed and sometimes the methionine itself, leaving serine or alanine as the terminal amino acid. There are two tRNAs for methionine, one antiparallel to GUG and capable of being formylated, the other, antiparallel to AUG and incapable of formylation. What is interesting is that these two tRNAs respond differently to their own code depending on where this code is located. If GUG is at the start of a mRNA, it is recognized by the anti-GUG tRNA, and formyl methionine is incorporated; but if GUG is in the internal message, another anti-GUG tRNA incorporates valine. If AUG is at the start of the code, it is recognized by the GUG tRNA, and formyl methionine is incorporated; if AUG is internal, it is recognized by AUG tRNA and methionine is incorporated. Thus the code for protein initiation is GUG. Similarly, the code for chain termination is UGA, UAA, or UAG, that of UGA being the code for normal termination and the others occurring in certain mutants.

If the tRNA for glycine (CGA) is treated with nitrous acid (to deaminate cytosine), it still reacts with the glycine-activating enzymes and still forms glycine-charged tRNA; but now it will incorporate glycine where glutamic should be and glycine where arginine should be. Of course, the glycine codon CGA is the codon on the mRNA, and the anticodon in the tRNA would be CCU; the cytosine could be deaminated by nitrous acid to uracil-yielding CUU (the anticodon for glutamic, GAA) or UCU (the anticodon for arginine, AGA).

Termination codons, which indicate when the peptide (or protein) is complete and is to be removed from the ribosome, are UGA and to a lesser extent UAG (amber mutants) and UAA (ochre mutants). Polynucleotide phosphorylase seems to be the agent breaking down mRNA when it has been adequately used. We have said nothing about the nature and assembly of ribosomes (for reviews, see 5,6). Any, indeed many, other interesting aspects, such as the factors required before the ribosome will work, have been omitted. But one should mention that there seem to be enzymes that destroy incomplete or malformed proteins (7).

We have presented an introduction and thus a somewhat elementary picture of this field. The development of knowledge just described is a remarkable accomplishment, and if one did not know that in fact it had developed in less than two decades, one would hardly believe it possible. This area of knowedge has become so vast and has attracted the attention of so many workers that it has become a specialized field of knowledge, molecular biology.

REFERENCES

Niederman, R. A. 1976. *Molecular Biology and Protein Synthesis.* Dowden, Hutchinson & Ross, Inc., Stroudsburg, Pa.

DNA

Boyer, H. W. 1971. DNA restriction and modification mechanisms in bacteria. *Ann. Rev. Microbiol.* *25*:153–176.

Goulian, M. 1971. Biosynthesis of DNA. *Ann. Rev. Biochem.* *40*:855–898.

Hayes, D. 1967. Mechanisms of nucleic acid synthesis. *Ann. Rev. Microbiol.* *21*:369–382.

Helinski, D. R., and D. B. Clewell. 1971. Circular DNA. *Ann. Rev. Biochem.* *40*:899–942.

Kjeldgaard, N. O. 1967. Regulation of nucleic acid and protein formation in bacteria. *Advan. Microbial Physiol.* *1*:39–96.

Klein, A., and F. Bonhoeffer. 1972. DNA replication. *Ann. Rev. Biochem.* *41*:301–332.

Koerner, J. F. 1970. Enzymes of nucleic acid metabolism. *Ann. Rev. Biochem.* *39*:291–322.

McCarthy, B. J. 1967. Arrangement of base sequences in deoxyribonucleic acid. *Bacteriol. Reviews* *31*:215–229.

Younghusband, H. B., and R. B. Inman. 1974. Electron microscopy of DNA. *Ann. Rev. Biochem.* *43*:605–619.

RNA

Burgess, R. R. 1971. RNA polymerase. *Ann. Rev. Biochem.* *40*:711–740.

Gauss, D. H., F. von der Haar, A. Maelicke, and F. Cramer. 1971. Recent results of tRNA research. *Ann. Rev. Biochem.* *40*:1045–1078.

Kozak, M., and D. Nathans. 1972. Translation of the genome of a ribonucleic acid bacteriophage. *Bacteriol. Reviews* *36*:109–134.

Littauer, U. Z., and H. Inouye. 1973. Regulation of tRNA. *Ann. Rev. Biochem.* *42*:439–470.

Ribosomes

Attardi, G., and F. Amaldi. 1970. Structure and synthesis of ribosomal RNA. *Ann. Rev. Biochem.* *39*:183–226.

Kelly, W. S. 1968. The "life cycle" of bacterial ribosomes. *Advan. Microbial Physiol.* *2*:89–142.

Pace, N. R. 1973. The structure and synthesis of the ribosomal ribonucleic acid of prokaryotes. *Bacteriol. Reviews* *37*:562–603.

Schlessinger, D. 1969. Ribosomes: development of some current ideas. *Bacteriol. Reviews* *33*:445–453.

Protein Synthesis

Attardi, G. 1967. The mechanism of protein synthesis. *Ann. Rev. Microbiol. 21*:383—416.
Fridkin, M., and A. Patchornik. 1974. Peptide synthesis. *Ann. Rev. Biochem. 43*:419—443.
Haselkorn, R., and L. B. Rothman-Denes. 1973. Protein synthesis. *Ann. Rev. Biochem. 42*:397—438.
Lengyel, P., and D. Soll. 1969. Mechanism of protein biosynthesis. *Bacteriol. Reviews 33*:264—301.
Lucas-Lenard, J., and F. Lipmann. 1971. Protein biosynthesis. *Ann. Rev. Biochem. 40*:409—448.

Papers

1. *Nature 249*:116 (1974).
2. *J. Bacteriol. 75*:102, 369 (1958).
3. *Biochim. Biophys. Acta 103*:466 (1965).
4. *J. Bacteriol. 100*:1350 (1969).
5. *Science 179*:864 (1973).
6. *Nature 253*:569 (1975).
7. *Nature 236*:143 (1972).

9
Energy—The Force of Life

Energy is the ability to do work. The transformations that any living cell makes in its nutrient are made for two reasons: to obtain the chemical structures from which it can build its substance, and to obtain the energy necessary for getting its work done. What work does a cell have to do?

A cell, once formed, needs a certain amount of energy to stay alive (energy of maintenance). It is thought that many of the compounds in the cell are unstable and must be continually resynthesized to maintain an adequate supply for particular functions. Some bacteria are motile; energy is expended in this motility, but bacteria do not move especially fast for their length. But the work done in motion uses very little energy, roughly 0.1 kg cal/g of bacterial nitrogen per day. Some energy is liberated as heat. In warm-blooded animals this is an important part of maintenance, but in the bacteria, where heat diffuses away so readily that the temperature of a colony is hardly detectably different from that of the environment, the heat produced represents a waste of energy, for heat per se is of no use to the bacteria.

A growing cell needs energy for the synthesis of new material, for the formation of new cells, and for all the processes of synthesis, cell division, and so on. A growing cell is expanding and increasing its surface, but the work required to do this against surface tension is very small (less than 0.1 kg cal/g of bacterial nitrogen) even though the new surface formed is very large. Most of the energy, then, is utilized to build new cell substance and to replace the vital parts that are continually breaking down.

The cell, then, growing or not, has a need for energy; it must continually do work. The cell obtains this energy by carrying out reactions yielding energy. It will now be our purpose to inquire how this is done. For the living cell is limited in what it can do and in what kinds of energy it can use. It is not a steam engine or a gasoline motor, and only in a remote sense is it an electric engine. The cell is a chemical engine and can use only chemical energy; it operates at a nearly neutral pH, relatively low temperatures, and with plenty of water.

The energy obtainable from a chemical reaction

$$A + B = C + D$$

is maximum work =

$$-\Delta F = \text{free energy} = RT \ln K - RT \ln \frac{(C)(D)}{(A)(B)}$$

$$-\Delta F = \begin{matrix}\text{concentration of}\\ \text{reactants and products}\\ \text{at equilibrium}\end{matrix} - \begin{matrix}\text{concentration of}\\ \text{reactants and products}\\ \text{at start of the reaction}\end{matrix}$$

Looking more closely at this equation, at equilibrium there is no energy: $\ln (C)(D)/(A)(B) = \ln K$. Therefore, the amount of energy one obtains depends on how far away one is from equilibrium. Thus "chemical energy" is a matter of *concentration* of substrates and products.

1. When ΔF is negative in sign, the reaction proceeds from left to right and is capable of liberating energy. Such reactions are called *exergonic*. Exergonic reactions (one should really say "exergonic conditions") release energy.

2. When $\Delta F = 0$, no energy is released or taken up. In this case

$$RT \ln \frac{(C)(D)}{(A)(B)} = RT \ln K$$

3. When ΔF is positive, the reaction does not proceed from left to right. Since the reaction does not proceed (from left to right), it does not exist. By definition, *endergonic* reactions *do not occur*, in biological systems or anywhere else. No one has ever found any reaction that requires energy (ΔF positive) going on in the living cell (or anywhere else). It never has; it never will.

However, when we see reactions that apparently require energy (e.g., the synthesis of starch from glucose), we find that the living cell has so adjusted the conditions that the reaction yields energy, that the reaction is not the one we suppose it to be. All chemical reactions that proceed at all in the living cell release energy. They must, or they would not proceed. The usual way of explaining the fact that organisms carry out reactions that, as written, have a $+\Delta F$ is that energy must be put into the reaction from some other reaction that is releasing energy. If we have reaction $A + B \rightleftarrows C + D$ ($\Delta F = -$) and reaction $E + F \rightleftarrows G + H$ ($\Delta F = +$), the way the second reaction can go is to use the energy released by the first. These are *coupled* reactions. However, there is *only one* known mechanism for the transfer of *chemical energy* from one reaction to another; the two reactions must have a common reactant. In the reaction above, in order for A + B to "supply" its energy to E + F, the real reaction must be

$$A + B + X \rightleftarrows C + D + X' \quad \Delta F = -$$

$$E + F + X' \rightleftarrows G + H + X \quad \Delta F = -$$

and both the reactions *must* have a $\Delta F = -$; otherwise, the reactions will not proceed. To "transfer" chemical energy from one reaction to another, there must be a reactant common to both. When the final reactions occur, they are exergonic. This is what is meant by saying that an exergonic reaction may be used to "drive" a normally endergonic reaction.

Biological systems employ a *common intermediate*, usually ATP (less often ADP). This substance is generated in certain (but not all) special exergonic reactions and enters into other reactions to make them exergonic. This is a rather convenient shorthand for the actual situation, which is that even syntheses are so arranged that they are exergonic reactions. This ATP is a common reactant of reactions yielding energy (exergonic) and those "requiring" it.

In order that an exergonic reaction be made to "drive" an endergonic one, the two reactions must be consecutive and have a common reactant. Yet many organisms build their entire cell substance from, say, inorganic salts and acetate. How can the acetate itself possibly enter into thousands of reactions as a common intermediate? This problem was solved by the cell by using a single reactant for "driving" the many apparently endergonic reactions involved in growth. This single common reactant is used in a large number of reactions, and, for reasons that have their basis in the chemical structure of molecules, is in the majority of the cases a compound of phosphorus, usually termed *energy-rich* phosphate. It is formed (usually from inorganic phosphate) in exergonic reactions. It enters as a common reactant into normally endergonic reactions (which do not proceed), and these thus become exergonic (and thus can proceed).

There are three types of energy-rich phosphate bonds:

1. -P-O~P bond, as in ATP (adenosine triphosphate).

2. $-\overset{\|}{C}-O\sim P$ (note that $-\underset{|}{\overset{|}{C}}-O-P$ bonds are not energy rich); in 1-3-diphosphoglyceric acid (the 1 bond is energy rich; the 3 bond is not).

3. $-N-\underset{\underset{N}{\|}}{C}-\underset{H}{N}\sim P$, as in phosphocreatine and phosphoarginine.

All three types may be regarded as "anhydride" bonds and written as formed by the elimination of water.

Type 1

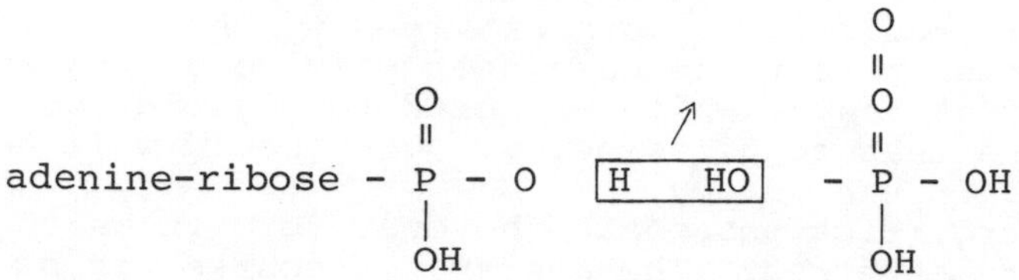

```
                                                  O
                                                  ‖
                     O              ↗             O
                     ‖                            ‖
adenine-ribose - P - O   [H    HO]   -  P - OH
                     |                            |
                     OH                           OH
```

to ADP

Type 2

```
CH3 - C = O
      |
      O    H
             ↗
               O
               ‖
     H - O - P - OH
               |
               OH
```

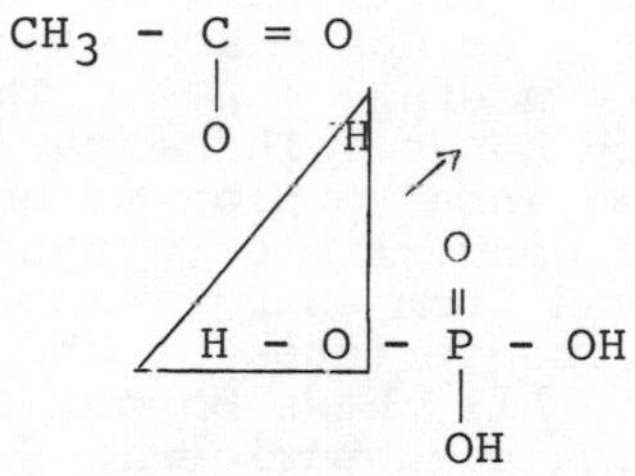

to acetylphosphate

Type 3

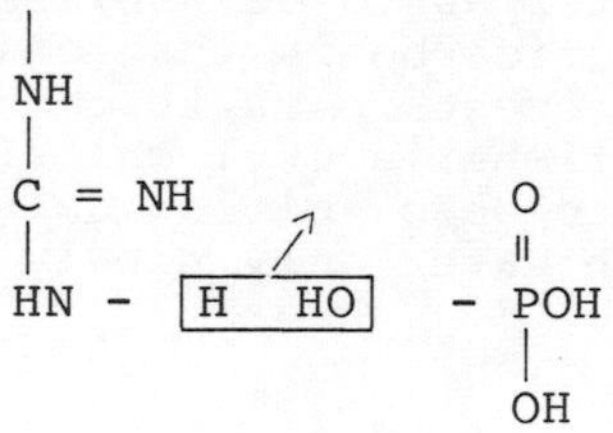

```
 |
 NH
 |
 C = NH            O
 |        ↗        ‖
 HN -  [H   HO]  - POH
                   |
                   OH
```

to phosphocreatine

Since such energy-rich bonds are formed only in certain reactions, these reactions become the energy generators for the cell. Indeed, the question of how energy is generated becomes then a question of how energy-rich bonds are formed. Actually, four reaction types generate the energy available to the cell, that is, energy-rich (phosphate) bonds, as follows:

1. Certain dehydrogenations.
2. Certain dehydrations.

3. The process of respiration.
4. The process of photosynthesis.

In a sense, these energy-rich bonds are the currency of the living cell. The work provided by the reactions that an organism runs to release energy is converted into *energy-rich bonds*, which are used to pay the cost of the reactions that require energy. They could well be called *biological free energy*.

Is energy-rich phosphate the only form in which energy may be transferred? The answer, of course, is no; there are other energy-rich bonds, not based on phosphorus, but all are convertible into energy-rich phosphate. For example, when pyruvate is oxidized by some enzyme systems, it forms an active acetyl group, which becomes attached to coenzyme A, in which form it may be transported to other reactions (e.g., the acetylation of amines and the acetylation of oxalacetate to form citrate) without the involvement of phosphorus. Yet there is an enzyme which converts acetyl CoA + H_3PO_4 to CoA + acetylphosphate, and acetylphosphate + ADP $\rightleftharpoons$ acetate + ATP. There are, then, certain other energy-rich bonds. These may be involved in reactions without the intervention of energy-rich phosphate, but they are all convertible into energy-rich phosphate and may be derived from energy-rich phosphate.

One further characteristic of the energy-rich bonds should be noted. All these bonds, or all the compounds containing them, are anhydrides and have a tendency to replace their anhydride bonds with water. It seems probable that anhydride bonds are used by nature for energy transport and accumulation, because the valence linkage per se is capable of storing energy. But the living cell must operate in a system in which there is a great deal of water, and most of the anhydride bonds known to the chemist (e.g., acetyl chloride and acetic anhydride) are so reactive with water as to be useless for biological purposes. The anhydride bonds that can be used in a water environment are thus restricted to those of phosphate and a few other combinations; acetyl chloride reacts too readily with water to be of biological use; acetyl coenzyme A reacts with water only slowly and may serve as an energy carrier.

REACTIONS RELEASING USABLE ENERGY

All reactions that proceed release energy, but only certain reactions release energy in a form available for use by the cell. That is, only certain reactions yield energy-rich phosphate or its equivalent.

It is convenient, indeed necessary, at this point to consider the Meyerhof—Embden system (which we assume is known to you, but to refresh your memory see Figure 9-1) from the viewpoint of energy transformations and energy release rather than from the viewpoint of carbon flux or hydrogen movement. Two reactions in this series (9-1A and 9-1B) warrant our particular attention, since they

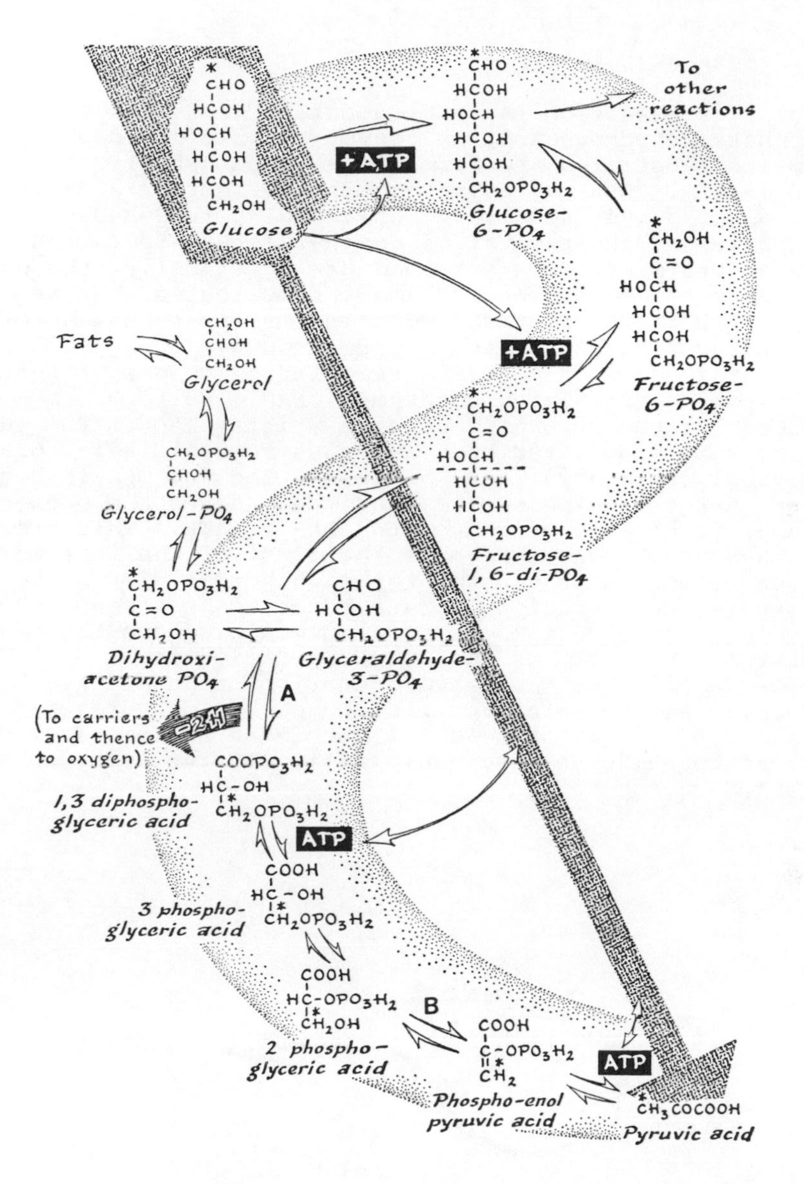

FIGURE 9-1. *The Meyerhof—Embden system — a primary pathway from glucose to pyruvate. From AN INTRODUCTION TO BACTERIAL PHYSIOLOGY, Second Edition, by Evelyn L. Oginsky and Wayne W. Umbreit. W. H. Freeman and Company. Copyright (c) 1959.*

represent two of the four principal mechanisms for the generation of that kind of energy which is available to the cell.

Dehydrogenation

The first reaction is carried out by glyceraldehyde-3-phosphate dehydrogenase, an enzyme capable of removing 2H from its substrates and forming 1-3-diphosphoglyceric acid (9-1A). One may represent this (Figure 9-2) as the removal of a hydrogen linked to the glyceraldehyde-3-phosphate and the removal of another linked to the enzyme with an enzyme product intermediate. Naturally, the detailed mechanism may be much more complicated, but at least this representation accounts for the formation of the energy-rich phosphate, represented by ~.

The hydrogen so removed is transferred to one of the nicotinamide-containing coenzymes, NAD^+, to form its reduced derivative, which we have written NADH. The substance remaining after such hydrogen removal is 1-3-diphosphoglyceric acid, and the upper, the "carbonyl" bond, is an energy-rich phosphate bond that, given the proper enzyme, is capable of converting ADP to ATP. This process of dehydrogenation represents the first of the four main processes for generating energy-rich bonds. For such substrate energy-rich phosphate bonds to be formed, the dehydrogenation must occur on a molecule so constituted that the net result of the dehydrogenation is a $-\overset{\|}{C}-O-PO_3H_2$. Thus the dehydrogenation of an aldehyde generally yields an energy-rich bond, but dehydrogenation of a carbon to carbon bond (as succinate to fumarate) does not, although subsequent manipulation of the hydrogen so released may do so.

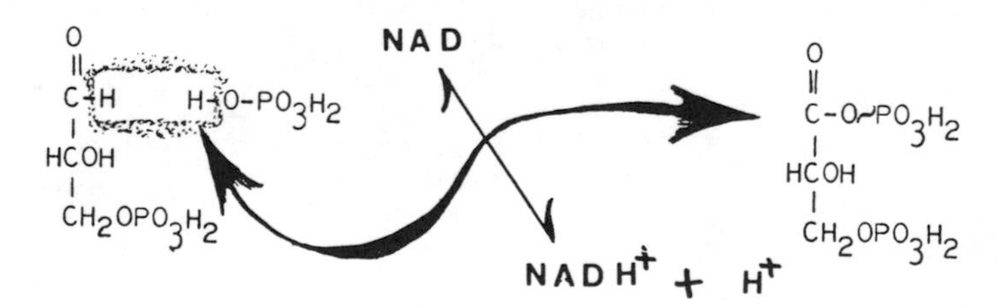

FIGURE 9-2. *Dehydrogenation of phosphoglyceraldehyde. From Umbreit, W., Metabolic Maps II, Burgess Publishing Co., Minneapolis, 1960.*

Dehydration

The second reaction (9-1B) of interest to us is that represented by the dehydration (removal of the elements of water) of the 2-phosphoglyceric acid to form enolphosphopyruvic acid (Figure 9-3). In this case the dehydra-

FIGURE 9-3. *Dehydration of 2-phosphoglyceric acid.*

tion forms a structure, $-\overset{\|}{C}-O-PO_3H_2$, comparable to that formed in reaction 9-1A by dehydrogenation and again resulting in an energy-rich phosphate bond. This is the second of the four main reactions for generating energy that is available to the cell.

Respiration

The third reaction type that yields usable energy to the cell is respiration. The earlier pictures of respiration (contact with oxygen) conveyed no real notion of the complexity of the situation. In bacteria we still have a very confused picture. There are, however, several substances involved, which require some discussion. The general idea is that these materials originate by reactions between iron (and copper) and proteins, and that these metal ligands generally develop into more complex structures. In the case of iron, some combined with porphyrins to form heme substances. Others simply retained the iron as an integral part of the protein structure. The heme proteins we know today are hemoglobin (a blood oxygen carrier), myoglobin (a muscle oxygen carrier), cytochromes (involved in electron transport not necessarily to oxygen), cytochrome oxidase (the enzyme that reacts with oxygen), the hydroperoxidases (catalase and peroxidase), and some oxygenases (as tryptophan oxidase). Oxygenases, or oxygen transferases, carry both atoms of the oxygen taken up into the substrate. All the above occur in bacteria except hemoglobin and myoglobin.

CYTOCHROMES

Cytochromes are heme proteins in which iron associated with the heme can exist in either the ferrous or ferric state. As such, they act as electron carriers. There are four general types:

1. Cytochromes of type a are *red* heme proteins having a formyl group on the porphyrin. There are several kinds (a, a_3, a_1, a_2, etc.) from different sources. Some may act as low potential electron carriers, but most react with oxygen, uniting electrons, oxygen, and H^+ from the milieu to form water. The cytochrome a proteins are frequently called *cytochrome oxidase.*

2. Cytochromes of type b are heme proteins with the heme a substituted protoporphyrin. They function as electron carriers at somewhat lower potential and do not usually react with oxygen. (Those from bacteria appear to react with oxygen.)

3. Cytochromes of type c are heme proteins with a covalent link between the heme and the protein. The heme is not split off by pyridine (and thus extractable with ether) as with the other cytochromes. The protein has a molecular weight of about 13,000 and is stable to acid, alkali, and to boiling. The cytochrome c's differ somewhat depending on origin. That from horse has absorption bands at 550 (α band), 522 and 415 nm when reduced (pink), and 530 and 400 nm when oxidized (orange), a molecular weight of 12,234, and a chain of 104 amino acids. That from *Pseudomonas aeruginosa* has its reduced α band at 551, a molecular weight of 8,000, and 82 amino acids in the chain. Although cytochrome from various mammals, and even from yeast, differ in amino acid sequence and in length of chain (yeast has five amino acids added at the amino terminous and lacks one at the carboxy), they all react with mammalian cytochrome oxidase (cytochrome a_3). But the bacterial cytochrome c's (normally of shorter chain length) do not react with mammalian cytochrome oxidase. Instead, bacteria have a variety of cytochrome oxidases, which differ in one or more aspects from those of the animal.

4. Cytochromes of type d are heme proteins in which one of the methylene bridges of the porphyrin is reduced so that the heme is a dihydroporphyrin. They seem to be rather less important than the others in electron transport and frequently are not found at all.

Bacteria differ in their cytochrome content. Cytochromes are absent from most lactics and most clostridia, although anaerobes may have heme proteins of a special nature; some even have cytochromes (1). *Streptococcus epidermis* normally lacks cytochromes but when grown with heme in the medium it does contain cytochromes (and develops a respiratory system). Most aerobes contain cytochromes, but the kinds and numbers may differ considerably. *Bacillus subtilis*, for example, contains a, a_3, b, c, and c_1; *Escherichia coli* contains a_1, a_2, b_1, and a soluble c (2). The passage of electrons in the mammalian system seems to be from cytochrome b to cytochrome c to cytochrome a (or a_3), the latter being cytochrome oxidase. It is clear that in many bacteria this cannot be the pathway, but it is not certain what replaces it.

NONHEME IRON PROTEINS

The nonheme iron proteins are the ferredoxins (electron transport at low potential), nitrogenase (nitrogen fixation), and succinic dehydrogenase (transport to the cytochrome system). There are others of more importance in higher tissues, such as hemoerythrin (oxygen carrier in invertebrates), transferrin (iron transport), ferritin (iron storage), xanthine oxidase (purine metabolism), lipoxidase, and ribonucleotide reductase.

FERREDOXINS (AND FLAVODOXINS) OF BACTERIA[1]

Ferredoxins and flavodoxins are the electron-carrying proteins found in many species of bacteria. They are classified by their electron-carrying prosthetic groups, the iron sulfide and flavin moiety, respectively. These proteins function in different physiological classes of bacteria: fermentative, aerobic, and photosynthetic.

It has been shown that several species of bacteria have two chemically distinguishable species of ferredoxin. Multiple ferredoxins have been found in both photosynthetic (*Rhodospirillum rubrum*) and obligate aerobic bacteria (*Azotobacter vinelandii*). Ferredoxin from *Clostridium*, *Chromatium*, and *Azotobacter* contains eight iron and eight sulfide atoms per molecule; those from *B. polymyxa* contain four atoms of iron and sulfide per molecule. The iron sulfide group functions as a center for accepting and donating electrons. Redox potentials vary from organism to organism (-90—455 mV).

Flavodoxins represent another major class of electron carriers and have been isolated from a number of bacterial sources. They have been isolated from several species of fermentative bacteria other than the clostridia, which include *Peptostreptococcus elsdenii*, *Desulfovibrio*, and *E. coli*. These flavodoxins are active as substitutes for ferredoxin in the phosphoroclastic reaction and *Desulfovibrio* flavodoxin (Fld) also functions in the sulfite reaction:

$H_2 \rightarrow$ hydrogenase $\rightarrow$ Fld $\rightarrow$ sulfite reductase $\rightarrow$ sulfite

[1]Yoch, D. C., and R. C. Valentine. 1972. Ferredoxins and flavodoxins of bacteria. *Ann. Rev. Microbiol.* *26*: 139—162. (Reviewed by Carol M. Krause.)

Axotoflavin functions as an electron carrier for N_2 fixation reactions by *Az. vinelandii*. Maximum flavodoxin production occurs on an iron-deficient medium. Amino acid compositions of flavodoxins show that all have a high percentage of acidic amino acids, accounting for their high affinity for DEAE cellulose.

Ferredoxins and flavodoxins are essential components of the electron transport chains of many bacterial types.

In fermentative bacteria, anaerobic bacteria that grow at the expense of a diverse group of natural products, such as sugars and organic nitrogen compounds, often depend on both proteins for redox reactions. They do not possess enzymatic activity in these reactions but, instead, interact with specific dehydrogenases and reductases that handle the substrates to be oxidized or reduced.

In aerobic bacteria, nitrogenase from all species of bacteria requires, along with ATP, a strong reductant for the reduction of nitrogen to ammonia. In *Az. vinelandii*, both ferredoxin and a flavodoxin-like molecule, azotoflavin, are present and are required for N_2 fixation when NADPH serves as an electron donor.

In photosynthetic bacteria, organisms such as *R. rubrum* are capable of producing strong reducing agents and contain two types of ferredoxin when grown photosynthetically in the light and only one ferredoxin when grown heterotrophically in the dark. This is the first report of one organism from which two chemically distinct types of ferredoxin have been isolated. These proteins are primarily membrane bound. It has been suggested that ferredoxin functions in a side reaction in photosynthetic bacteria and that an unknown compound, *X*, functions between ferredoxin and reaction center chlorophyll. These flavodoxins and ferredoxins are extremely hard to isolate and study because they are tightly membrane bound.

In addition, the ferredoxins are acidic proteins (isoelectric point below 4) with a molecular weight of about 6,000, and have five molecules of nonheme iron and five molecules of *labile sulfide* (H_2S released on acidification). These proteins are reduced (in the presence of appropriate enzymes) by hydrogen, pyruvate, hypoxanthine, formate, and may be oxidized by NAD^+, $NADP^+$, methylenetetrahydrofolate, and some other substances. They are involved in nitrogen fixation and photosynthesis, and serve as electron (or hydrogen) carriers of low oxidation reduction potential. They have not been found in lactobacilli, pseudomonads, bacilli, hydrogen bacteria, *Azotobacter*, and enterics (*Aerobacter*, *Escherichia*), although strains of these may yet be found to possess them. These organisms undoubtedly also require carriers of low-oxidation—reduction potential and this suggests that other carriers are involved. A cytochrome c_3 has been described in *Desulfovibrio* (which is low in ferredoxin), and vitamin E has been found in *Cl. nigrificans* (which should probably not be classified with other clostridia) and hydrogen bacteria, both of which lack ferredoxin.

Succinic dehydrogenase is a complex enzyme that contains nonheme iron, FAD, and coenzyme Q reductase (see later), and it may be tightly coupled with a cytochrome oxidase system; the combination, which may separate as a single large particle, is called succinic oxidase, or *succinoxidase*.

COPPER ENZYMES

Along with iron, there seems to have been a similar development with copper, except that the copper porphyrins did not appear. The bacterial copper proteins are superoxide dismutase (protection against superoxide; see later) and cytochrome oxidase (the terminal enzyme in respiration, i.e., the enzyme that reacts with oxygen), which are the most important; but azurin (an electron carrier) and galactose oxidase (which also contains pyridoxal) are found. Other copper proteins are hemocyanin (an oxygen carrier in invertebrates), plastocyanin (electron carrier in plants), ceruloplasmin (ferro-oxidase; Fe^{2+} to Fe^{3+} in blood), lysine oxidase (important in cross linking in collagen and elastin), tyrosinase and dopamine β-hydroxylase (adrenaline synthesis), and ascorbate oxidase (terminal oxidase in some plants).

In higher organisms the typical pathway of respiration generating energy-rich phosphate begins with the dehydrogenation of the substrate, the passage of the hydrogen via carriers to flavoproteins which convert it into hydrogen ions and electrons. The electrons are then passed through a series of heme-like pigments (the cytochromes) and eventually passed to an oxidizing enzyme (cytochrome oxidase), which unites electrons, hydrogen ions, and oxygen to form water. This pathway is represented diagramatically in Figure 9-4 which is somewhat more complex than the usual pathway given, since it recognizes that succinic dehydrogenase can put electrons into the pathway without involving NAD or NADP. Thus those carriers which can connect with succinic dehydrogenase can find an outlet. The composition of each complex is given in Table 9-1.

COENZYME Q OR UBIQUINONE

Coenzyme Q is a lipid-soluble quinone structure (see Figure 9-5) and somewhat resembles the carrier lipid (Figure 3-7). The composition varies somewhat from organism to organism. The usual bacterial coenzyme Q has a side chain of eight isoprenoid units, and the usual animal type has ten units in the side chain. Virtually all the bacterial coenzyme Q is found in the bacterial membrane.

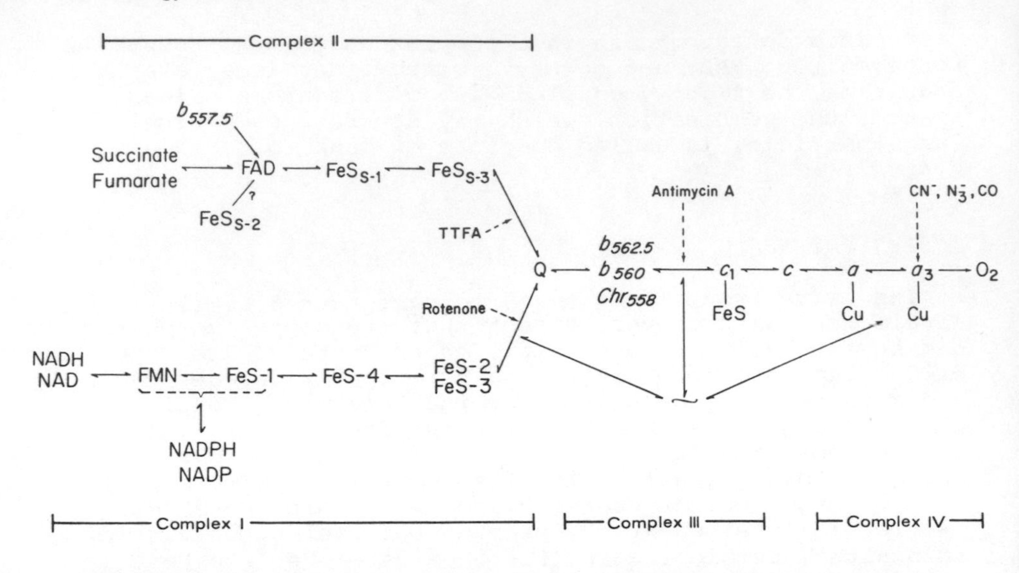

FIGURE 9-4. *Typical cytochrome pathway in mitochondria. FeS are iron–sulfur proteins; TTFA is thiophenyltrifluoroacetate; the remainder are conventional designations. From Hatefi, Y., et al. Reprinted from FEDERATION PROCEEDINGS 34:1699–1706, 1975.*

OXYGEN IS TOXIC

All living things are faced with the problem that the oxygen which usually supports their lives is, in fact, toxic, and their survival depends upon an elaborate system of defenses. The margin of safety is small, and they (and we) can be killed by exposure to five to ten times the usual oxygen pressure (which is 0.2 atm). Obligate anaerobes do not possess these defenses and are killed by exposure, even to air.

The reduction of O_2 to H_2O requires four electrons, and this is what the cytochrome oxidase system, a highly interrelated enzyme system, accomplishes; but if one looks at the stepwise reduction of oxygen (Figure 9-6), one finds two free-radical substances that can be produced. One is superoxide (O_2^-), the other the hydroxyl radical ($OH\cdot$) (3,4,5).

Until recently, these substances were of interest mostly to radiation chemists, but it is now known that numerous spontaneous as well as enzymatic oxidations do, in fact, generate substantial amounts of superoxide. The autooxidation of flavins, hydroquinones (like CoQ), catechol amines, ferredoxins, xanthine oxidase, aldehyde oxidases, and others generate superoxide. We therefore conclude that some fraction of the oxygen uptake of the cell unavoidably generates superoxide.

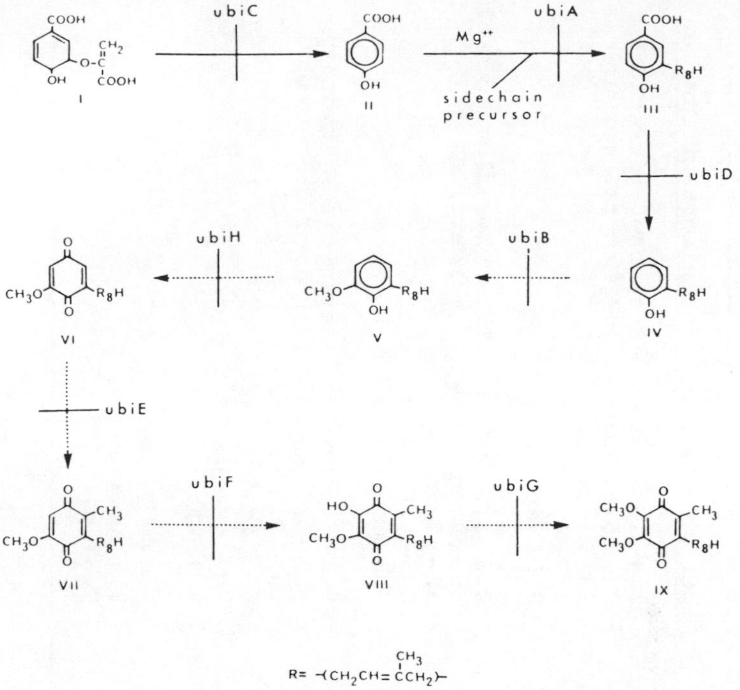

$R = -(CH_2CH{=}C(CH_3)CH_2)-$

FIGURE 9-5. *Pathway for the synthesis of coenzyme Q (ubiquinone) in Escherichia coli. The mutations in the genes controlling the synthesis are also shown. From Young, I. G., P. Stroobant, C. G. MacDonald, and F. Gibson, J. Bacteriol. 114:42 (1973). Used with permission.*

a. $O_2 + e^- \rightarrow O_2^-$

b. $O_2 + 2e^- + 2H^+ \rightarrow H_2O_2$

c. $O_2 + 3e^- + 3H^+ \rightarrow H_2O + OH\cdot$

d. $O_2 + 4e^- + 4H^+ \rightarrow 2H_2O$

superoxide dismutases

catalases and peroxidases

FIGURE 9-6. *Steps in the reduction of oxygen. O_2^- represents superoxide and $OH\cdot$ represents hydroxyl radical. From Fridovich, I., American Scientist 63:65–69 (1975). Used with permission.*

TABLE 9-1. *Components of the Cytochrome Electron Transport System*

	Complex	Component	Concentration per mg Protein	Inhibitors of Complex
I.	NADH-Q reductase	FMN[a]	1.4—1.5 nmoles	Rotenone, piericidin A, amytal, mercurials
		nonheme iron[b]	23—26 ng-atoms	
		labile sulfide	23—26 nmoles	
		ubiquinone	4.2—4,5 nmoles	
		lipids	0.22 mg	
II.	Succinate-Q reductase	FAD[c]	4.6—5.0 nmoles	Thenoyltrifluoracetone, mercurials
		nonheme iron	36—38 ng-atoms	
		labile sulfide	32—38 nmoles	
		cytochrome b	4.5—4.8 nmoles	
		lipids	0.2 mg	
III.	QH_2-cytochrome c reductase	cytochrome b	8.0—8.5 nmoles	Antimycin A, hydroxyquinoline-N-oxide, SN-59-49
		cytochrome c_1	4.0—4.2 nmoles	
		nonheme iron	10—12 ng-atoms	
		labile sulfide	6—8 nmoles	
		ubiquinone	2—4 nmoles	
		lipids	0.4 mg	
IV.	Cytochrome c oxidase	cytochromes a, a_3	8.4—8.7 nmoles	CN^-, N_3^-, CO
		copper	9.4 ng-atoms	
		lipid	0.35 mg	

[a]Complex I contains no FAD.
[b]Heme iron is less than 0.1 nmoles/mg protein.
[c]All the flavin of complex II is succinate dehydrogenase flavin.

After Hatefi *et al.*; see Figure 9-4.

Superoxide itself is quite unstable and decomposes rapidly without itself causing much damage. But it evidently reacts with hydrogen peroxide:

$$O_2^- + H_2O_2 \rightarrow OH^- + OH\cdot + O_2$$

It is the hydroxyl radical ($OH\cdot$) that is evidently capable of attacking any of a wide variety of materials found in cells. It accounts for a large part of the biological damage caused by ionizing radiation. Indeed, $OH\cdot$ is one of the most toxic substances known.

The cell, in order to live in air, must have a way of preventing $OH\cdot$. Enzymes that might attack $OH\cdot$ are not yet known. Enzymes attacking peroxide have been known for a long time (catalase, peroxidase), but only recently have enzymes been found that attack superoxide (O_2^-). These are called superoxide dismutases and at least four are already known. Two kinds, one from *E. coli* membranes and another from chicken liver mitochondria, contain manganese. A third is found in the periplasmic space of *E. coli* and contains iron. A fourth, seemingly characteristic of eucaryotes, is found in bovine erythrocytes and in the cytoplasm of chicken liver; it contains copper and zinc.

Obligate anaerobes do not have either catalase or superoxide dismutase, hence oxygen is toxic. Indifferents, such as streptococci and lactobacilli, do not have cytochromes or catalase, but they do have superoxide dismutase and can thus live in air, although they do not benefit from it and may indeed find it somewhat inhibitory. Too much oxygen may overwhelm even aerobes, and thus become toxic to them.

GENERATION OF ENERGY

In animal mitochondria (which may, after all, have been derived from bacteria), the energy of substrates is converted into phosphate-bond energy in small particles attached to mitochondrial membranes, called electron-transport particles (ETP). These are thought to carry out the reactions indicated in Figure 9-4. From the animal ETP, four enzymatically active proteins (complexes I through IV) have been separated in addition to a structural protein. These do not account for all the protein of the particle, and the remainder is thought to be involved in oxidative phosphorylation. In addition to the protein, the particle contains phospholipids (whose removal results in the loss of activity), coenzyme Q, and cytochrome c, all of which seem to be located in contact with the enzymes. The usual respiratory pathway is NADH to reduced coenzyme Q to reduced cytochrome to oxygen (complexes I, III, IV). The enzyme succinoxidase (or, as its first step, succinic dehydrogenase) reacts with CoQ without the intervention of the niacin coenzymes, and thus proceeds, CoQ, reduced cytochrome to oxygen (complexes II, III, IV). The cytochrome pathway is thus highly organized in a particle containing the components in a physical structure in which the components are spatially related. When properly prepared, this particle is capable of generating energy-rich phosphates. The P/O ratio (i.e., energy-rich phosphates formed per oxygen taken up) in mitochondrial preparations can approach 3; that is, three energy-rich phosphates can be generated from 2H passed through the system. But only two are found for succinate; that is, one energy-rich phosphate is generated by NADH-CoQ reductase (complex I), which is not formed by succinate-CoQ (complex II).

At present there are two major theories as to how energy-rich phosphate is formed, The first, the chemical theory, conceives of the formation of a high-energy chemical intermediate during the passage of electrons through the respiration chain. This intermediate reacts with inorganic phosphate and ADP to form ATP. This is, in principle, very similar to the generation of energy-rich phosphate via glyceraldehyde phosphate dehydrogenase (except that the enzymes are attached to mitochondrial membranes and that some exchange between them occurs through the phospholipid components). Indeed, it has proved possible to resolve the phosphorylation process into separate components by fractionation of electron-transport particles and then to reconstitute the oxidative phosphorylation by recombination of certain of the fractions. However, successful reconstitution has always produced a reasonably intact membrane structure. The second hypothesis, the chemiosmotic, postulates no high-energy chemical intermediate, but that the membrane is a necessary element in energy coupling. Such reconstituted membranes are capable of concentrating ions and can serve as a proton pump. Just how the membrane so operates is not clear at the moment.

GENERATION OF ENERGY BY BACTERIAL MEMBRANES[1]

1. Energy in bacterial membranes, derived from the electron-transport system, is conserved as phosphate bond energy of ATP or substrate transport across membranes.
2. Reactions of metabolism are oriented with respect to the membrane; enzymes are organized within membranes. Vectorial processes are intimately related to the generation and utilization of energy in bacterial membranes.
3. In principle, the relationships of energy-conservation and energy-utilization reactions in bacterial membranes are believed to be the same as those in mitochondrial membranes.

[1]Harold, F. M. 1972. Conservation and transformation of energy by bacterial membranes. *Bacteriol. Reviews* *36*:172—230. (Reviewed by Richard A. Bronen.)

Mitochondrial Membranes

Theories of energy conservation. Oxidation of electron donors (NADH) are catalyzed by a multienzyme chain that is associated with the inner membrane of mitochondria. Energy is conserved at specific coupling sites by synthesis of 3 ATP/NADH oxidized. Certain inhibitors [oligomycin, rutamycin, dicyclohexylcarbodiimide (DCDD)] block ATP synthesis and ATPase activity, and also block ATP-induced transhydrogenation and ion transport. These inhibitors do not affect these reactions (transhydrogenation ion transport) when reduced substrates donate energy.

Conclusion: There is an energized state that is a common link between different energy-dependent functions (and oligomycin inhibits the linkage between the energized state and ATP formation).

The following are three theories to explain the energized state:

1. *Chemical coupling hypothesis:* Energy released by redox reactions is conserved in the form of high-energy intermediates. Mechanism is similar to substrate phosphorylation. No high-energy intermediate has yet been found, possibly because the intermediate is stable only in the hydrophobic environment of lipid mitochondrial matrix.

2. *Conformational coupling:* Electromicrographs point to an energized structural state of mitochondria, which is supposed to be responsible for ATP synthesis, pH gradients, and ion transport.

3. *Chemiosmotic:* It is thought that in mitochondria the inner membrane is impermeable to most ions (including H^+ and OH^-); thus it has low electrical conductivity. There is a vectorial organization of the respiratory chain across membrane. Oxidation of a substrate results in translocation of protons from the interior to the exterior of the mitochondria. Oxidation of NADH results in six protons being transported; this results in a pH gradient and an electropotential across the membrane. These two gradients together result in a force (the proton motive force) that pulls protons across membrane into mitochondria. The pH and electrical potential gradients result in a reversal of direction of the ATPase reaction, so as to bring about net synthesis of ATP.

Usual reaction:

$$ATP + H_2O \xrightleftharpoons{ATPase} ADP + Pi$$

Mitchell's reaction:

$$ATP + H_2O + H^+_{(in)} \xrightleftharpoons{ATPase} ADP + Pi + H^+_{(out)}$$

$$Keq' = \frac{(ADP)\ (Pi)}{(ATP)} \frac{(H^+)_{out}}{(H^+)_{in}}$$

ATPase (F, "coupling factor") may be situated in membrane such that ATP, ADP, Pi, and H^+ have access to the active site only from inside of mitochondria, OH^- only from outside the membrane, and water has no access at all. Synthesis of ATP occurs by extraction of elements of water from ADP + Pi; H^+ is pulled inward by OH^- and negative potential, while OH^- is pulled outward. With removal of water in the form of $(H^+)(OH^-)$, the ATPase equilibrium is shifted toward production of ATP.

It is an accepted fact that the mitochondrial membrane is impermeable to small ions, including protons and OH^- ions. There is evidence of membrane polarity and vectorial organization of respiratory catalysts. Stalked spherical particles are attached to the matrix side of the membrane. When the membranes are disrupted, closed vesicles of two types result, one with the particles facing toward the matrix and one with the particles facing the medium. This latter type translocates protons and ions in the opposite direction of the former (or right-side-out type).

Most evidence has confirmed the postulate that proton extrusion occurs and that a proton motive force is created. Proton extrusion should lead to a negative mitochondrial interior, and the mitochondria should absorb cations and expel anions (the reverse should occur in inside-out vesicles). In fact, respiring mitochondria accumulate Ca^{2+} and K^+ (in the presence of lipid-soluble valenomycin).

The coupling device, which mediates the energy transfer reaction that results in ATP synthesis, is a complex enzyme system (which includes ATPase of F_2 particle). Studies of inhibitors of this complex reveal that the coupling device is quite separate from the oxidative chain, although the mechanism of ATP synthesis is unknown. The chemical coupling hypothesis involves the familiar "discrete steps and common intermediate" mechanism, although none has been found. The chemiosmotic hypothesis is supported by the following: (1) imposition of a pH or K^+ (w/valinomycin) gradient on chloroplasts or mitochondria results in ATP synthesis; (2) the addition of ATP to anaerobic mitochondria results in hydrolysis and subsequent proton ejection.

Typically, uncouplers block phosphorylation but stimulate respiration. Mitchell proposed that uncouplers act as circulating carriers which conduct protons across the osmotic barrier, thus dissipating the proton gradient. Experimental evidence indicates that uncouplers are proton conductors. However, another explanation is that uncouplers catalyze the acid hydrolysis of energy-rich intermediates (in the hydrophobic matrix).

Bacterial Membranes

The bacterial membrane carries out interconversion of chemical, mechanical, and osmotic forms of energy.

Stalked particles occur on the inner surface of plasma membrane, which contain Ca^{2+}-activated ATPase. (The ATPase of the homofermentative *Str. faecalis*, which does not carry out oxidative phosphorylation, appears to lack a visible stalk.) Functional proteins are associated with the bilayer of phospholipid molecules that make up the membrane. The membranes reseal after injury. Disruption of bacterial membranes results in several types of particles: fragments, right-side-out vesicles, inside-out vesicles, and patchwork (hybrid) vesicles (based on the side on which stalks appear).

Bacterial respiratory chains are more flexible than mitochondrial chains. A wider range of substrates and electron acceptors is used. Soluble coupling factors are required for oxidation and phosphorylation. ATPase is one factor common to all phosphorylating membranes. It is believed that the phosphorylating particles from bacterial membranes are resealed vesicles similar to the mitochondrial particle vesicles. Evidence indicates that the coupling of respiration to phosphorylation occurs indirectly, via one or more nonphosphorylated energy-rich intermediates. This is revealed, as in mitochondria, by the effects of inhibitors on various systems of the bacterial membrane. Studies of *Micrococcus denitrificans* suggest oxidative phosphorylation requires a proton-motive force across membrane; translocation of H^+ is associated with oxidative phosphorylation.

Photosynthetic bacteria can use either light or ATP as an energy source to generate ATP or reducing power. Illumination of chromatophores results in accumulation of H^+. Vesicles are thought to be inverted during preparation.

Bacterial membranes are largely impermeable to the majority of metabolites. Uptake is related to transport, not permeability; it depends on the interaction of substrate with a membrane component that specifically recognizes it and translocates it across the membrane. Bacterial transport systems can achieve large concentration gradients (1:10^6 K^+ gradient in *E. coli*). Some type of energy (from the metabolism) is needed to do this work. A well-known mechanism of substrate transport in bacteria is group translocation. A substrate is transported across the membrane after or during being chemically modified. Certain sugars are known to be transported in this way, and it is believed that uptake of purines, pyrimidines, and fatty acids may also occur in this fashion. However, it is difficult to differentiate between group translocation and translocation followed by chemical modification. It appears that phosphorylation occurs concomitantly with and is required for the translocation process itself. One experiment showed that the internal glucose-6-P was almost entirely derived from ^{3}H-glucose added externally and that ^{14}C-glucose already present in the vesicles was not phosphorylated.

In the β-galactoside permease system (in *E. coli*), there is no evidence that chemical modification of β-galactoside occurs during translocation; the coupling of transport to metabolism occurs with the carrier rather than the substrate. Most of the evidence indicates that β-galactoside transport does not require metabolic energy; only when it is accumulated against a concentration is there a need for the coupling to metabolism. Metabolic inhibitors and uncouplers stop accumulation and dissipate the accumulated substrate, but do not inhibit translocation and equilibration of substrate. A mutant deficient in the energy-coupling step could not accumulate β-galactoside. Kepes concludes that a high-affinity permease becomes a low-affinity permease when it becomes energized on the interior side of the membrane. The view that energy is required only for accumulation but not for translocation is highly disputed.

It is supposed that there is a link between ion gradients, transport, and metabolism. The primary ion gradient is the driving force for secondary carriers, which have affinity sites for both the secondary ion and the accumulated nutrient. The most likely coupling ion systems in bacteria are Na^+, K^+, and H^+. But no Na^+ has been directly implicated as the source of energy for bacterial transport. All bacteria, however, accumulate K^+, and they might use some of the potential energy stored in this gradient to accumulate other metabolites. At present, no evidence exists to support this.

OXIDATIVE PHOSPHORYLATION IN BACTERIA

Studies of oxidative phosphorylation in bacteria have yielded a variety of results such that no single unifying picture is as yet possible. It is therefore necessary to list several kinds of systems separately, because we do not know whether or not they are related. For example, *Alcaligenes faecalis* contains particles, apparently membrane fragments, that carry out oxidative phosphorylation with a P/O ratio of about 0.3 starting with NADH. After precipitation with ammonium sulfate, this activity is lost, but can be restored by the addition of a tetranucleotide containing one uridylic, two adenylic, and one guanylic acid, and an additional protein factor. The system is relatively insensitive to uncoupling by dinitrophenol. From *Az. vinelandii*, particles have been obtained acting on a variety of substrates and requiring a protein cofactor. The P/O ratio was usually about 0.3 to 0.5. Similar particles have been obtained from *E. coli* with P/O ratios from 0.3 to 1.0; they also require a protein cofactor. *Mycobacterium phlei* yields a particle relatively more efficient in oxidative phosphorylation (i.e., P/O of 1.2—2). The oxidative phosphorylation activity is sensitive to light, and it appears that vitamin K_1 is required. A soluble system has been obtained that contains a protein coupling factor, phospholipid, vitamin K_1 reductase (as well as other factors). From *M. lysodeckticus*, a system can be prepared with a P/O (from NADH) of about 0.5 that is almost insensitive to dinitrophenol, oligomycin, KCN, and so on. Many such bacterial systems (from *E. coli*, pseudomonads, micrococci, etc.) can also transfer electrons to nitrate (to form nitrite) with P/NO_2^- ratios of 0.2—0.5. Several autotrophic bacteria have yielded preparations capable of oxidative phosphorylation. But it is obvious that so far P/O ratios are low, and that much research is needed on this subject before any clear picture is obtained. We also do not know at this juncture whether all these systems capable of oxidative phosphorylation are composed of membrane particles or reconstituted membranes.

LOCATION OF RESPIRATORY ENZYMES

The content of respiratory enzymes and their location within the cell may be varied considerably by conditions of cultivation. For example, addition of oxygen to anaerobically growing *Staphylococcus aureus* results in cells that contain 15 times as much cytochrome a and 55 times as much cytochrome oxidase (cytochrome a). The electron-transport system is found in the membrane, and other components of the membrane (vitamin K_2 and phospholipids) change in response to aerobic conditions.

OTHER RESPIRATORY PATHWAYS

The cytochrome respiratory pathway so far discussed is not the only pathway to oxygen. At least three other major systems react with gaseous oxygen. These are the oxidases, the hydroxylases, and the peroxide formers; each is discussed below. All these have Michaelis—Menten constants with respect to oxygen of about 2×10^{-5} or even 10^{-4} *M* compared to the constant obtained with the cytochrome system of 2×10^{-6} *M*. That is, they require 10 to 100 times as much oxygen as the cytochrome system; thus when the oxygen concentration becomes limiting (as in a colony or at high cell concentration), they cannot compete with the cytochrome systems. Furthermore, it is not at all clear that these other systems generate energy-rich phosphate. In most cases they evidently do not, but certain of them might be capable of such energy-rich bond formation. Clearly, however, it is the cytochrome system that generates most of the energy derived from respiration.

Peroxide Formers

In peroxide-forming reactions the oxygen taken up does not appear in the product derived from the substrate but appears as hydrogen peroxide (H_2O_2). The cytochrome system is similar except that the product is H_2O. These peroxide-forming enzymes have been called *two electron transfer oxidases* (the cytochrome system is a *four electron transfer oxidase*). All are flavoproteins; some have metals associated with them. A glycerol oxidizing system obtained from *Str. faecalis* permits a rather rapid rate of respiration in cells completely lacking in the cytochrome system (6). The dihydroxyacetone phosphate formed can be metabolized via the normal metabolic routes. In most organisms, glycerol phosphate is oxidized via NAD^+ linked dehydrogenases, but this particular organism has this rather unusual system. Some peroxide-forming enzymes can use other electron acceptors, but others must have gaseous oxygen.

Hydroxylases

Hydroxylases have also been called "mixed function" oxidases or "mono oxidases" and incorporate only one atom of oxygen per molecule of substrate. The other atom of oxygen is reduced to water. In some cases the substance itself serves as the endogenous electron donor; in others (the majority) an external electron donor is required. Reduced niacin coenzymes, reduced cytochrome, sometimes metals (Fe, Cu), ascorbate, and pteridine derivatives may serve as the donor depending on the enzyme.

Such hydroxylases are found in many bacteria. These enzymes are also found in animal cells in the supernate from which mitochondria have been removed. Here they are associated with the endoplasmic reticulum, and, given the proper substrate, oxygen uptake may be as rapid as that of the mitochondrial fraction. In animal and microbial cells there are enzyme systems of this type (steroid hydroxylases, phenolases, fatty alcohol oxidases, etc.), and in the animal they attack *xenobiotic* substances (i.e., drugs and the like, foreign to the metabolic network of the cell).

Oxidases

Oxidases have been called "oxygen transferases" or "dioxygenases," but the term "oxidase" is more reasonable. However, many enzymes were called "oxidases" (e.g., amino acid oxidase) long before this category of enzymes was delineated; therefore, the term "oxidase" is not too precise. However, the enzymes referred to here are those which add oxygen to the substrate molecule, and both atoms of oxygen taken from the gas phase unite with the substrate. A good example of an oxidase is tryptophan oxidase (pyrrolase) (Figure 9-7) in which the oxygen taken up appears directly in the *N*-formyl-kynurenine formed. Enzymes of this type are all metalloproteins, are activated by reducing agents and inhibited by metal ion binding agents, and require a metallic ion (usually ferrous ion) for activity. Oxygen appears directly in the product and does not exchange with water. It appears that the enzyme unites with oxygen, and the oxygen-carrying enzyme reacts with the substrate to form the product.

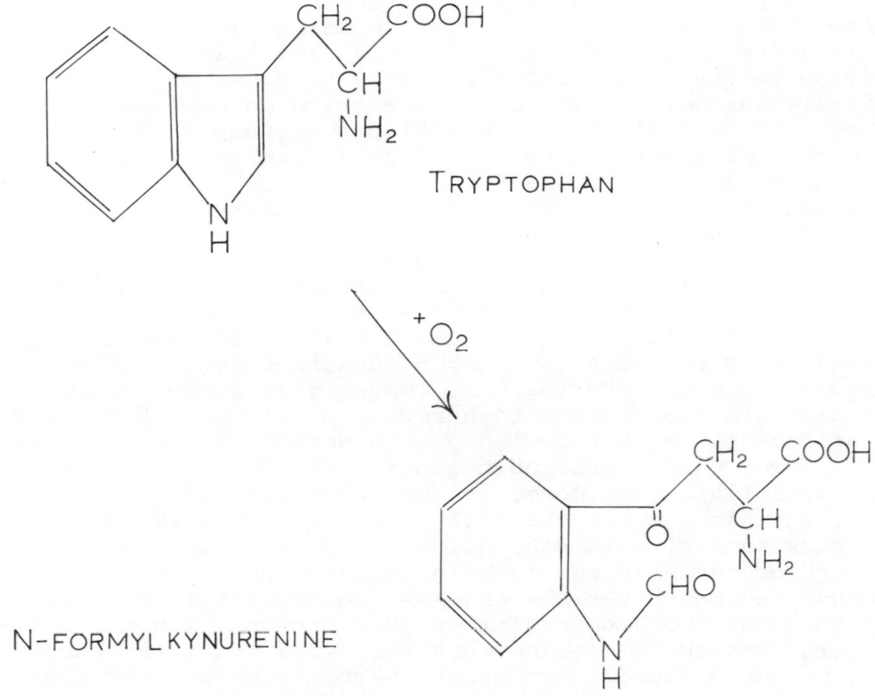

CH_2 COOH
CH
NH_2
N
H
TRYPTOPHAN

+ O_2

CH_2 COOH
O
CH
NH_2
CHO
N
H
N-FORMYLKYNURENINE

FIGURE 9-7. *Tryptophan oxidase.*

Oxidases are particularly prevalent among the pseudomonads and have been crystallized from some. They are characteristically involved in the oxidation of aromatic rings.

ORGANISMS POSSESSING PATHWAYS TO BOUND OXYGEN

Some organisms can grow in the absence of air if a bound form of oxygen is present. The most prominent substance serving as a source of bound oxygen is nitrate (sometimes nitrite). It is true that certain strict aerobes can grow somewhat under anaerobic conditions if supplied with external bound-oxygen sources (or electron acceptors) such as ferricyanide, but in most of these cases the electron acceptor is capable of reacting at some step with the normal pathway to oxygen. The use of nitrate as a hydrogen acceptor, however, does not appear to involve the normal pathways to oxygen.

There seem to be two situations. The first is that of several common facultative organisms (*Aerobacter aerogenes*, *E. coli*, etc.) which can use nitrate as a terminal hydrogen acceptor under anaerobic conditions, producing nitrite. For example, *A. aerogenes* can utilize nitrate as the sole source of nitrogen in a minimal medium under both aerobic and anaerobic conditions. We can call this *assimilatory nitrate reductase* (abbreviated ANR) and recognize that the process normally produces not only nitrite but further reduces the nitrite to ammonia. Under anaerobic conditions, nitrate also functions as a terminal hydrogen acceptor, which we can call *respiratory nitrate reductase* (abbreviated RNR).

It turns out that ANR is very sensitive to sonic disintegration and other homogenization procedures, is relatively indifferent to oxygen, but is very low in cells grown anaerobically (in some organisms it is repressed by ammonium ion). RNR, on the other hand, is not inhibited by sonic disintegration (it may actually be stimulated), is strongly repressed by oxygen, and is inactivated by oxygen *in vitro*. Furthermore, when cells grown anaerobically are transferred to aerobic conditions, there is a considerable lag before log growth occurs, indicating that RNR, formed under anaerobic conditions, does not serve in the assimilatory process. After resumption of growth in air, the nitrate reductase (to nitrite) activity gradually becomes sensitive to sonication, suggesting that ANR is being formed while RNR is being diluted out. These results suggest that ANR and RNR are two different enzymes, but both have similar pH functions, kinetic parameters, and the like; and when purifying RNR from anaerobic cultures, certain fractions become sensitive to sonication and thus resemble ANR.

Although the reduction of nitrate as a respiratory device (RNR) is quite widespread among bacteria, the reduction of sulfate (to H_2S) and carbon dioxide (to CH_4) is restricted to a few very highly specialized organisms.

REACTIONS RELEASING USABLE ENERGY: PHOTOSYNTHESIS

The fourth reaction yielding energy available to the living cell is photosynthesis. Of course, this source of energy is available only to photosynthetic organisms, which have the necessary apparatus to utilize the light. These are the photosynthetic bacteria and green plants.

Green plants have been the prime exploiters of the energy of light, but photosynthetic bacteria capable of using light have also evolved. These bacteria possess a chlorophyll that differs from plant chlorophyll only in minor chemical details. Bacterial chlorophyll is green. In some photosynthetic bacteria that are purple, the color is caused by an abundance of red or brown pigments, which mask the green color of the bacterial chlorophyll. But this system seems to be incapable of producing oxygen. The typical photosynthetic bacterium does not produce oxygen, but the procaryotic blue-green algae, which are otherwise identical to the bacteria, are characterized as "algae" because they produce oxygen.

Although there is much yet to be learned about the photosynthetic mechanism, a unifying viewpoint is now possible. This viewpoint has as its basis the assumption that the primary photochemical event in photosynthesis is the ejection of an electron from the chlorophyll complex. This leaves a charged chlorophyll complex plus an electron. We may distinguish three types of results depending upon what happens to the electron and the charged chlorophyll complex subsequent to the photochemical event.

As a matter of convenience, we shall consider the chlorophyll system as the proverbial "little black box" upon which light falls and which then ejects an electron (Figure 9-8). Just how the light ejects the electron we shall not consider. The charged chlorophyll system (the black box without its electron) cannot again react with light until the charge is somehow neutralized. There are at least two ways of doing this in photosynthetic bacteria that do not produce oxygen.

In the first case (Figure 9-8A), the electron moves back to the charged complex through a system of heme pigments resembling cytochrome (and involving ferredoxin), and in the process generates energy-rich phosphate. The energy-rich phosphate may, of course, be used by the cell directly, and such a cell uses the photosynthetic systems directly as a source of ATP. The process is called *photophosphorylation* or *cyclic photophosphorylation*. One can obtain enzyme systems, especially from bacteria, that utilize light solely for the production of ATP.

In the second case (Figure 9-8B), the electron moves into other parts of the cell where it may become involved (via carriers) in the reduction of hydrogen ion (to hydrogen gas), the reduction of NADP, and even in nitrogen fixation (N_2 to NH_3). But this process would soon stop, because it leaves a charged chlorophyll complex that cannot respond to light. This charge is removed by some other electron donor — thiosulfate, sulfur, succinate, etc. This is called noncyclic photosynthesis.

Indeed, photosynthetic bacteria not producing oxygen may be grouped into three general categories as follows

I. Those using inorganic materials as a source of electrons to neutralize charged chlorophyll (*Thiorhodaceae*). The organisms are strictly anaerobic, obligately photosynthetic, and use H_2S, thiosulfate, and other reduced sulfur compounds. They are also called the *purple sulfur bacteria*.

II. Similar to I except the cells are green, since the green color of chlorophyll is not masked by other pigments (*Chlorobacteriaceae*, or *green sulfur bacteria*).

III. Those using solely organic sources of electrons to neutralize chlorophyll (*Athiorhodaceae*, or *purple nonsulfur bacteria*). They may use a variety of organic materials (succinate, malate, even glucose). A good example is the use of isopropanol, which is converted into

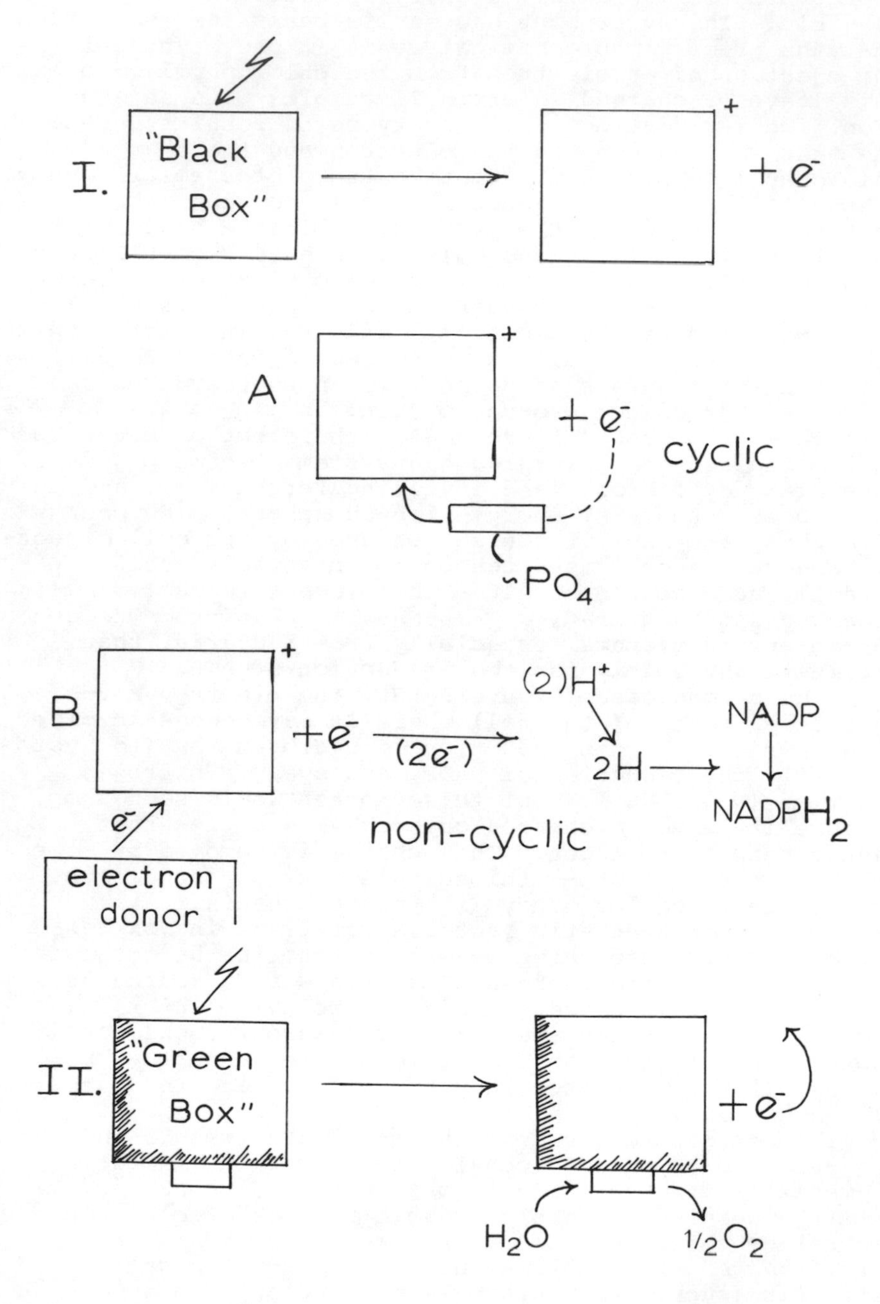

FIGURE 9-8. *Schematic diagram of photosynthesis.*

acetone as follows:

$$CO_2 + 2\underset{\displaystyle CH_3}{\overset{\displaystyle CH_3}{|}}\!\!\!\!CH{-}OH \xrightarrow{\text{light}} (CH_2O) + 2\underset{\displaystyle CH_3}{\overset{\displaystyle CH_3}{|}}\!\!\!\!C{=}O + H_2O$$

It is evident that the isopropanol is used merely as a source of hydrogen (which is capable of neutralizing the charged chlorophyll) and not as a source of carbon, since the acetone is not assimilated.

Mention should also be made of the purple color. The actual photosynthesis is carried out by bacterial chlorophyll, which is green like plant chlorophyll. But the red pigment absorbs longer wavelengths of light, converts this energy to forms usable by chlorophyll, and thus permits these organisms to grow in layers (e.g., in polluted water) below the algae layers and into which light of longer wavelengths can penetrate.

However, in the photosynthetic bacteria that do produce oxygen (the blue-green algae) and in algae and higher plants, a second photosynthetic system, a "little green box," has developed. This green-box system can absorb light and split water to form oxygen and electrons (Figure 9-8), the electrons being moved by a series of complex carriers to neutralize the charge on the "little black box," which is present in all photosynthetic organisms whether they produce oxygen or not.

One may also note that in eucaryotes the chlorophyll system is organized differently (in chloroplasts). There is also a certain amount of evidence that chloroplasts, like mitochondria, may have developed from procaryotic elements.

This undoubtedly is an oversimplified picture of the photosynthetic process (and we have deliberately refrained from speculating on *how* the light ejects an electron from the chlorophyll complex); nevertheless, we have found this picture useful, and until a clearer hypothesis is available, it can be used to explain and correlate a great deal of otherwise divergent information.

The principal, but not the sole, pathway of carbon in photosynthesis, plant or bacterial, is via CO_2 addition to ribulose-diphosphate.

BACTERIAL PHOTOSYNTHESIS[1]

Bacterial photosynthesis begins with the absorption of light by light-harvesting antenna, bacteriochlorophyll (BCHL). Most if not all of the antenna pigments probably occur noncovalently bound to proteins. The best known is

[1]Parson, W. W. 1974. Bacterial photosynthesis. *Ann. Rev. Microbiol.* *28*:41—59. (Reviewed by Charles C. Amilo.)

the one isolated from the green bacterium *Chlorobium*. This is a water-soluble protein, with a molecular weight of 152,000 which can be crystallized after purification without the use of detergents. It contains four subunits, each with five BCHL molecules.

Extraction of pigment—protein complexes from the purple photosynthetic bacteria requires detergents or organic solvents. On disruption of *Rhodopseudomonas spheroides* chromatophores with detergents, the antenna BCHL emerges in association with a polypeptide with a molecular weight of about 10,000; using gentler means, aggregates with a molecular weight exceeding 100,000 have been found, which dissociate into smaller units upon boiling with 1.5 percent SDS. Other organisms yield BCHL protein on the order of 100,000. Evidently, the antenna pigments in the *Thiorhodaceae* and *Athiorhodaceae* occur on hydrophobic polypeptides with molecular weight on the order of 10,000, with each polypeptide holding a small number of pigment molecules. These complexes associate to form larger assemblies, bringing many molecules of pigment into interaction in a manner that remains to be elucidated.

The electron donor in the primary electron-transfer reaction is a complex called P_{870} for a wavelength at which bleaching of a major absorption band occurs when the complex loses an electron. The oxidation of P_{870} also results in other absorbance changes throughout the UV, visible, and IR regions of the spectrum. Although the positions of the maxima and minima in the absorption spectrum vary among different species of photosynthetic bacteria (especially if one turns to species that contain BCHL b), the basic properties of the reactive complex probably are much the same, and the name P_{870} is useful in a general sense.

Reed and Clayton succeeded in purifying a protein—pigment complex that contained P_{870} but was free of antenna BCHL. This product exhibits all the spectral properties that characterize P_{870} in chromatophores and intact cells. The P_{870} in the preparation responds to illumination by undergoing photooxidation. On polyacrylamide gel electrophoresis with 1 percent SDS, the particle dissociated into three proteins with molecular weights of approximately 21,000, 24,000, and 28,000. All the pigments dissociate from the proteins under these conditions. On analysis it was found that the amino acid compositions of all three of the reaction-center proteins are strongly hydrophobic; approximately 70 percent of the amino acids are nonpolar.

P^{+}_{870} extracts an electron from a c-type cytochrome. All species examined contain at least two pools of different types of cytochromes. One of these invariably is a c-type cytochrome with a midpoint potential (Em) on the order of 0.3 V, which appears to react directly with P^{+}_{870} in all cases. Many species also contain a c-type cytochrome with a midpoint redox potential near 0.0 V, which has also been shown to react directly with P_{870}. Some species contain b-type cytochrome. These do not transfer electrons directly to P_{870}^{+}, but react instead with the high-potential c-type cytochrome.

Illumination results in the oxidation of several different cytochromes. In *Ch. vinosum* each photosynthetic unit contains two copies of the high-potential cytochrome c_{555} and two (or possibly three) of a low-potential cytochrome c_{552}. Both types of cytochrome are bound tightly to the chromatophore membrane. On disruption of the membrane with SDS, the two emerge together as part of a reaction-center particle that contains P_{870} and some antenna BCHL. Further dissolution with deoxycholate solubilizes the cytochromes, but still does not afford a separation of the two types.

The identity of the component *X* that accepts an electron from P_{870} in the primary photochemical reaction continues to enjoy spirited debate. Two principal contenders for this position are nonheme iron and ubiquinone (UQ). Chromatophores contain both of these components, and various reaction-center preparations contain one or the other or both. Dutton and Leigh have made a strong case for a nonheme iron complex, which they call photoredoxin.

The picture that emerges from the review is that light pumps electrons from two or more types of cytochromes, through P_{870} and nonheme iron, to coenzyme Q (UQ). The cytochrome pools supply P_{870} with a large reservoir of electrons, and UQ may provide a reservoir for the storage of electrons at the opposite pole of the system. Photosynthetic bacteria evidently meet their need for reduced pyridine nucleotides by energy-linked reverse electron flow, rather than by direct photoreduction. ATP or pyrophosphate can replace light in driving the reduction of NAD^{+} by succinate, and uncouplers prevent the reduction whether it is energized by light or by ATP or pyrophosphate.

REFERENCES

Electron Transport Pathways

Baltscheffsky, H., and M. Baltscheffsky. 1974. Electron transport phosphorylation. *Ann. Rev. Biochem.* *43*:871–897.

Bartsch, R. G. 1968. Bacterial cytochromes. *Ann. Rev. Microbiol.* *22*:181–200.

Benemann, J. R., and R. C. Valentine. 1971. High-energy electrons in bacteria. *Advan. Microbial Physiol.* *5*:135–172.

Forrest, W. W., and D. J. Wlker. 1971. The generation and utilization of energy during growth. *Advan. Microbial Physiol.* *5*:213–274.

Gottlieb, S. G. 1971. Effect of hyperbaric oxygen on microorganisms. *Ann. Rev. Microbiol.* *25*:111–152.

Green, D. E., editor. 1974. The mechanism of energy transduction in biological systems. *Ann. N. Y. Acad. Sci. 227*:5—680.
Henderson, P. J. F. 1971. Ion transport by energy-conserving biological membranes. *Ann. Rev. Microbiol. 25*:393—428.
Horio, T., and M. D. Kamen. 1970. Bacterial cytochromes. II. Functional aspects. *Ann. Rev. Microbiol. 24*:399—428.
Hughes, D. E., and J. W. T. Wimpenny. 1969. Oxygen metabolism of microorganisms. *Advan. Microbial Physiol. 3*:197—232.
Kamen, M. D., and T. Horio. 1970. Bacterial cytochromes. I. Structural aspects. *Ann. Rev. Biochem. 39*:673—700.
Neims, A. H., and L. Hellerman. 1970. Flavoenzyme catalysis. *Ann. Rev. Biochem. 39*:867—888.
Orme-Johnson, W. H. 1973. Iron-sulfur proteins: structure and function. *Ann. Rev. Biochem. 42*:159—204.
Payne, W. J. 1970. Energy yields and growth of heterotrophs. *Ann. Rev. Microbiol. 24*:17—52.
van Dam, K., and A. J. Meyer. 1971. Oxidation and energy conservation by mitochondria. *Ann. Rev. Biochem. 40*:115—161.
White, D. C., and P. R. Sinclair. 1971. Branched electron-transport systems in bacteria. *Advan. Microbial Physiol. 5*:173—212.

To "Fixed" Oxygen

LeGall, J., and J. R. Postgate. 1973. The physiology of sulfate-reducing bacteria. *Advan. Microbial Physiol. 10*:82—135.
Payne, W. J. 1973. Reduction of nitrogenous oxides by microorganisms. *Bacteriol. Reviews 37*:409—452.

Photosynthesis

Bishop, N. I. 1971. Photosynthesis: the electron transport system of green plants. *Ann. Rev. Biochem. 40*: 197—226.
Gest, H. 1972. Energy conversion and generation of reducing power in bacterial photosynthesis. *Advan. Microbial Physiol. 7*:243—282.
Kirk, J. T. O. 1971. Chloroplast structure and biogenesis. *Ann. Rev. Biochem. 40*:161—196.
Lascelles, J. 1968. The bacterial photosynthetic apparatus. *Advan. Microbial Physiol. 2*:1—42.
Lascelles, J. 1973. *Microbial Photosynthesis*. Dowden, Hutchinson & Ross, Inc., Stroudsburg, Pa.
Pfennig, N. 1967. Photosynthetic bacteria. *Ann. Rev. Microbiol. 21*:285—324.

Vernon, L. P. 1968. Photochemical and electron transport reactions of bacterial photosynthesis. *Bacteriol. Reviews 32*:243—26.

Walker, D. A., and A. R. Crofts. 1970. Photosynthesis. *Ann. Rev. Biochem. 39*:389—438.

Papers

1. *J. Bacteriol. 122*:325 (1975).
2. *Arch. Biochem. Biophys. 150*:459, 473, 482 (1972).
3. *Proc. Natl. Acad. Sci. 68*:1024 (1971).
4. *J. Bacteriol. 117*:456 (1974).
5. *Amer. Scientist 63*:54 (1975).
6. *J. Bacteriol. 98*:1063 (1969).

10
Enzymes: Their Nature and Control

GENERAL PROPERTIES

As you know, the reactions of the cell are carried out by enzymes. Enzymes are surface catalysts. The nature of their surface is such that certain molecules are adsorbed to the surface and once on it are oriented into particular positions so that a reaction can occur. One of the earliest scientific concepts was that of Michaelis, who said in a statement still valid today that "When each molecule of two reacting substances must collide in solution, the chance of a reaction between them is small and the chance of collision is small. But when held (on a surface) close together in appropriate juxtaposition and orientation with respect to each other, the substances remain in this special spatial relationship for a time long enough for the reaction between them to occur." It is curious how much of life is dependent upon these "weak" surface adsorption forces (the van der Waal forces) rendered highly specific.

Today almost 1,000 enzymes are known and over 100 have been crystallized. They all are protein; most are globular protein composed of at least 100 amino acids. One, ribonuclease, has been chemically synthesized (Figure 10-1). The amino acid sequence in enzymes, as in all native protein, is precisely defined and the chain is folded in a complex fashion so that each enzyme has a unique and reproducible three-dimensional structure. The interactions of the structure of amino acids, the components of the peptide bond $(-\overset{\text{H}}{\overset{|}{\text{N}}}-\overset{\text{O}}{\overset{\|}{\text{C}}}-)$ linking them together, and finally the side chains in the amino acids determine the folding of the chain and the three-dimensional structure. The genetic code has only to determine the linear sequence of the amino acids (termed the *primary structure* of the protein), whose further interaction after incorporation into the peptide chain will determine the three-dimensional structure. The three-dimensional structure is dependent upon the structure of the amino acid (e.g., proline versus alanine), the α-helix brought about primarily by the peptide bond and the side chains of the amino acids, some of which permit cross linking.

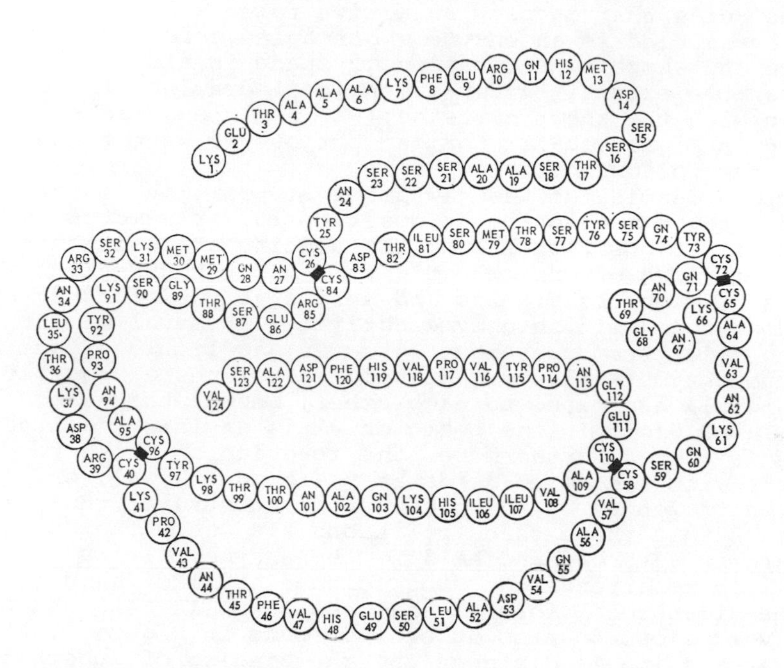

FIGURE 10-1. *The amino acid sequence and cross linking found in ribonuclease from bovine pancreas. From Anfinson, C. B., Science 181:233–230 (1973). Copyright 1973 by the American Association for the Advancement of Science.*

The substrate does not bind randomly to the enzyme but only at specific sites; there is one or at most only a few such binding sites per molecule of enzyme. The binding sites are highly specific and are able to distinguish among closely similar substances.

Close to this is a *catalytic site*, which actually effects the reaction. That the binding site and catalytic site are different is known because it is possible to inhibit or destroy the catalytic property of an enzyme without greatly affecting the ability of the substrate to adsorb to it.

Most substrates are much smaller than the enzyme attacking them, and where the substrate is large (as cellulose being hydrolyzed by a cellulase) the enzyme attaches to only a small part of the larger substrate molecule. Only a fraction of the enzyme protein surface is in contact with the substrate. Thus, of the amino acid sequences of the enzyme protein, only a few amino acids in the peptide chain are in direct contact with the substrate, although some others may play a role in the enzyme activity (e.g., at the catalytic site). Probably the majority of the amino acids in the enzyme protein play no role at all in the enzyme activity per se, but rather are involved in the positioning in space of those amino acids that do play an active role.

Ribonuclease is an enzyme whose amino acid sequence is known and whose configuration in space is also known to a reasonable extent. A two-dimensional drawing of the chain with the known cysteine (—S—S—) cross links is shown in Figure 10-1. For our purpose we want to note only the following:

If one deamidates the glutamine at position 11, or oxidizes the methionine at position 13, or removes the aspartic at position 121 or the alanine at position 122, there is loss of enzyme activity. However, breaking the bonds between the alanine and serine at positions 20 and 21 does not diminish enzyme activity. Removal of the amino acids from positions 14 to 20 has little effect upon the enzyme activity. Histidine residues at positions 12 and 119 are close to each other, and if they are alkylated or if lysine position 41 is arylated, all activity is lost. Furthermore, the reaction of these reagents with the amino acids at positions 12, 41, or 119 can be prevented by the presence of substrate; thus, it appears that these three positions are where the substrate is adsorbed and held to the enzyme. It is possible to arrange the ribonuclease peptide chain in a three-dimensional form so that positions 12, 41, and 119 are very close to each other (not done in Figure 10-1) and at a suitable distance for the binding of substrate. In this case, histidine and lysine seem to be parts of the active site. In other enzymes, histidine, serine, tyrosine, cysteine, and lysine have been found at the active sites.

Ribonuclease has been chemically synthesized by the chemical addition of the proper amino acid, one at a time, to provide the proper sequence, or by the joining of two larger peptides that had been built up by chemical addition of one amino acid at a time in the proper sequence. The synthetic protein thus obtained had the enzyme activity comparable to the isolated enzyme. It is evident, therefore, that the substrate adsorbs onto the enzyme at a specific site and that at this point, or close within a small area on the enzyme surface, certain amino acids in the peptide chain have a particular significance in the action of the enzyme. Similar data are available for other enzymes (e.g., glutamic dehydrogenase, coenzyme binders (1)). The enzyme, of course, concentrates the substrate molecules, and this could account for perhaps a doubling of the reaction rate; yet the enzyme causes perhaps a million-fold increase in such a rate. A much more important factor is that specific adsorption to the enzyme surface can orient the reacting molecules so that groups that can react are close to each other. This *orientation and juxtaposition* was early recognized as one of the principal features of enzyme action. However, this is still not enough to account for the enormous catalytic activity of the enzymes, and a likely theory is that there is not an exact fit between the substrate molecule and the enzyme, as between pieces of a jigsaw puzzle, but that on adsorption at the active site the substrate induces a conformation change in the enzyme protein which leads to the appropriate alignment of the catalytic groups. In one sense, this is the "strain theory" of enzyme action in modern terms backed by modern isolation methods and kinetic analysis.

RATE CONSTANTS IN ENZYME STUDY

There are certain constants that are used either to characterize or to describe particular enzymatic activity; although these are readily calculated, we shall take time to explain their meaning.

The substrate reacts with the enzyme to form an enzyme-substrate complex, which breaks down into enzyme and products. If S is substrate, and E is enzyme, and P is products, then we may write

$$S + E \underset{}{\overset{(1)}{\rightleftarrows}} ES \underset{}{\overset{(2)}{\rightleftarrows}} E + P \qquad [10\text{-}1]$$

Reaction 1 is a reaction between two molecules (a bimolecular reaction), whereas reaction 2 is a breakdown of one (albeit a complex) molecule (a monomolecular reaction). Since in any enzymatic action these two reactions are always proceeding in both directions, the amount of ES (enzyme—substrate complex) is dependent on (1) the amount of enzyme (E), the concentration of the substrate (S), and the affinity of E and S (i.e., the strength of the forces that adsorb the substrate and hold it to the enzyme), and (2) the relative stability of the ES combination (i.e., the rate at which ES breaks down to E + P or to E + S, and the rate at which E + P form ES).

When one measures the conversion of substrate into products in the early stages (when E and S are the materials supplied), the rate of reaction will be largely dependent on the substrate concentration and the stability of ES; later, when products accumulate, the reaction may slow up as the reverse reactions assume more importance. If either the substrate concentration is very low or the ES is very unstable, thus resulting in relatively little ES, the reaction E + S becomes the rate-limiting one, and the rate of the overall reaction is proportional to the substrate concentration over a wide range. When the substrate concentration is high or the affinity of E for S is high or ES has a high stability, most of the enzyme is in the form of ES, and the reaction rate is independent of the substrate concentration. By determining the rate of the reaction under different conditions, one may obtain an estimate of the relative stability of the ES complex. If all the enzyme present in a given solution were in the form of enzyme—substrate complex (ES), the rate at which products were formed would be dependent upon the rate at which the ES broke down (to E + P). One could obtain an estimate of the ability of ES to break down by having all the enzyme present in the form of ES. The speed of the reaction would then be the greatest that one could obtain for a given quantity of a given enzyme solution (we shall call it V maximum, or V max = maximum velocity). The number of molecules of substrate converted into product per second per molecule of enzyme (at V max) is called the turnover number. If we happen to be working with an enzyme that has been prepared pure and its turnover number is determined, the V max will determine how many molecules of enzyme we have in any solution we may measure. In most cases the turnover number has not yet been determined; so V max is merely an estimate, in terms of reaction velocity, of the relative amount of enzyme in a given preparation.

When not all the enzyme is in the form of ES, the formation of products is proportional to the amount of enzyme that is in the form of ES; that is, V (velocity of reaction at any instant) = kES (k is called the velocity constant). The reaction between a substrate and an enzyme (E + S $\rightleftarrows$ ES) may be written in the form

$$K_S = \frac{\text{free enzyme x substrate}}{\text{ES}} \qquad [10\text{-}2]$$

in which K_S is the equilibrium constant of this reaction (usually called the dissociation constant, and sometimes written K_m for Michaelis constant). Equilibrium is actually never reached in most circumstances because ES breaks down to E + P; unless this is reversible under practical conditions, one cannot *experimentally* obtain a direct measure of the E + S $\rightleftarrows$ ES equilibrium. However, the free enzyme present would be the total enzyme supplied (let us call it Σ) less that in the form of ES; thus

$$K_S = \frac{(\Sigma - \text{ES})(\text{S})}{\text{ES}} \qquad [10\text{-}2a]$$

Solving for ES, we have

$$\text{ES}K_S = (\Sigma)\text{S} - (\text{S})(\text{ES})$$

$$\text{ES}K_S + \text{S}(\text{ES}) = (\Sigma)(\text{S}) \qquad [10\text{-}3]$$

$$\text{ES} = \frac{(\Sigma)(\text{S})}{K_S + \text{S}}$$

Now the velocity at any instant is $V = k(\text{ES})$; so ES = V/k. Substituting in equation 10-3, we have

$$V = \frac{k(\Sigma)(S)}{K_S + S} \qquad [10\text{-}4]$$

The greatest velocity possible (V max) will occur when all the enzyme is in the form of ES. Thus V max, in any preparation, will be proportional to the amount of enzyme supplied (Σ); so V max = $k(\Sigma)$. Hence substitution in equation 10-4 yields

$$V = \frac{(V \text{ max})(S)}{K_S + S} \qquad [10\text{-}5]$$

Equation 10-5 is known as the Michaelis—Menten equation; it relates the dissociation constant (i.e, a measure of the affinity of the substrate to the enzyme; actually $1/K_S$ is a physical measure of breakdown) to three factors that may be measured: rate of reaction (V), substrate concentration (S), and amount of enzyme (expressed in terms of something, V max or the maximum possible rate, that can also be measured if necessary). Equation 10-5 may be written

$$K_S = S\ \frac{V \text{ max}}{V} - 1$$

which is the equation of a rectangular hyperbola.

Equation 10-5 may be tinkered with mathematically in three ways, as follows [in analytic geometry the slope-intercept equation of a straight line is $Y = aX + b$, where a is the slope and b the intercept on the Y (ordinate) axis]. Take the reciprocal of equation 10-5:

$$\frac{1}{V} = \frac{K_S + S}{(V \text{ max})(S)} = \frac{K_S(1/S)}{V \text{ max}} + \frac{1}{V \text{ max}} \qquad [10\text{-}6]$$

If one plots $1/V$ against $1/S$, a straight line whose slope is K_S/V max and whose ordinate intercept is $1/V$ max results.

Multiply equation 10-6 by S:

$$\frac{S}{V} = \frac{S}{V \text{ max}} + \frac{K_S}{V \text{ max}} \qquad [10\text{-}7]$$

If one plots the substrate concentration divided by the rate of reaction at that concentration (S/V) against the substrate concentration (S), the slope is $1/V$ max and the ordinate intercept is K_S/V max.

Multiply equation 10-5 by K_S + S:

$$VK_S + VS = (V \text{ max})(S) \qquad [10\text{-}8]$$

Transpose terms:

$$VS = -VK_S + (V \text{ max})(S)$$

Divide by S:

$$V = \frac{-V}{S} K_S + V \max$$

If one plots the velocity (V) against the velocity divided by the substrate concentration (V/S), a straight line results in which the slope is K_S, the ordinate intercept is V max, and the abscissa intercept is V max/K_S.

If one then determines the initial rate of reaction at several substrate concentrations, one can calculate the dissociation constant (K_S) of the enzyme—substrate complex and the V max. The first is a measure of the affinity of the substrate and the enzyme, the latter (when related to the number of molecules of enzyme present, which is not always known) is a measure of the stability of the enzyme—substrate complex.

One note regarding K_S is of interest. If half the enzyme supplied is in the form of enzyme—substrate complex (ES), E (or Σ - ES) = ES, and equation 10-2a becomes K_S = S; that is, K_S = the substrate concentration at which the enzyme is half in the form of ES and half free, or is operating at half its maximum possible velocity.

These equations apply to an enzyme operating under conditions where it reacts with its substrate and where the enzyme—substrate complex is broken down to products, the enzyme being released to participate further in the reaction. Under these conditions (which apply to a wide variety of enzymes), the equations hold. However, one may have a situation in which they do not hold; for example, a percentage of the enzyme may become inactive during the reaction, and in this case straight lines are not obtained. Such anomalies serve as clues to further information on the mechanism of particular reactions. The same type of treatment may be applied to the dissociation constants of coenzymes (in this case substrate is provided at maximum concentration, and the rate becomes dependent upon the amount of coenzyme). This type of analysis is also used to determine the nature of the action of inhibitors on enzymes, and to distinguish between competitive and noncompetitive inhibition. In *competitive inhibition* the inhibitor unites with the enzyme (rendering it inactive) but also dissociates from the enzyme, yielding active enzyme again. Inhibitor and substrate (or coenzyme) can thus compete for active spaces on the enzyme surface; hence, the name. In *noncompetitive inhibition* the inhibitor unites with the enzyme and remains at its surface, preventing substrate (or coenzyme) combination with the enzyme.

ENZYME NOMENCLATURE

Rapid growth in our knowledge of enzymes has led to difficulties in nomenclature. By international agreement, six major types of enzymes are recognized and a numbering system has been devised for more precise designation. Note that the first number corresponds to the major type and subsequent numbers refer to subdivisions within the type.

1. *Oxidoreductases.* All enzymes catalyzing oxidoreductions with the names formed on the pattern donor—acceptor—oxidoreductase. An example is glucose-6-phosphate oxidoreductase (number 1.1.1; the 1.1.1 means oxidoreductases acting on a CHOH group with NAD as the acceptor). Similarly, pyruvate dehydrogenase (or pyruvate oxidase) should be known as pyruvate oxidoreductase. Its number would be 1.2.1. One can refer to a key provided by the International Union of Biochemistry (and normally available in handbooks published after 1960) to obtain the proper number.

2. *Transferases.* Enzymes transferring one carbon group (2.1), acyl group (2.3), nitrogenous groups (2.6), phosphorus (2.7), and others.

3. *Hydrolases.* Enzymes acting on ester bonds (3.1), on glyceryl compounds (3.2), on peptide bonds (3.4), and others.

4. *Lyases.* Enzymes that remove groups from substrates nonhydrolytically, leaving double bond. Prefixes may be used to designate the group removed (as l-malate-hydrolyase) to designate an enzyme removing the elements of water from l-malate resulting in the product fumarate, 4.2.1.2. The trivial name of this enzyme is "fumarase."

5. *Isomerases.* These enzymes catalyze the inversion of asymmetric groups, and are termed *racemases* if the substrate has one center of asymmetry (as lactic acid) or *epimerases* if the substrate has more than one asymmetric center (as tartaric acid).

6. *Ligases.* Enzymes that catalyze the linking together of two molecules coupled with the breakdown of ATP. These were formerly called synthetases, which use should be discouraged.

Along with systematic nomenclature, trivial and less logical systems of nomenclatures continue to exist, and one must examine the details of the reaction catalyzed rather than rely solely upon the name to designate specific enzymes. As with many investigators, we shall continue to use the most common name while attempting to identify the precise reaction carried out.

Not only do enzymes work in teams (one enzyme taking up where the other leaves off), but enzymes themselves are frequently multicomponent systems. For example, threonine deaminase of *Escherichia coli* exists in two forms, a monomer (molecular weight of 40,000, 3.2S) and a tetramer (molecular weight of 160,000, 8S), and conversion of the monomer to the larger form is enhanced by AMP (2,3). In *Salmonella*, the same enzyme isolates as a single protein but can be dissociated into four subunits (4).

Sometimes the subunit components of enzymes are surprising. For example, in lactose synthesis the enzyme carrying out the reaction,

UDP-galactose + glucose = UDP-lactose

can be separated into two subunits, A and B. It turns out that B is lactalbumin, a common protein of milk which was not known to have enzymatic function (5). It also turns out that component A by itself, carries out the reaction:

UDP-galactose + *N*-acetyl-glucosamine =

N-acetyl-lactosamine + UDP

which reaction is inhibited by lactalbumin (6). Thus the presence of lactalbumin changes the specificity and the product of another enzymatic protein. This might be of considerable importance in the mammary gland for the control of milk production.

Some enzymes are exceedingly complex. For example, pyruvic dehydrogenase of *E. coli*, although isolating as a unit, can be dissociated into four separate enzymes:

1. Pyruvic decarboxylase (pyruvate = CO_2 + acetyl-2H-enzyme).
2. Lipoic reductase (lip + acetyl-2H-enz = lip-2H + acetyl-enzyme).
3. Reduced lipoic dehydrogenase (lip-2H + NAD = lip + NAD-2H).
4. Transacetylase (acetyl-enz + CoA = enzyme + acetyl CoA).

The isolated complex contains 16 molecules of enzyme 1 and 8 molecules of enzyme 3 in a matrix of 64 subunits of 2 and 4; together they carry out the reaction

pyruvate + NAD + CoA = acetyl-CoA + NAD-2H + CO_2

Sometimes such complexes are so internally structured that they not only isolate as a unit but actually crystallize. An example is yeast fatty acid synthetase, which consists of seven enzymes that crystallize as a single unit.

Another situation is well illustrated in the lactic dehydrogenase of muscle and heart. Two types of protein are seen in gel electrophoresis; one is predominant in muscle (called M), the other, predominant in heart (called H). Crystalline lactic dehydrogenase consists of four molecules of protein. Lactic dehydrogenase of muscle (M_4) and of heart (H_4) are each composed of identical (although different) subunits, but the enzymes from other tissues may be M_3H, M_2H_2, and so on (these are all called *isozymes*). Why do we have isozymes, especially in the same organism? The H enzyme is not inhibited by pyruvate, whereas the M type is; hence, one may have enzymes with different degrees of inhibition by the reaction product. In the animal, for example, heart can keep going even though enough pyruvate has been formed to inhibit muscle.

INDUCTION AND REPRESSION OF ENZYMES

The enzymes present in a cell are dependent upon its genetic constitution, the environment in which it was grown, and, to a lesser extent, what happened to it subsequent to growth. The present picture of the genetic control mechanism is the following (updated in 7):

1. The DNA of the gene carries out its function by serving as a template for the synthesis of mRNA. The protein-synthesizing machinery translates the mRNA into polypeptide chains, which fold up to form enzyme proteins.
2. Messenger RNA cannot be started at any point on the DNA strand, but only at certain points. These initiation points are called *operators*, and they may control several adjacent structural genes. Such groups of genes are called the *operon*.
3. Associated with the operator is a *promotor* gene (P). Transcription is actually initiated at the promotor, and mutations at the promotor site result in a coordinate change in the level of all the products of the operon.
4. Transcription of the operon is mediated by a *repressor*, which binds specifically and tightly to the operator, thereby preventing transcription of the operon controlled by it. Mutations in the operator site result in constitutive synthesis of the products of the operon. The repressor is a protein that is coded by another portion of the DNA, which may be located some distance from the operator and its structural genes, and this portion of DNA is called the repressor gene.

The repressor—promotor—operator system works in two ways as follows:

Inducible Enzymes

Inducible enzymes are not present unless and until the substrate appears in the medium. Such a substrate is called an *inducer*, and may not be metabolized by the cell at all. The idea is that the repressor gene is continually making repressor mRNA, which is making repressor (protein), which combines with the operator and prevents it from acting so that it cannot permit the transcription of the genes under its control; thus no enzyme proteins controlled by this operator can be made. When the inducer is added, it combines with the repressor protein, freeing the operator from repressor and permitting the synthesis of the *adaptive* enzymes or *inducible enzymes*.

For example, *E. coli* normally does not possess the enzymes (Figure 10-2) for metabolizing lactose if it is grown in the absence of lactose. When grown with lactose, the enzymes for its utilization appear. There are three enzymes involved: (1) a galactoside (lactose) permease (the *y* gene), (2) a β-galactosidase, breaking the lactose to galactose and glucose (the *z* gene), and (3)

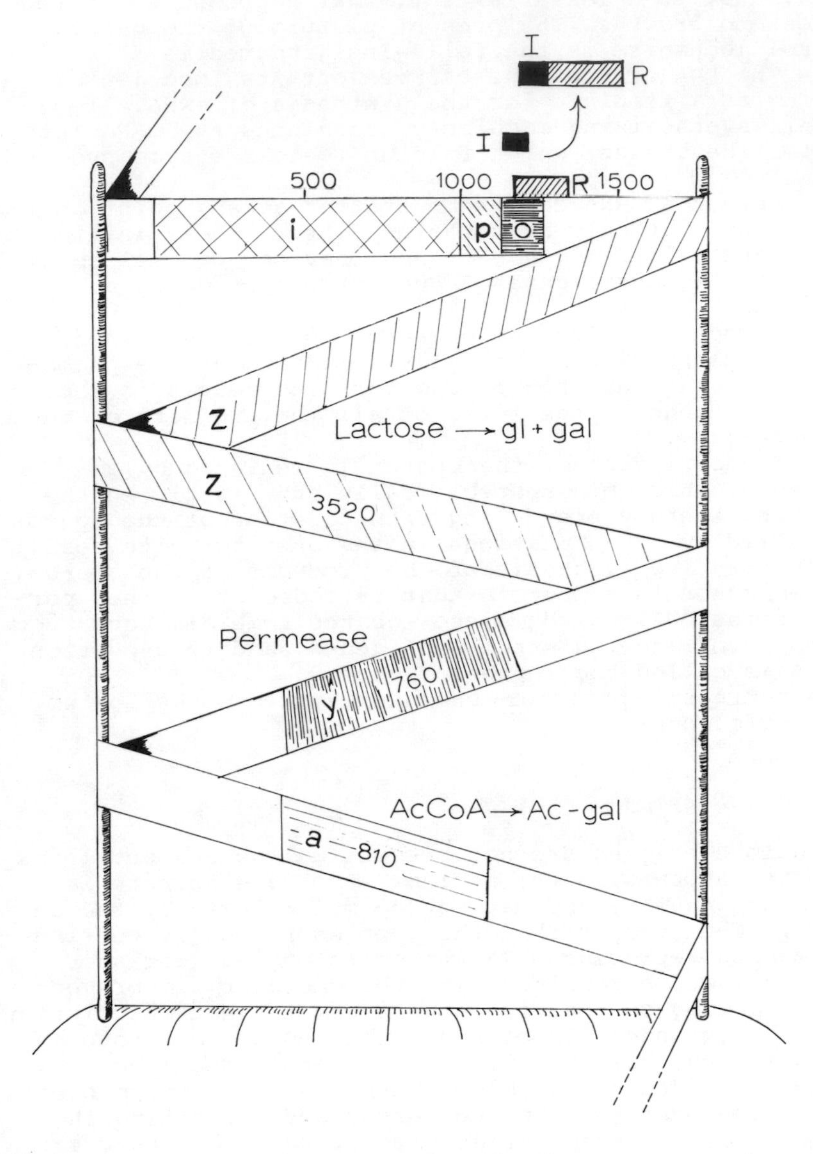

FIGURE 10-2. *A description of the lac operon. The lactose operon, according to the drum model of cellular DNA (Figure 8-1). The repressor binds to operator sequence. When inducer is added repressor is converted into inactive form and dissociates from operator. Length of genes drawn roughly to scale.*

galactoside acetylase (*a* gene), all three of which are under the control of a single operator, promotor, and inducer gene, the whole complex being called the *lac operon*. When grown on nutrient broth, repressor is formed, the operator is inhibited ("switch closed"), no mRNAs for the three enzymes are made, and hence the enzymes themselves are not made. When lactose is added, it combines with the lac-repressor protein, the operator is thus freed from repressor ("switch open"), mRNAs for the three enzymes are made, and their proteins are synthesized. The situation here is that, in the absence of utilizable substrate, no enzymes for its utilization are made. Why should they be? There is nothing for them to metabolize.

Repressible Enzymes

In the case of repressible enzymes, the repressor gene codes for the synthesis of a repressor protein but the latter is inactive, or at least it does not combine with the operator to inhibit it. The presence of the repressor substance (the *effector*, a *corepressor*) permits combination of repressor protein and effector, and this complex inhibits the operator so that the proper proteins are not made.

For example, *E. coli* growing in a mineral salts—glucose medium is synthesizing tryptophan; that is, it has the enzymes for manufacturing tryptophan (Figure 10-3). But as soon as tryptophan is added, the synthesis of these particular enzymes ceases. After all, why should a cell make enzymes to synthesize something of which it has plenty? In this case tryptophan (the effector) combines with an inactive repressor protein; the combination now shuts off the operator and no further enzyme is made.

By the use of these two mechanisms, induction and repression, and their many varients, a cell can control what enzymes it possesses. However, other mechanisms for induction and repression are now being discovered (8).

ALLOSTERIC ENZYMES

But this is not enough. A cell must clearly also be able to control the operations of the enzymes it does have. The genetic mechanism provides the pattern of enzymes with which a cell must work, but there must be in addition a minute by minute control of the enzymes already present. This is taken care of by means of allosteric proteins subject to *feedback inhibition*. It seems easiest to discuss an example, since in this way the principles can be developed. For our purpose we will choose threonine deaminase (or threonine dehydrase) and follow the reaction series shown in Figure 10-4.

Threonine deaminase first removes water (thus its other name, threonine dehydrase) and forms an unstable inter-

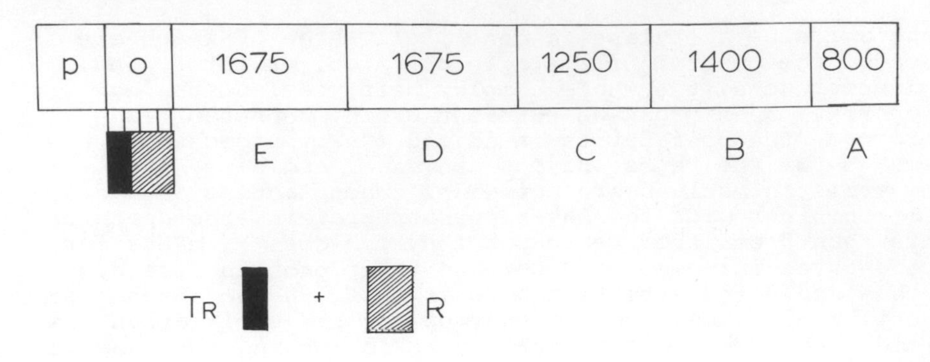

FIGURE 10-3. *The tryptophan operon of* _Escherichia coli_. *Repressor (R) is inactive until it combines with corepressor (tryptophan, tr). Reactions are: (E,D) chorismate to anthranilate to phosphoribosyl anthranilate to (C) carboxy phenyl amino deoxyribosyl phosphate to (B) indole glycerol phosphate to (A) tryptophan.*

mediate, which chemically forms α-ketobutyric acid, which condenses with pyruvate to form intermediate compounds, finally forming isoleucine. Isoleucine, the end product of this reaction series, inhibits threonine deaminase. This is called *feedback* inhibition and can serve to control the activities of the threonine enzyme. When isoleucine is adequate, the enzyme involved in forming it is inhibited and less is made. When isoleucine is used up, the inhibition is released and more is made. The inhibition is reversible and competitive, the latter meaning that it depends upon the ratio of threonine to isoleucine. At a given level of isoleucine, it takes x moles of isoleucine to inhibit, but if the threonine is increased, the enzyme can proceed again, and so on.

Inhibition of enzymes in a reversible and competitive manner is not due to destruction of the enzyme or its active sites but rather to the adsorption of the inhibitor on the substrate adsorbing site, or on the catalytic site, or indeed at some other site close at hand in which the inhibitor molecule overlaps the active site, thus preventing the substrate from adsorbing or being acted upon. Such substances would be expected to bear some structural relation to the substrate of the enzyme inhibited, and one can see a relationship in this sense between threonine and isoleucine. In isoleucine, a methyl group and an ethyl group replace the hydroxyl and methyl of threonine.

However, this kind of explanation is not really adequate since α-ketobutyrate, which seems to be more closely related structurally, does not inhibit, and nor-

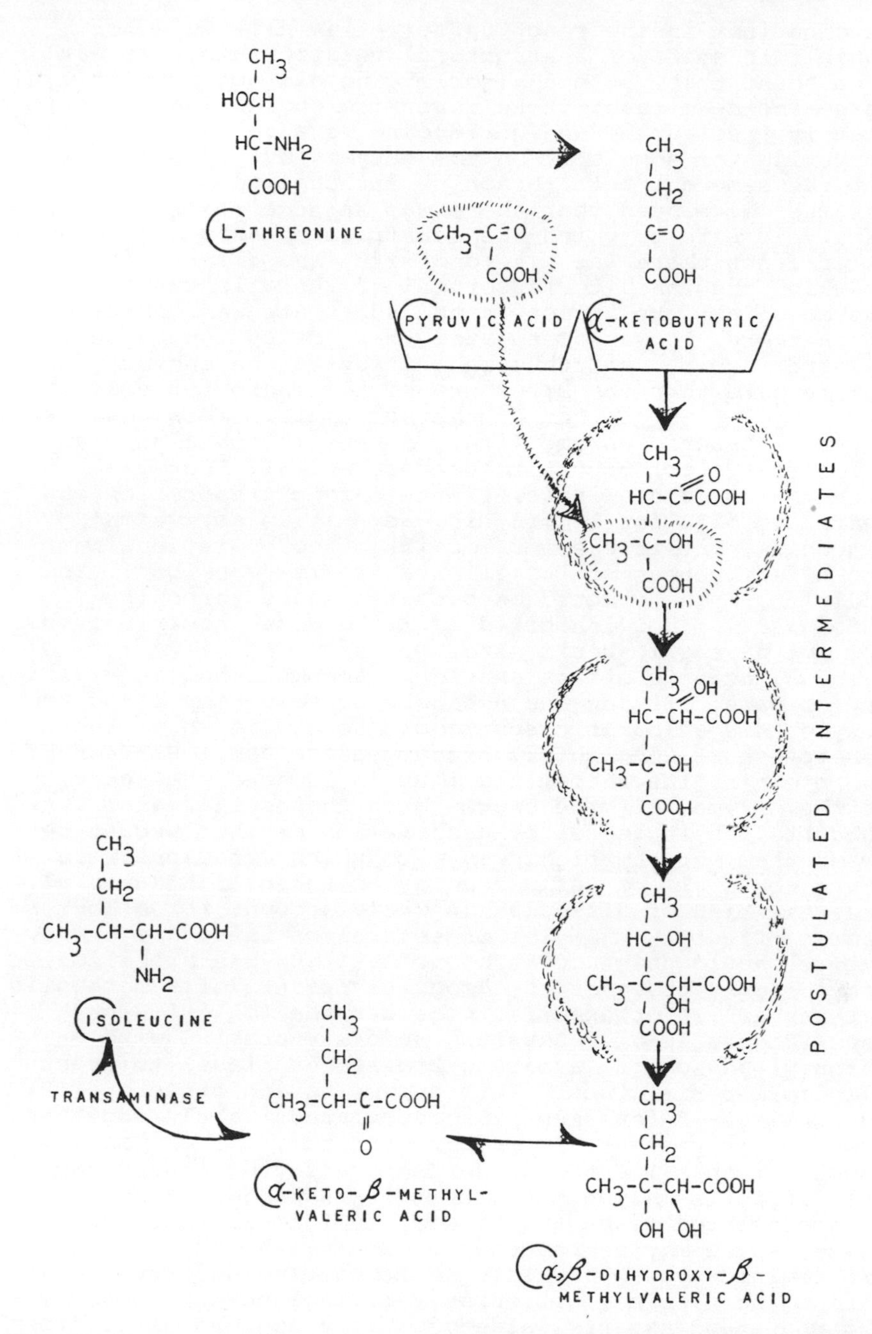

FIGURE 10-4. *Pathway from threonine to isoleucine. From Umbreit, W., Metabolic Maps, Vol. II, Burgess Publishing Co., Minneapolis, Minn., 1960.*

leucine (not in the reaction series at all) does not inhibit in spite of a structural relationship. It has been found that, although norleucine did not inhibit threonine deaminase, it relieved isoleucine inhibition; that is, isoleucine and norleucine were interacting and competing for some spot on the enzyme, but this spot was not the same one that threonine attached to. The idea was then developed that there was an adsorption site and catalytic site to which the substrate adsorbs, but that in addition there was a second site, the *allosteric site*, to which the inhibitor adsorbed; if the allosteric site was occupied, the action on the substrate was inhibited. Furthermore, it was postulated that adsorption to the allosteric site, and thus inhibition of the enzyme, did not require that the structure of the inhibitor bear a direct relation to that of the substrate. This postulate of an allosteric enzyme (i.e., an enzyme containing an allosteric site) proved to be very useful. For example, with threonine deaminase, if the enzyme preparation was heated to 55°C for 10 min, it was just as active on threonine, but it was not now inhibited by isoleucine; the effectiveness of the allosteric site was lost. In addition, mutants could be obtained whose threonine deaminase was not inhibited by isoleucine; their enzymes did not have allosteric sites.

The allosteric effect should be accomplished in a variety of ways. If one had a single protein comprising the enzyme, the allosteric spot could be one in which the adsorption of the inhibitor changed the configuration of the protein (its shape) and thus influenced the activity of the enzyme. If the enzyme were composed of associated subunits, their degree of association or the spacing between them, or the tightness with which cofactors were bound could all be influenced by an inhibiting material, and this kind of dissociation could account for allosteric effects. Examples of several of these are known. But we should point out that not all enzymes are allosteric; only those at the "control" points of a metabolic pathway are. For example, consider the following pathways (Figure 10-5). Aspartic acid may combine with carbamyl phosphate (aspartic transcarbamilase) to eventually form pyrimidines. This enzyme is inhibited by CTP. It consists of four subunits associated tightly together. When CTP is present, in proper quantity, the enzyme "swells"; that is, the subunits move farther apart and this inhibits enzyme activity (9). Aspartate also is phosphorylated (aspartyl kinase) to form aspartyl phosphate, which may be reduced to aspartyl aldehyde (aspartic semialdehyde), and this to homoserine and thus to serine, methionine, threonine, and isoleucine. Aspartyl phosphate and aspartyl aldehyde may also condense to form (in *E. coli*) diaminopimelic acid and thus lysine. In this complex pathway there are several allosteric enzymes. We have already discussed the reaction pathway threonine to isoleucine, and its control by isoleucine. Threonine inhibits allosterically the first enzyme of the

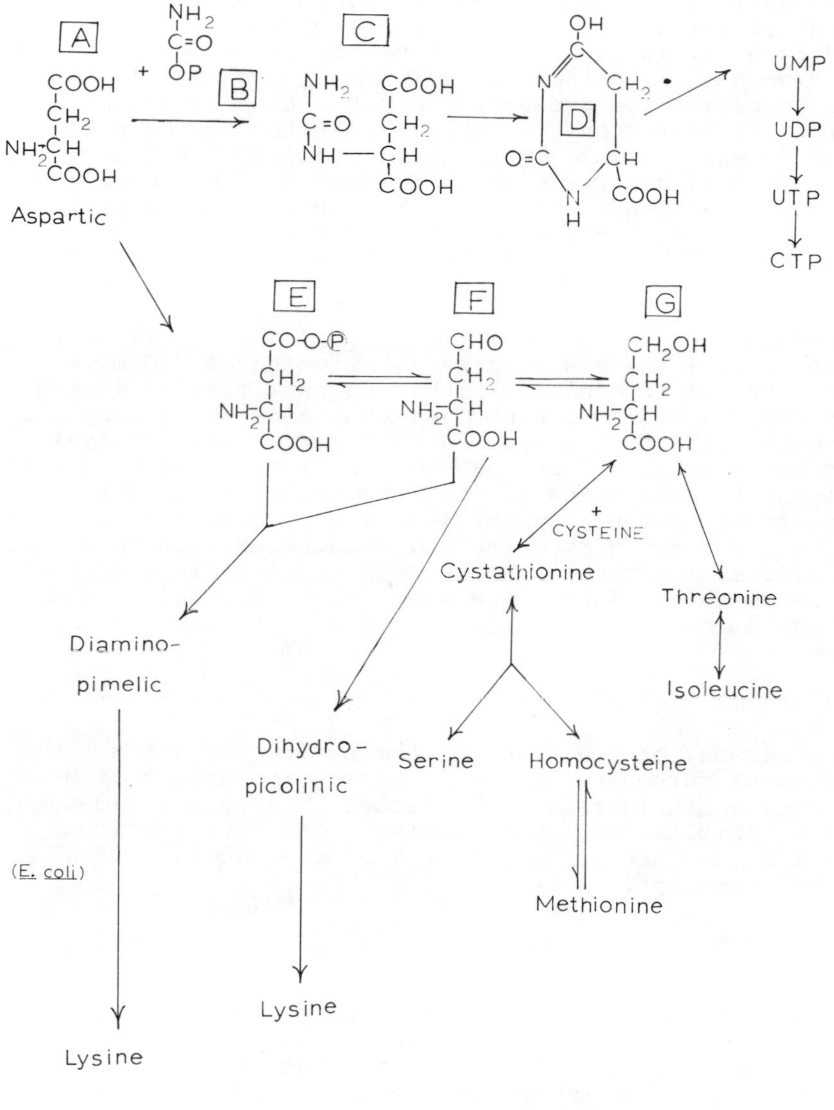

FIGURE 10-5. *Amino acids, etc. derived from aspartic acid. Substance A is aspartic, B is carbamyl-phosphate, C is carbamyl aspartic, D is dihydroorotic, this series forming the pyrimidines. E is aspartyl phosphate, F is aspartyl aldehyde and G is homoserine.*

pathway homoserine to threonine, and also the pathway aspartyl aldehyde to homoserine. Methionine inhibits the pathway homoserine to cystathionine. Serine inhibits the pathway aspartic aldehyde to homoserine, and aspartyl aldehyde inhibits the pathway aspartic to aspartyl phosphate. This latter enzyme is of even more interest because it introduces a new principle. This enzyme, aspartyl kinase, is inhibited, as mentioned, by aspartyl aldehyde; it is not inhibited by the end products, lysine, methionine, threonine, or isoleucine; but if *both* lysine and threonine are present, the inhibition shows up.

The allosteric enzymes thus permit the feedback inhibition to exert minute by minute control over the level of the critical materials in the cell. But the controls are even more refined. In fact, there appear to be at least six control methods, which of course employ allosteric enzymes. These are as follows (10):

Isoenzymatic Control

In *E. coli* there are three aspartokinases. One is sensitive to threonine, another is sensitive to lysine, and repressible by lysine, and the third is repressible by methionine. When growing in a medium high in lysine and methionine, *E. coli* has less aspartokinase than when growing in mineral salts, but the enzyme left is sensitive to end-product inhibition (i.e., threonine). Also, *E. coli* has two homoserine dehydrogenases, one threonine sensitive, one methionine sensitive. The distribution of isozymes mentioned here seems to be typical of the enterics.

Sequential Control

An excess, or addition, of isoleucine increases the amount of threonine, which increases the amount of homoserine, which increases the amount of aspartic aldehyde, which inhibits the aspartokinase. One may sometimes actually inhibit or stop growth by the addition of a single nutrient.

Linked Pathways

Cysteine is a strong inhibitor of homoserine dehydrogenase (on the path to methionine). If cysteine is high, formation of methionine stops, in spite of the fact that cysteine is involved in methionine synthesis. But if threonine is also high, less homoserine is used, more accumulates, and cysteine inhibition can be overcome.

Concerted Feedback Inhibition

We have already examined a case of this kind where more than one substance is necessary for allosteric inhibition; that is, aspartokinase is inhibited only when both lysine and threonine are present.

Compensatory Feedback Control

Again, we have seen an example in the effect of norleucine on isoleucine inhibition of threonine deaminase; that is, allosteric spots can adsorb substances which are not themselves inhibitory but which prevent the inhibitory substance from acting. Norleucine is not a microbial metabolite, but a more physiological example is the observation that inhibition of homoserine dehydrogenase by threonine is reversed by isoleucine and methionine.

Cumulative Feedback Inhibition

An example is glutamine synthetase (ATP + glutamic + ammonia = glutamine + ADP + Pi). This enzyme is inhibited by the following substances, and inhibition is additive: alanine, glycine, tryptophan, histidine, CTP, AMP, carbamyl phosphate, and glucosamine phosphate. With all eight present, one obtains complete inhibition, and the evidence seems to indicate that there are separate allosteric sites for all eight substances. However, there is also some interaction between sites, since AMP and histidine are synergistic in *Bacillus* (11) and alanine and carbamyl phosphate are synergistic in *Neurospora* (12).

EXTERNAL CONTROLS

In addition to these internal control methods, there are a group of external controls other than end product inhibition, allosteric enzymes, or induction-repression. One, of course, is pH; in general, if the pH of the cell is at the optimum for the enzyme, one presumably has maximum activity. If the pH is somewhat different than the optimum for the enzymes, some enzymes increase in amount so that the total enzymatic activity remains the same; other enzymes do not (13).

A second control is the level of inorganic phosphate, ADP, and others, since many enzymes require ATP/ADP/AMP/Pi in their reactions; the level of each tends to influence the reaction rate. Third, the level of coenzymes will affect the rate of reaction. Fourth, several enzymes have cofactors that arise from other phases of metabolism. For example, lactic dehydrogenase (*Streptococcus lactis*) requires for its optimal activity the presence of fructose-1-6-diphosphate (14). Fifth, enzymes may associate with other proteins, and this may change their specificity and rate. We have cited the example of the UDP-galactose enzyme association with lactalbumin. Sixth, there is recent evidence that cyclic AMP affects the activity of various enzymes.

CYCLIC AMP

In addition to its function as an energy carrier, ATP can be converted by an enzyme (adenyl cyclase) to cyclic AMP (i.e., adenosine-3'5'-monophosphate). This substance has become of considerable interest in endocrinology because of evidence that many hormones and neurohormones produce their effects by changing the concentration of cyclic AMP in the target tissues. It also appears to be the substance that initiates or signals the roundup of individual cells of slime molds to form colonies. Studies on bacteria are only beginning, but here too it seems to play an important role. The synthesis of β-galactosidase in *E. coli* is inhibited (repressed) when glucose is present. This has been called *catabolite repression*, and the evidence is that glucose represses the synthesis of the mRNA specific for β-galactosidase. Cyclic AMP stimulates the formation of β-galactosidase, overcomes the catabolite repression due to glucose, and acts at the level at which DNA is transcribed to RNA. Cyclic AMP also stimulates tryptophanase production in cells induced with tryptophan. From these small beginnings it seems reasonable to expect that cyclic AMP may play an important role in bacterial cells. Cyclic AMP is destroyed by a specific phosphodiesterase which converts it to adenosine-5'-phosphate.

SIMULTANEOUS ADAPTATION

Another phenomenon, simultaneous adaptation, needs to be mentioned. Figure 10-6 shows the pathway of breakdown of aromatic substances by pseudomonads. There are essentially two pathways, depending upon whether the ring has a side chain, but the enzymes in the two paths, although certainly analogous, are distinctly different; that is, the enzyme attacking catechol does not act on protocatechuic. To take just one series (mandelic to benzaldehyde to benzoic to catechol), if the organism is grown on mandelic it will oxidize all these substances. The enzymes have all been induced simultaneously. If grown on benzoic, the organism will not oxidize mandelic or benzaldehyde but only benzoic and catechol. That is, the only enzymes appearing are those for the substance supplied and for the further substances in the series, but not for those substances preceding it in the reaction series. Presumably, when supplied benzoic, the enzyme for catechol is also induced because catechol is present due to the oxidation of benzoic. But how is *this* managed? It must be something different from the operator—structural gene

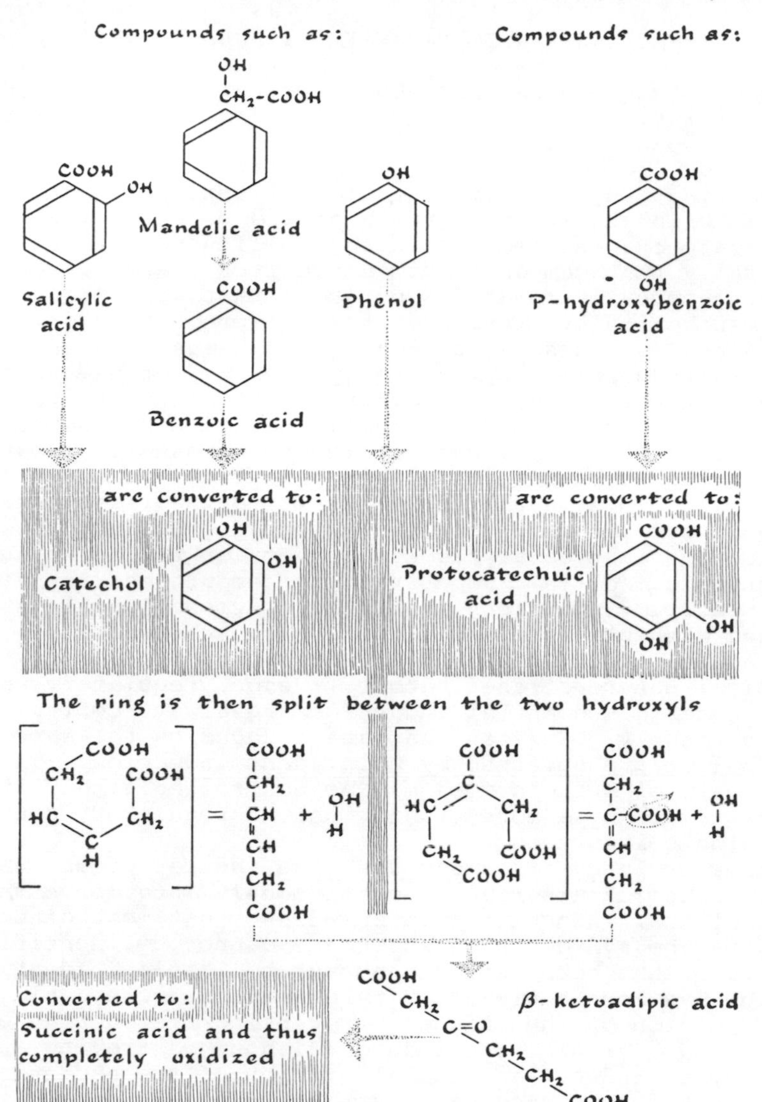

FIGURE 10-6. *Pathway of the breakdown of aromatic substances by many pseudomonads. From AN INTRODUCTION TO BACTERIAL PHYSIOLOGY, Second Edition, by Evelyn L. Oginsky and Wayne W. Umbreit. W. H. Freeman and Company. Copyright (c) 1959.*

sequence, or each enzyme must have its own individual operator, which seems rather unlikely (but might, nonetheless be the case). We know of no adequate explanation of the experiments on simultaneous adaptation in terms of the genetic mechanisms involved.

REGULATION OF CATABOLIC PATHWAYS IN PSEUDOMONAS[1]

Pseudomonas species possess extraordinary nutritional versatility: one strain of *P. multivorans* for an extreme example, can utilize up to 108 of 146 diverse organic compounds proffered as growth substrates.

The first physiological studies of catabolic pathways in bacteria showed that growth at the expense of a single substrate endowed cells with the ability to oxidize specifically that compound and catabolites thereof. As a rule, chemically related substances that were utilized via metabolically distinct pathways were not oxidized immediately. This was called *simultaneous adaptation*; that is inducible enzymes are synthesized at high rates only when they serve a necessary function during growth. Each enzyme mediating a catabolic reaction is induced by its substrate. Thus, when cells are exposed to a potential growth substrate, the chemical elicits the synthesis of only the enzyme that converts it to the first catabolite in the pathway. The newly formed compound induces the enzyme that catalyzes its conversion to the subsequent intermediate. Repetition of the process results in the specific induction of all the enzymes directly associated with utilization of the growth substrate. A sequential induction is characterized by a shift in the chemical nature of the inducer. Hence, regulatory units that undergo sequential induction are always controlled independently. Enzymes that are induced by the same metabolite are governed by coincident induction. The synthesis of such enzymes may be controlled either independently or by a more tightly united regulatory control, coordinate induction.

Under ordinary circumstances, the inducer of an enzyme is a member of a series of continuously interconverted metabolites. Hence, the intermediate that most directly elicits the synthesis of an enzyme cannot be identified before it has been rendered nonmetabolizable. Identification has been achieved by three methods: chemical modification of the inducer, preventing the *in vivo* enzymatic hydrolysis of the inducer without destroying its ability to induce; genetic alteration of the bacterial strain; and physiological restriction of the metabolism of the cell.

In pseudomonads, enzymes that catalyze neighboring catabolic reactions are frequently governed as units of metabolic function subject to coincident inductive control exerted by a single metabolite. Therefore, a sequential inductive step often initiates the synthesis of a group of related enzymes. On the average, between two and three enzymes were subject to coincident control, and in some instances a single metabolite induces as many as five enzymes. Some enzymes were induced by their

[1]Ornston, L. N. P. 1971. Regulation of catabolic pathways in *Pseudomonas*. *Bacteriol. Reviews* *35*:87–116. (Reviewed by Edwin R. Guzman.)

substrates and others by their products.

Two counterbalancing selective factors appear to have directed the evolution of semisequential inductive control. One is the economy of protein synthesis that limits the formation of inducible enzymes only to the presence of their substrates. The other is the information permitted by the use of a common regulatory mechanism to govern the synthesis of several enzymes. By uniting the inductive control over enzymes catalyzing neighboring reactions, cells restrict the amount of regulatory information required to achieve the full regulation of enzyme synthesis.

To elicit the synthesis of a functional enzyme, an inducer must bind to a specific recognition site within the cell. The only known property of *Pseudomonas* recognition sites is their ability to bind metabolites. Chemical and physiological factors appear to have played a determining role in the selection of metabolites as inducers for metabolic function. Chemical instability of a metabolite would restrict its use as an inducer. Some enzymes that participate in specialized catabolic sequences are under highly specific inductive control. Within *Pseudomonas*, some functional units are governed by product induction; the primary substrate must be metabolized by at least one enzyme in the functional unit before induction commences. The specific control of induction decreases the possibility of the bacteria initiating a catabolic pathway that they cannot complete.

REFERENCES

Abou-Sabé, M. 1973. *Microbial Genetics*. Dowden, Hutchinson & Ross, Inc., Stroudsburg, Pa.

Atkinson, D. E. 1969. Regulation of enzyme function. *Ann. Rev. Microbiol. 23*:47–68.

Calvo, J. M., and G. R. Fink. 1971. Regulation of biosynthetic pathways in bacteria and fungi. *Ann. Rev. Biochem. 40*:943–968.

Cohen, G. N. 1965. Regulation of enzyme activity in microorganisms. *Ann. Rev. Microbiol. 19*:105–126.

Daigen, K., and B. Williams. 1970. Catabolite repression and other control mechanisms in carbohydrate utilization. *Advan. Microbial Physiol. 4*:252–324.

Ginsburg, A., and E. R. Stadtman. 1970. Multienzyme systems. *Ann. Rev. Biochem. 39*:429–472.

Haschemeyer, R. H., and E. de Harven. 1974. Electron microscopy of enzymes. *Ann. Rev. Biochem. 43*:279–301.

Hegeman, G. D., and S. L. Rosenberg. 1970. The evolution of bacterial enzyme systems. *Ann. Rev. Microbiol. 24*:429–462.

Jost, J.-P., and H. V. Rickenberg. 1971. Cyclic AMP. *Ann. Rev. Biochem. 40*:741–774.

Kirsch, J. F. 1973. Mechanism of enzyme action. *Ann. Rev. Biochem. 42*:205–234.

London, J., and K. Kline. 1973. Aldolase of lactic acid bacteria: a case history in the use of an enzyme as an evolutionary marker. *Bacteriol. Reviews 37*:453—478.

Mildvan, A. S. 1974. Mechanism of enzyme action. *Ann. Rev. Biochem. 43*:357—399.

Rickenberg, H. V. 1974. Cyclic AMP in prokaryotes. *Ann. Rev. Microbiol. 28*:353—369.

Sanwal, B. D. 1970. Allosteric controls of amphibolic pathways in bacteria. *Bacteriol. Reviews 34*:20—39.

Schlesinger, M. J., and C. Levinthal. 1965. Complementation at the molecular level of enzyme interaction. *Ann. Rev. Microbiol. 19*:267—284.

Srere, P. A., and K. Mosbach. 1974. Metabolic compartmentation: symbiotic, organellar, multienzymic, and microenvironmental. *Ann. Rev. Microbiol. 28*:61—83.

Papers

1. *Nature 252*:543 (1974).
2. *J. Biol. Chem. 243*:167 (1968).
3. *Science 167*:75 (1970).
4. *J. Biol. Chem. 243*:178 (1968).
5. *J. Biol. Chem. 242*:1391 (1967).
6. *Proc. Natl. Acad. Sci. 59*:491 (1968).
7. *Science 187*:27 (1975).
8. *Science 189*:22 (1975).
9. *J. Biol. Chem. 244*:1846, 1860, 1869 (1969).
10. *Science 165*:556 (1969).
11. *Arch. Biochem. Biophys. 118*:736 (1967).
12. *J. Bacteriol. 93*:1045 (1967).
13. *Bacteriol. Reviews 7*:139 (1943).
14. *Science 146*:775 (1964).

11
The Environment

A living cell is a remarkable achievement. There is much in its environment that is against its development; in fact, Jeremy Taylor in 1650 remarked that "to preserve a man alive in the midst of so many chances and hostilities, is as great a miracle as to create him." This is the idea we wish to express here. Cells live and interact with their environment in such a way that, however unlikely, if it is possible, life will survive. This interaction with the environment has several aspects. It may relate to the populations possible in a given situation — the kinds of organisms, their sequences, their numbers per cubic inch. Or it may relate to the distribution of the ability to grow at high temperatures or to the survival of some types of cells under conditions decidedly toxic to others. One may further extend this consideration of environment to ways in which man has exploited the environment to his own advantage and the methods by which he is able to control microorganisms. We have grouped these all together since we sometimes cannot make meaningful distinctions between "chemical" or "physical" factors, and sometimes practical application and fundamental physiology are but different facets of the same phenomenon. We shall therefore let the subject flow, to some extent at will, among the varied aspects of the interaction of microbial cells and their environment.

We should first observe that the environment surrounding a microorganism may be enormously varied, but certain characteristics are always there; they are unavoidable. These are temperature, water, and hydrogen ion. No life is possible without water and where there is water there, too, are hydrogen ions. But other aspects of the environment do not always have to be present: oxygen, inorganic salts, organic chemicals, or radiation, for example. We shall look first at those aspects of environment that are always and unavoidably present and then at those conditions and materials which may be introduced, starting with the more common and progressing to the more exotic, with explorations of a few byways and cul-de-sacs along the way.

EFFECTS OF TEMPERATURE

It is taken as axiomatic that the bacterial cell has no way of maintaining itself at temperatures lower or higher than those of its environment. This defect is imposed both by the aqueous medium and by the structure of the cell itself. Heat generated by various metabolic processes can be lost by radiation or convection or by both. The data obtained on the growth or the survival of bacteria at different temperatures are in agreement with the premise of uniform temperature in both medium and cell. The two types of studies are distinguished from each other by the same criteria applied in the study of the chemical environment: first, the ability to grow, meagerly or well or not at all, at specific low or high temperatures; second, the ability to maintain viability and survive exposure to extremes of temperature for definite short intervals. A great deal of the methodology is the same, and, superficially at least, the results are the same — inhibition or death of the cell. The overall range of temperatures known to permit the growth of bacterial cultures is from about -10 to 90°C. Growth at -10°C, of course, requires the addition of antifreeze to the medium. A minimum, an optimum, and a maximum temperature exists for each particular species, and even for a particular strain; the optimum temperature is ordinarily taken to be that at which the organism grows most rapidly over short incubation periods (18—24 h), although a greater total number of cells is produced over longer incubation periods at temperatures somewhat lower than the "optimum." The net effect of increasing temperature is the speeding up of chemical reactions. At limiting low temperatures, the cells grow very slowly because their enzymatic processes are operating at a very low rate. Unless the temperature drops below freezing, there is little disruption of cellular organization. If freezing occurs rapidly enough to induce the formation of small, rather than large, ice crystals, the integrity of the cell is retained; such rapidly frozen and dried (lyophilized) cultures remain viable for years. At limiting high temperatures, the metabolic functions of the cell are markedly accelerated. Growth at such temperature denotes that all the vital functions are proceeding in reasonable balance, and that the rate of synthesis of cell material — the processes of repair — is somewhat higher than the rate of destruction. Survival at temperatures above the growth limits implies that not enough destruction has occurred to damage permanently the processes of repair, so that growth is possible when the organism is returned to its normal temperature.

The bacteria can be classified into three general

groups on the basis of their temperature requirements: (1) the *psychrophilic* or cold-loving, (2) the *mesophilic*, and (3) the *thermophilic* or heat-loving.

Psychrophiles and Psychrotropes

For many years psychrophiles were considered to be organisms that grew in 1—2 weeks at 0—3°C but grew better at 10°C and even better at 20°C, but did not grow at 37°C. But in 1964 bacteria were found by several investigators that grew well at 0°C with a generation time of 80—90 min, compared to several hours for the first group. Indeed, some had an optimum for cell yield at 4°C and in 3 days in rich media could attain an O.D. of 1.4. Others had optima of 10 or 15°C but grew much less, if at all at 20°C. It has therefore been proposed to call the latter group psychrophiles (cold-loving) and the first group psychrotropic. Most of the studies in the literature are on psychrotropic organisms (and this is the term we shall use, even though the papers reporting the studies call them psychrophiles).

It turns out that psychrophiles must be kept cold constantly. Pour plates cannot be used (melted agar, even at 42°C, kills them). All pipettes, media, and so on must be kept cold at all times. The organisms are only found where the environment is cold all the time, but since 14 percent of the land is in the polar regions and 71 percent of the earth's surface is ocean, of which 90 percent is below 5°C, there seems to be adequate place for them. Evidently, they are not found in water that may be warmed to over 20°C even if for only part of the year. After isolation they must be transferred frequently, since the refrigerator is their incubator rather than a storage area.

The psychrophiles seem to be extraordinarily sensitive to increased temperature. The difficulty seems to be not only in the temperature sensitivity of their enzymes, but their cell wall seems to disintegrate rather readily.

Psychrotropes are found among the bacteria, molds, yeasts, and streptomyces; most of the former seem to be highly oxidative pseudomonads, although anaerobic psychrotropes have been found (1). Two general theories are current for explaining psychrophilic and psychrotropic growth. One points to evidence that their enzymes are different; they have a lower optimum temperature and a lower temperature of inactivation than similar enzymes from mesophiles. The other suggests that psychrotropes lack an adequate feedback control mechanism, and as the temperature rises the reactions get out of synchrony and unbalanced growth results.

With respect to enzymes, there are adequate data to show that some psychrotropic enzymes are more heat labile. For example, in a psychrotrope whose maximum growth was at 35°C, the oxidative and fermentative enzymes were destroyed in 1 h at 46°C, while the comparable enzymes from mesophilic *Escherichia coli* were not affected (2). Not all the enzymes from the psychrotrope were so heat labile. But it is not only the heat inactivation of the enzymes already present, but the synthesis of new enzymes that is important. Indeed, using cell-free systems for protein synthesis, a factor P has been reported (3) that enables ribosomes from psychrotropes to make protein at 0°C. This factor may be extracted from the ribosomes by washing them with 1 *M* NH_4Cl, a process that tends to remove protein cofactors; this washing does not destroy the ability to form protein at 25°C, but protein can no longer be made at 0°C. Addition of P restores the ability to function at 0°C. Ribosomes from *E. coli* are practically inactive at 0°C, but when washed and treated with P they can more than double their rate of synthesis at this temperature. A mutant of the psychrotrope that no longer grows at room temperature but can now grow at 37°C does not contain P, but its ribosomes, while having low activity at 0°C, can be activated at 0°C by the addition of P. The P factor itself is extremely heat sensitive and is rapidly inactivated at 35°C. One may, on the basis of these data, conclude that psychrotropes possess a mechanism which permits protein synthesis to occur at temperatures where it is stopped in mesophilic organisms. Although not true of microorganisms, it appears that certain fish living in waters actually below 0°C contain a special glycoprotein which acts as an antifreeze (4).

Mesophilic mutants of a psychrotrope seem to have the same enzymatic complement and even the same ribosomes as the psychrotrope except for the temperature of growth (5).

Mesophiles

Mesophiles seem to separate into two groups, both having a minimum of about 15°C and a maximum of 50—55°C. The first group has an optimum of from 25—30°C, The second an optimum from 37—45°C. When mesophiles are grown at different temperatures within their growth range, the composition, both chemical and enzymatic, differs. For example, at lower growth temperatures *E. coli* tends to have more unsaturated fatty acids than when grown at higher temperatures. It has been thought that the maximum growth temperature is just below that of the minimum temperature for enzyme inactivation. Early data showed such a relationship, but there are obviously certain difficulties with this hypothesis: we cannot study *all* the enzyme of a cell; perhaps many could be inactivated and still permit growth; and perhaps it is not entirely enzymes that are involved in the inactivation process, but possibly one might denature DNA, RNA, ribosomes, or control mechanisms. One can conceive of higher temperature damaging separate functions differently; indeed, within the growth range, growth at different temperatures may result in different reactions. For example, *Lacto-*

bacillus plantarum needs phenylalanine for growth at 37°C but not for growth at 26°C. Metabolic pathways may differ and different products may be formed. One can readily see how even minor damage by too high a temperature might prevent growth. Many temperature-sensitive mutants have been isolated from bacteria in which the wild-type organisms are capable of growth at both temperatures but the mutant does not grow at the higher temperature. Most frequently, such mutants are auxotrophic for a particular metabolite at the higher but not at the lower temperature.

Another real problem, however, is why mesophiles do not grow at low temperature. *Escherichia coli* grows slowly at 8°C, but it does not grow at all at 6°C no matter how much time is allowed. Data are available to show that certain enzymes are simply not made although others are (6). A cold-sensitive mutant of *Pseudomonas putida* simply does not make the enzyme that converts muconic acid to its lactone (7), although the enzyme, once made (at 30°C), operates perfectly well at low temperature, nor is it cold sensitive at the moment of synthesis. It is not, therefore, that the mesophile simply runs more slowly at the lower temperatures, but that certain enzymes are simply not manufactured. Although this might appear to be related to an imbalance of reaction rates or some lack of synchrony of the cell functions (which do indeed have different temperature coefficients, i.e., DNA synthesis may be stopped by low temperature sooner than RNA or protein synthesis), the evidence is that the factor involved is genetic. Similar results suggest that psychrotropes and thermophiles are also restricted to their characteristic range by means of genetic facters.

A possible mechanism may be suggested as to why mesophilic enzymes may be inactive at low temperatures. For example, a mitochondrial ATPase was inactive at 5°C because it dissociated from 28×10^4 to 4.5×10^4 daltons (8). That is, low temperatures can influence the dissociation of proteins and thus can alter either their activity or their allosteric response.

Thermophiles

Thermophiles are those organisms having a minimum of 37—40°C, an optimum in the 50—60°C range, and a maximum of 90°C (9,10). There are two principal theories to explain their properties: one that the proteins are more heat stable, the other that they have a more rapid resynthesis of protein to compensate for its more rapid breakdown. It has been shown that the flagella and various enzymes are indeed more heat stable when prepared from thermophilic bacteria (11). Moreover, it has been possible genetically to transfer the ability to grow at 55°C from thermophiles to a mesophilic strain of *Bacillus subtilis* (12), which suggests that there must be a *thermophilic factor* common to the entire cell rather than that each protein is individually more thermoresistant.

The triose phosphate isomerases from psychrotropic, mesophilic, and thermophilic clostridia have been isolated and, although very similar, the proteins do have different physical properties (13).

Thermophiles are not limited to bacteria, since there are thermophilic fungi and algae, and even thermophilic phage. Their full potential has yet to be exploited; for example, thermophiles growing on hydrocarbons might be a good source of protein. At high temperatures these organisms grow rapidly, and turn over a rather large amount of substrates in unit time.

At first sight it seems remarkable to find organisms living at 90°C, but, after all, the cell wall is heat stable and certain proteins are heat stable. For example, luciferase in Sephadex still retains 40 percent of its activity after exposure to 135°C for 36 h. Why shouldn't there be organisms whose proteins are more heat stable than the average? Of course, it has to be everything — not only protein, but RNA and DNA (these are normally heat stable, but "melt" at higher temperatures), phospholipids, and so on, and also the organization — the ribosomes, the membrane, the nuclear structure. The ribosomes, for example, *are* more stable, and there is a rough correlation between the maximum growth temperature and the stability of the ribosome (14).

Cell-free protein synthesis has been carried out at 65 and 70°C using enzymes and ribosomes from thermophilic organisms. Thermophiles tend to have DNA with a higher GC ratio when compared to their mesophilic counterparts.

In some cases enzymes isolated from thermophilic organisms have proved to be no more heat stable than their mesophilic counterparts, although as a general rule those from thermophiles are more heat resistant. The fact that they do contain heat-sensitive enzymes suggests that conditions or organization inside the cell are such that the enzyme is stable, or such heat lability may account for increased nutritional requirements at more elevated temperatures. But lest one be overconfident that an explanation for the thermophilic character is now available, we should point out that we do not yet know *why* the ribosomes are more heat stable or the enzymes or other structures more resistant.

Elevated temperature has another effect of psychrotropes, mesophiles, and thermophiles; when it exceeds the maximum, it becomes a toxic factor in the environment (15). It seems that one of the first effects is damage to the membrane. Some strains of *E. coli*, for example, grow at 28°C but not at 40°C. When placed at 40°C the cell membrane becomes fragile, RNAse (and other enzymes) are released in the surrounding fluid, and ribosomal RNA begins to be degraded and leak out. *Staphylococcus aureus* grows at a slightly higher temperature, but at 50°C it shows the same membrane damage, leakage, and ribosomal breakdown (16). A psychrotrope shows the same leakage response at 20°C (17). Cells so injured by heat may die, but if placed in a suitable environment they may repair the injury and recover (18). Such an environment usually consists of rich media to replace the materials that leak from the cell. The more drastic the heat treatment, the longer will be the subsequent lag before recovery. Of course, the environment in which the cells are overheated influences the damage inflicted. Dried cells of mesophiles in high vacuum can survive 65°C for 5 days; spores are more resistant in oil than in water; the ionic and organic composition of the heating environment influences the result obtained. All these factors are reflected in the death time and temperature data for which conditions, both of heat application and media, and conditions for estimating subsequent viability must be specified.

PROTEINS FROM THERMOPHILIC MICROORGANISMS[1]

Microorganisms growing at temperatures around the boiling point of water must contain unique mechanisms for survival since proteins, nucleic acids, and enzymes might be denatured at high temperatures.

Interest in thermophilic microorganisms is due to several factors:

1. Thermophily might be a primitive characteristic from an evolutionary point of view.
2. Thermal pollution, due to increased industrial activity, increases thermophilic growth.
3. Enzymes from thermophilic microorganism may have applications in industrial processes.

Thermophilic microorganisms are found everywhere in nature, ranging from thermal pools to desert soils. Thermophiles can exist at elevated temperatures as either obligate, facultative, or thermotolerant. Most bacterial genera contain thermophilic microorganisms that resemble their mesophilic counterparts. Data suggest that evolution proceeded from an environment considerably warmer than the present; hence mesophiles would have to originate from the thermophiles.

The mechanism of survival can be classified into three general categories:

1. Stabilization may be achieved through lipid interaction.
2. Heat-denatured cellular components may be rapidly resynthesized.
3. Thermophilic organisms may contain macromolecular complexes with an inherent heat stability.

Lipids

It has been observed that thermally stable organisms had lipids with higher melting points than did their meso-

[1]Singleton, R., Jr., and R. E. Amelunxen. 1973. Proteins from thermophilic microorganisms. *Bacteriol. Reviews 37*:320–342. (Reviewed by Jenny Ng.)

philic counterparts, which suggests that the temperature at which cellular lipids melted might set an upper limit for cellular growth.

Differences in the cell membrane and cell wall do occur between thermophiles and mesophiles. The percentage of saturated and branched-chain fatty acids increases as the growth temperature increases. These changes might provide the organisms with a more stable membrane.

Rapid Resynthesis

In the genus *Bacillus*, smaller organisms tend to have a greater heat resistance than the larger. Smaller cells should have higher metabolic rates owing to the greater ratio of surface to volume, which would facilitate the rapid transport of substrates and waste products into and out of the cell; thus growth at elevated temperatures could be the result of rapid resynthesis of heat-denatured cellular components.

Thermally Stable Macromolecules

Three possible mechanisms can explain survival in molecular terms:

1. Thermophiles may contain factors that increase the stability of their components to high temperatures.
2. Mesophiles may contain factors that increase the lability of their components to high temperatures.
3. Cellular components of thermophiles may have an inherent heat stability, independent of exogenous factors.

Mixing cell-free extracts from thermophilic microorganisms with extracts prepared from mesophiles shows that the thermophilic extracts possessed a marked degree of heat stability, which was not transferable to the mesophilic extract. The mesophilic extract did not cause a loss in thermal stability in the thermophilic extract. Thus transferable factors are not related to the thermal stability of the thermophiles, although it is conceivable that a stabilizing factor may be very tightly bounded and not be transferred. The evidence supports the hypothesis that thermophilic microorganisms possess an intrinsic thermostability which is independent of any transferable, stabilizing factors. This may be a more stable membrane, more rapid growth, some type of structural stabilization, or the inherent heat stability of the cellular protein.

Studies show that flagella from thermophiles have a higher degree of heat resistance than flagella from mesophiles. Thermophilic flagellin molecules have fewer charged groups upon titration than mesophilic flagellin. The smaller numbers of charged groups might cause a decreased dissociation.

Comparisons of the ferredoxin from thermophiles and mesophiles of *Clostridia* show a significant physicochemical difference between the ferredoxins due to the presence of histidine in the thermophilic proteins. The thermophilic proteins are considerably more heat stable. They lost only 5—10 percent of their activity upon heating at 70°C for 1 h, whereas the mesophilic proteins lost 70 to 75 percent.

It has been demonstrated that the binding of Mg^{2+} to the yeast tRNA specific for phenylalanine resulted in increased thermostability. Perhaps the major difference in proteins from thermophilic and mesophilic sources is the ability to bind certain ions more tightly and thereby fold into a more stable conformation. There appears to be an increased level of hydrophobic amino acids in proteins from thermophilic sources when compared with their counterparts from mesophiles. Proteins from thermophilic sources seem somewhat impervious to reagents that disrupt hydrophobic bonds.

It appears that thermophilic microorganisms synthesize thermostable proteins, and, except for some rather minor points, these proteins appear to be physicochemically similar to their mesophilic counterparts. Their points of similarity include (1) molecular weight, (2) subunit composition, (3) allosteric effectors, (4) amino acid composition, and (5) primary sequences.

OTHER RELATIONS TO TEMPERATURE

All organisms when growing or metabolizing are producing heat; *E. coli* at very high aeration rates produces 0.1 to 1.0 cal/s/10^{12} cells (19). Thermophiles produce 0.6 to 1.25 cal/s/10^{12} cells. Normally, this heat is rapidly dissipated, although one could assume that a colony on a plate is warmer by some millionths of a degree than the media on which it grows. But for all practical purposes the organism is at the temperature of its environment. However, under conditions of restricted heat flow the temperature may rise considerably, enough to heat a bale of wool to 70°C. Certainly, microorganisms play a key role in spontaneous combustion, but presumably the temperature could not reach the ignition point solely by microbial action. The heat given off during growth can be accurately measured, and microcalorimetry has been used by some in attempts to study metabolic processes, with indifferent success.

Another aspect of temperature relations to be noted is *cold shock* or *heat shock*, resulting from the *rapid* change from one temperature to another. For example, an organism growing at 37°C is plated using dilution water blanks at 10°C, and only 10 percent as many cells are found compared to dilutions made at 37°C. It appears that such cold-shocked cells are more permeable, that they have lost certain periplasmic enzymes and peptides, and that a certain percentage can be restored to viability by supplementary nutrients in the plating media or by the addition of the *shockate*, the fluid in which the first shock occurred. Such sensitivity to temperature shock (without freezing) is much greater during the exponential growth phase than it is in the lag or stationary phase, and the same is true for other lethal agents (toxic materials, high temperatures, etc.). Many years ago this was called "physiological youth" upon the assumption that in the exponential phase the cells were no more than one generation time since the last division (and most were less than one generation old); this rather correlated with the more general observation that the young of any species tended to be more sensitive to harmful influences than the mature individuals. With respect to freezing, however, this relation does not exist, and freeze-kill of cells from the lag or from the stationary phase are essentially similar.

EFFECT OF PRESSURE

Most bacteria live at atmospheric pressure, and the variation in this pressure from day to day has little effect upon them. It takes several hundred atmospheres of pressure to inhibit growth, and in general the higher the temperature, the more pressure is required. Several bacteria can grow at pressures as high as 600—800 atm, but at these pressures they tend to grow in the form of filaments; that is, the process of cell division is inhibited. Cell-free protein synthesis is also inhibited by 200—800 atm, but the degree of inhibition depends upon the origin of the ribosomes. Ribosomes from pseudomonads appear to be more resistant, but the ribosomes from a pressure-resistant, deep-sea pseudomonad did not appear to be better adapted to higher pressure than the ribosomes from common terrestrial forms (20).

Sonic waves, especially ultrasonic sound, may be regarded as rapidly altering pressure waves, and, of course, these are capable of disrupting bacteria. Applications of ultrasonic sound may therefore decrease the number of viable bacteria, but since it also breaks up bacterial clumps it may, on occasion, increase the microbial count.

EFFECTS OF OSMOTIC PRESSURE

The effects of osmotic pressure on bacterial cells are of a different sort. Most bacteria except marine bacteria will grow well at very low salt concentrations, and most will survive in water. The movement of solutes out of the cell and the diffusion of water inward seem to be sharply restricted by the cell membrane, and this produces a higher osmotic pressure in the cell than in the medium. In the reverse situation, when bacteria are placed in a solution of high solute concentration (50 percent or greater), their growth is markedly inhibited. The concentration required for inhibition depends on the particular solute (i.e., whether salt or carbohydrate) and on the organisms tested, since some are much less sensitive than others. In media of growth-inhibitory osmotic pressure the cytoplasm becomes dehydrated and contracts away from the cell wall. Such *plasmolyzed* cells may be quite capable of growth when returned to a medium of normal osmotic pressure.

Halophilic organisms are characterized by their ability to grow in media containing concentrations of NaCl that usually inhibit the multiplication of nonhalophilic forms. To the best of our knowledge, the ability to grow in high (10—15 percent) NaCl concentrations is not the same as the ability to grow in high sugar concentrations. Furthermore, only a few cultures are obligate halophiles, that, will grow *only* when large amounts of salt are present. Since "large amounts" is ambiguous, it has been suggested that the halophilic bacteria be divided into two groups: *moderate* halophiles, which grow in from 1—2 percent to as high as 20 percent NaCl, and *extreme* halophiles, which require no less than 15 percent and can grow in as much as 31 percent NaCl.

The *intracellular* salt concentrations of the extreme halophiles are very high, of the order of 10—20 percent. Potassium is the major component of the internal salt and is evidently required to maintain the structural integrity of the ribosomes. Added potassium salts are required to obtain the maximal rate of protein synthesis in *in vitro* systems, and the enzymes obtained from the cells are adapted to function at high salt concentrations. Extremely halophilic bacteria contain a second ("satellite") DNA comprising 11—36 percent of the total DNA. This material has a different density than the major DNA, although its base composition and amount vary from one strain to another.

The growth of extremely halophilic bacteria is generally slow, and they have been thought to be unable to attack carbohydrates and not to require external vitamins. However, further study showed that these organisms have a specific potassium requirement, and when potassium is adequate (about 1 mg/ml) both vitamins and carbohydrates contribute to growth. The high potassium requirement is reflected in the high potassium content of the cells. *Halobacterium salinarium*, growing in nearly saturated NaCl, contained 4.7 *M* K^+, a concentration greater than that found in saturated KCl.

WATER AVAILABILITY

When a solute is dissolved in water, the solution has a lower freezing point, a higher boiling point, and a lower vapor pressure. The solute has associated with it some of the water, so not all the water is available for other purposes. It has been found convenient to define a value called *water activity* or A_w as follows:

$$A_w = \frac{\text{vapor pressure of solution}}{\text{vapor pressure of solvent}}$$

$$A_w = \frac{\text{number of moles of solvent } (N_2)}{\text{moles of solute } (N_1) + \text{moles of solvent } (N_2)}$$

$$A_w = \frac{N_2}{N_1 + N_2}$$

$$A_w = \frac{\text{relative humidity}}{100}$$

This value (A_w) is related to osmotic pressure as follows:

$$\text{osmotic pressure} = \frac{-RT \ln A_w}{V}$$

Where V is the partial molar volume of water at 25°C. A 1-*M* solution of an unionized material has an A_w of 0.9823 and an osmotic pressure of 22.4 atm. Typical minimum values for water availability are the following:

	A_w
Normal bacteria	0.91
Yeasts	0.88
Molds	0.80
Halophilic bacteria	0.76
Halophilic fungi	0.65
Osmophilic yeasts	0.60

This means that most bacteria have a minimum water requirement of $A_w = 0.91$ and will not grow when the A_w is less than this point. A few common values for A_w are helpful:

	A_w
30% Glucose	0.964
1% Glucose + 20% glycerol	0.955
1% Glucose + 40% sucrose	0.964
Saturated NaCl	0.78
Saturated $CaCl_2$	0.30
Saturated $MgCl_2$	0.30
Saturated LiCl	0.11

One may first note that sugars do not decrease water availability as greatly as one might suppose. In fact, considering A_w only, most bacteria should be able to grow at the concentrations of sugars listed. Their effect, therefore, must be on some factor other than water availability. But salts decrease the water activity markedly. Lithium, especially, owing to high hydration of the lithium ion, can produce a very low A_w, and a saturated solution of LiCl has the highest known osmotic pressure. A bacterium has been found that will grow in saturated LiCl. The organism appears to be much like *B. megaterium* but has a thick cell wall and is gram-negative. If it is diluted to about one third saturated (which is still much above the A_w of the extreme halophiles), the cytoplasm begins to clear from the cell walls inward and the cells become optically empty, but they do not swell or burst. If diluted with saturated NaCl, which changes the osmotic pressure but not the ionic strength, the cytoplasm clumps. The extreme halophile, *Halobacterium cutirubrum* grows in saturated NaCl (about 30 percent) but not in half-saturated (15 percent). When it is transferred to water it does not swell up and burst, but rather disintegrates into lipoprotein droplets.

All this information suggests that ability to grow at high salt concentrations, and indeed the requirement for high salt before growth is possible, is not due to a requirement for high osmotic pressure but rather for the ions per se; presumably, this ionic requirements results from the requirement for cell membrane and ribosomal stability. As with many types of observations, it appears that more than one critical factor is in operation and not all such factors are evident in all organisms. Halophilism seems to depend upon the following properties:

1. Salt is required for metabolic activity in part to protect the cell against internal osmotic pressure.
2. Salt penetrates in some cases to levels higher than the external environment; in other cases it is partially excluded.
3. Enzymes of halophiles may require higher salt concentration for optimal activity.
4. The cell surface of some halophiles has exterior sites specifically requiring sodium for their integrity.

SALT-DEPENDENT PROPERTIES OF PROTEINS FROM EXTREMELY HALOPHILIC BACTERIA[1]

The cytoplasmic fraction of whole cells of various halophiles exhibit an excess of acidic groups as high as 17 to 18 mole percent, unlike those of representative nonhalophilic cells. The cell envelopes of *Halobacterium* are highly charged, the excess of acidic groups amounting to 19 to 20 percent. Some cellular structures of extreme halophiles have been separated into fractions by lowering the salt concentration. *Halobacterium cutirubrum* ribosomal proteins, more acidic than those of *E. coli* were fractionated by exposure to buffers that lacked the high

[1]Lanyi, J. K. 1974. Salt-dependent properties of proteins from extremely halophilic bacteria. *Bacteriol. Reviews 38*:272–290. (Reviewed by Beatrice F. Minassian.)

KCl concentrations found in the cells, but which contained 0.05 *M* $MgCl_2$. A considerable fraction of the proteins was removed by this treatment; the detached proteins were found to be more acidic than the original ribosomes, whereas the residual bound proteins were less acidic.

The effect of salt on the binding of substrates has been investigated for *H. salinarium* isocitrate dehydrogenase. The binding of isocitrate in the *H. salinarium* dehydrogenase showed a maximum at 0.75 *M* NaCl or at 1.5 *M* KCl, whereas nicotinamide adenine dinucleotide phosphate binding decreased continually with salt concentration. Similarly, the K_m for the substrate increased with salt concentration for the lactic acid dehydrogenase of *H. salinarium*, even though the maximal rate (*V* max) increased under these conditions. In these systems the binding of substrates is affected by the salt concentration. The salt response of other halophilic enzymes was found to be influenced by the binding of substrates.

When the effective concentration of salt is small, the binding of ions by a macromolecule may be thought to be specific, and it may be possible to identify a binding site. However, when a larger concentration of salts is required to affect the proteins, they act in a less specific manner and exert their effects also through changing the structure of the solvent.

MARINE MICROORGANISMS

In addition to the extremely halophilic bacteria, there exists another type of salt-requiring organism, usually marine forms. Some of these are so highly adapted to seawater that "synthetic" seawater will not suffice, but most can grow on the proper salt mixtures. These organisms require at least 1 percent salt, and usually of the order of 5—10 percent. Below 0.5 percent salt they are lysed but can be protected by sucrose. Evidently, NaCl, LiCl, and sucrose do not penetrate the cell. They differ from the halophiles in that they do not swell before bursting. Halophiles placed in conditions of inadequate salts change from rods to spheres and eventually burst; but the lower limit of salt required can be extended (from about 0.5 to 0.3 *M*, i.e., about 4 to 2 percent NaCl) by the addition of very small quantities of calcium and magnesium. It appears that in these cases there is indeed an effect of ions on the cell wall stability, but that the external environment must be in osmotic relation to the cell contents. If the osmotic pressure externally is too low, water enters the cell, the cell wall is weak and unable to stand the pressure of the membrane against it, and the cells swell and burst; indeed both cell wall and membrane may rapidly disintegrate, owing to these osmotic forces. Halophilic bacteria and some of the "lesser halophiles" (but still salt-requiring forms) seem to have lower muramic acid content, and some none at all. Thus there is evidence that the obligate halophiles have a weakened cell wall structure such that, when the organism is placed in water or dilute salt, the cell's swelling simply breaks the cell wall and the cell lyses. But this is not the whole story, since isolated, cell-free, protein-synthesizing systems from halophiles require a higher salt content *in vitro* than do comparable preparations from more usual cells.

With osmophilic yeasts, for example, whose resistance to the osmotic effects of high sugar concentrations is well known, the mechanism appears to be different. These organisms will not tolerate high salt concentrations, and the A_w values of 0.60 must be achieved by way of sugar, not inorganic ions. Most organisms when placed in 30 percent glucose plasmolyze readily, and the protoplasm shrinks away from the cell wall. They may still metabolize, but growth seems to be impossible. Clearly, the water has been drawn out of the cell. Indeed, if such plasmolyzed cells are suddenly placed in distilled water, most will burst; the penetration of water into the internal regions (possibly in the area *between* the cell membrane and the cell wall) generates expanding forces that the cell wall is unable to tolerate. But not the osmophiles. They remain in an unplasmolyzed state and continue their metabolism without appreciable change. When placed in distilled water, they do not swell or burst, and most that we are familiar with experimentally will grow at lower levels of sugar (e.g., 1 percent glucose). They do not *require* high sugar content; they are simply not affected by it.

EFFECTS OF pH

There is another component of the physical environment that exerts decisive influence on whether an organism will be able to grow in a given medium. This is the concentration of hydrogen ions, which is customarily designated by the term pH [log $1/(H^+)$].

In one sense, hydrogen ion is as toxic as mercury, for at a concentration of 10^{-4} *M* (pH 4) hydrogen ion stops the growth of most bacteria. But hydrogen ions are a normal part of any environment; they cannot be excluded. Even at pH 14, there is a 10^{-14} *M* concentration of H^+. Each organism has a pH range, a minimum, an optimum, and a maximum. Below the minimum pH, there is first inhibition of growth, but as the pH drops farther below the minimum, killing of the cell may begin, but not as drastically as one might suppose. Cells whose growth is stopped at pH 4.5 may be exposed to 1 *N* HCl (pH about 0) for 24 h without appreciable drop (less than one log) in count, but longer exposure will accentuate death. One should remember that a drop of one log in numbers represents a 90 percent kill. At the other end of the scale, high pH is normally a more drastic killing agent, largely because cell walls tend to dissolve and cell membranes become damaged. Too much hydrogen ion (low pH) or too little (high pH) may stop growth and may kill preformed cells. But within the range where growth is possible, the external pH profoundly influences the enzymatic composition of the cells.

The pH limits for the growth of any particular organism are a reflection of the pH limits for the activity of the enzymes with which that organism synthesizes new protoplasm and divides into daughter cells. There is considerable variation in the minimum, optimum, and maximum pH values for growth from one genus to another, but for the majority of bacteria the figures range from a minimum pH of 4.5—5.0 to a maximum pH of 8.0—8.5. The optimum pH is generally within 0.5 pH unit of neutrality, that is, 6.5—7.5.

The metabolic activity of the bacterial culture may produce from the components of the medium a variety of end products, some acidic, some basic. Even if the medium were initially at pH 7.0, the release of such substances during growth would shift the pH above or below the outside limits for growth, and thereupon cell reproduction would become impossible. It is therefore necessary to include as a component of a culture medium a chemical that will act as a hydrogen-ion buffer over the pH range at which the organism can grow. The most widely used buffers are K_2HPO_4 and KH_2PO_4, separately or in combination.

Temperature, the concentrations of materials in solution (and thus osmotic pressure), and pH are always present surrounding a living cell. There is no avoiding them; they are the givens, and must always be considered as the environment, because in any conceivable environment they are, of necessity, present. But other environmental factors can be introduced. Of these, we shall consider only radiation.

EFFECTS OF RADIATION

The effects of radiation on living cells as well as on microorganisms is a vast field of investigation that can only be summarized briefly here. Only that radiation which is absorbed by the cell can be effective, although radiation of the media may produce substances that subsequently affect the cell. Such absorbed radiation may have no effect, it may result in damage or death, or it may result in mutation. Certain types of radiation may be used as sources of energy if the photosynthetic apparatus is present, and some radiations may give off light of altered wavelength (fluorescence). However, the radiations we shall discuss are those causing damage or death.

Irradiation effects on bacterial cultures can be produced by electromagnetic radiation: ultraviolet rays of 210—310 nm wavelength (with maximal effect generally at 265 nm) or short X rays (γ rays) of .005—1.0-μm wavelength; or by particle radiation: β rays (high-speed electrons), α rays (helium nuclei), or neutrons. The absorption of the radiant energy by the bacterial cell has two major demonstrable results: either *death of the cell*, indicated by its inability to form colonies, or *mutation* to a genetically altered pattern.

The effects of radiation on cells are brought about by two general mechanisms. One, an indirect but nonetheless lethal effect arises because the radiation results in ionization of the intracellular water and oxygen molecules along the path the radiation takes through the cell, and that the particles thus produced react with the cell constituents. Water ionization due to radiation produces hydrogen atoms and hydroxyl radicals; the interaction of these particles with each other and with oxygen under irradiation is not known with certainty, but it was early postulated that H_2O_2 (hydrogen peroxide) was formed. However, the addition of small amounts of H_2O_2 to the medium, equivalent to those produced by effective radiation doses, does not produce equivalent results. For this and other reasons, H_2O_2 per se is not now considered to be the mechanism by which radiation produces its effect, although it is possible that organic peroxides are involved. It is more likely, however, that the more potent superoxide and OH radicals are concerned by their reaction (primarily oxidation) with the molecules of cytoplasm and nucleus. There is evidence that ionizing radiation (i.e., X rays) can inactivate enzymes obtained in the cell-free state, and that the inactivation is due mainly to oxidation of enzyme protein sulfhydryl (—SH) groups to disulfide (—S—S—) groups. Compounds containing sulfhydryl groups have been shown to furnish protection against the effects of ionizing radiation on enzymes, bacteria, and even animals, presumably by competition for the injurious reactive substances formed by the radiation. The lethal effect of X rays and γ rays is markedly dependent on the presence of oxygen. This oxygen dependence has been demonstrated in a variety of ways: by replacement of the air environment with gases such as nitrogen or helium, by the presence of reducing compounds such as sodium hydrosulfite, by metabolic exhaustion of the environmental oxygen through oxidation of substrates as succinate or pyruvate, or even by endogenous oxidation of heavy bacterial suspensions. Chemical substances may be added to the medium that mimic the effect of radiation. These are called *radiomimetic* materials and are generally agents that react with DNA, are highly oxidative, react with —SH groups, or generate reactive free radicals.

It would be oversimplification, however, to assume that the only effect of radiation on bacteria is via the indirect mechanism of oxidation by such substances as OH radicals. There is a direct effect of radiation, especially upon the nuclear material of the cell. This effect is particularly related to the DNA rather than RNA, and killing results, in a sense, from the production of lethal mutants.

Protection against ionizing radiation by oxygen removal,

whether by physical, chemical, or metabolic means, is still only partial protection; the rate of killing is slower, but there is killing nevertheless. The effects on the nuclear material may result not only in the death of the cell but also in surviving mutants. Since lethal doses of irradiation produce depolymerization and a decrease in the viscosity of desoxyribonucleic acid *in vitro*, one might suppose that, if similar changes occur *in vivo*, a profound disorganization of the controlling genetic processes would result. However, it appears that the effect of radiation is much more specific.

There seem to be several processes involved. Effective radiation causes the formation of thymine dimers; when there are two adjacent thymines on a DNA strand, the hydrogen bonds between the strands may be broken and dimers formed as follows:

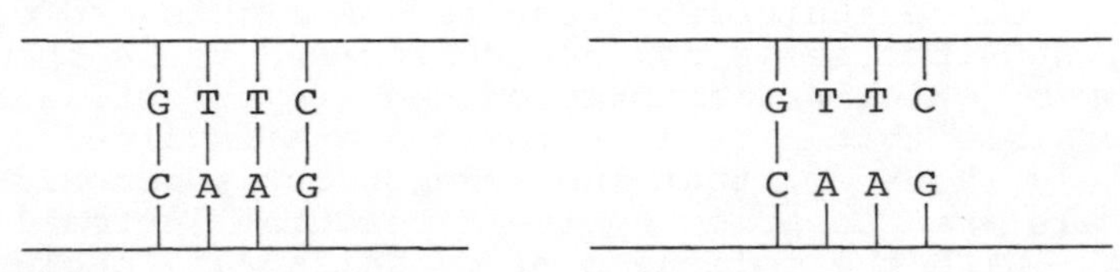

One dimer in a DNA strand 350 μm long is sufficient to block further synthesis. Ultraviolet radiations can cause other dimer formations involving deoxycytidine, and can also alter the thymine itself, forming, for example, dihydrothymine (*E. coli*). Furthermore, radiation can inactivate an inducer of cell division. All these effects are on DNA, but it has also been shown that ultraviolet radiation can inactivate RNA (at least for β-galactosidase); so the effect of radiation, although predominantly on the DNA, can be exerted on RNA as well.

There also exist repair mechanisms, one of the most interesting being *photoreactivation*. Operationally, when a suspension of bacteria is irradiated with ultraviolet light and plated at intervals, it shows the usual log death curve due to irradiation. But if an aliquot of the same suspension is subsequently exposed to visible light (3,200—4,500 Å), there is a markedly greater number of survivors. Indeed, if two sets of plates are made and one kept in the dark and the other exposed to visible light, the number of colonies found increases markedly on the plates exposed to visible light. It is not necessary to have the whole cell and it is possible to inactivate and photoreactivate transforming principle, although a final enzymatic step involving the intact cell may be involved. What appears to happen is that visible light activates an enzyme which splits the thymine dimer, and the hydrogen bonding across to the adenine is restored. Visible light, in addition to its photoreactivating effect has other influences on nonphotosynthetic microorganisms. It stimulates hydrogenase production by *Proteus* and increases mutation to T5 resistance in *E. coli*, but the mechanisms are not clear.

Restricting our attention to the effect of ultraviolet radiation on *E. coli*, most strains of this organism are equally susceptible to pyrimidine dimer formation (the primary dimer is thymine—thymine). Mutations that increase sensitivity to ultraviolet light may do so by reducing the ability to repair the damage or by reducing the ability to tolerate unrepaired damage. The most sensitive strains can neither repair nor tolerate more than a few pyrimidine dimers per cell. The most resistant can repair thousands of dimers and can tolerate as many as 100 without repair. Three repair mechanisms are known. Two (photoreactivation and excision) eliminate the thymidine dimers. The third (postreplication repair) acts on single-stranded gaps in replicating DNA, not on the DNA actually exposed to radiation.

Photoreactivation is caused by an enzyme activated by visible light, which splits or monomerizes pyrimidine dimers and thus restores normal DNA structure. It is the only repair mechanism that requires visible light; the primary damage is reversed and the entire reversal takes place in a single enzymatic step.

Excision repair occurs in the dark; the dimers are not altered chemically but are removed physically. This is not just removal of the two dimer nucleotides in one strand of the DNA. Rather there is a cutting out of a longer nucleotide strand including the dimer with a small number of nucleotides on either side of it, the total nucleotides removed being about 5 to 8 (21). The gap produced in one strand of the DNA is evidently first enlarged, and then nucleotides are repolymerized using the opposite unbroken strand as template. This type of repair is not restricted to ultraviolet damage (and thus to pyrimidine dimers) but can repair DNA damaged by nitrogen mustard, 4-nitroquinoline, mitomycin c, and nitrous acid.

Postreplication repair seems to be a variation of recombination involving the pairing of two daughter strands that may contain gaps, but never at the same locus. Postreplication repair seems to be caused by factors in episomal form.

Each of these processes is enzymatic and thus under genetic control, and therefore mutants exist deficient in one or more of the repair mechanisms. But of perhaps more physiological interest, inhibitors of the repair processes are known, some of which (e.g., caffeine or acriflavin) increase the yield of ultraviolet-induced mutations, especially at low radiation doses. Because repair of radiation damage is dependent upon enzymatic activity, because the occurrence of adjacent thymines and their dimerization may affect the synthesis of various necessary substances, the temperature of irradiation, the temperature range used to cultivate the irradiated cells, the media employed, the age of the culture when irradiated, the nature of the radiation, the medium in which it occurs, and doubtless other factors influence the results obtained.

There exist mutants markedly resistant to radiation; *E. coli* B has such a mutant (*E. coli* B/r), but in part *E. coli* B is more sensitive to ultraviolet radiation than most other *E. coli* strains. However, the B and B/r strains differ but little in the number of thymine dimers produced by ultraviolet radiation, in the rate of excision of thymine dimers, in the ability to promote host cell reactivation of ultraviolet-inactivated bacteriophage, or in the recovery time for resynthesis of DNA after irradiation. Evidently, the resistance to radiation in B/r is not to be found in any fundamental alteration in these factors. In *E. coli* B/r, resistance to radiation seems to be related to an inducer of cell division, which is destroyed by ultraviolet radiation. Some mutants of *E. coli* B form long filaments after exposure to ultraviolet radiation, but the growth of cell mass is essentially unharmed. A purified extract of normal *E. coli* B added to such cultures permits cell division; that is, a chemical factor appears to be involved in cell division. One could conceive of radiation resistance as being due to an altered DNA (e.g., one with fewer adjacent thymines), even though the cell content of DNA would be essentially similar. One might also suppose that some organisms might be protected by pigments which absorb the radiation before it can reach the DNA, and indeed many organisms that survive in air (where they might be exposed to the ultraviolet in sunlight) are pigmented. But radiation-resistant mutants do not seem to differ from the wild-type in nature or extent of pigmentation or carotenoid content.

Production of Light

Relatively few microorganisms are capable of producing visible radiation, but they are rather interesting. In the firefly and higher organisms there is a system that oxidizes a benzothiazole derivative (called luciferin), according to the following reactions (this system is used to measure small amounts of ATP):

$$\text{reduced luciferin } (LH_2) + ATP \rightarrow LH_2 - AMP + \text{pyrophosphate}$$

$$LH_2 - AMP + O_2 \rightarrow L - AMP + 2H_2O + \text{light}$$

In bacteria the situation is somewhat different, and the reactions appear to be

$$\begin{matrix}\text{reduced} \\ \text{flavoprotein}\end{matrix} + \begin{matrix}C_1 \text{ to } C_8 \\ \text{organic} \\ \text{aldehydes}\end{matrix} + O_2 \rightarrow$$

$$\begin{matrix}\text{oxidized} \\ \text{flavoprotein}\end{matrix} + \begin{matrix}\text{organic} \\ \text{acid}\end{matrix} + \text{light}$$

REFERENCES

Temperature

Castenholz, R. W. 1969. Thermophilic blue-green algae and the thermal environment. *Bacteriol. Reviews 33*: 476–504.
Farrell, J., and L. L. Campbell. 1969. Thermophilic bacteria and bacteriophages. *Advan. Microbial Physiol. 3*:83–110.
Farrell, J., and A. Rose 1967. Temperature effects on microorganisms. *Ann. Rev. Microbiol. 21*:101–121.
Friedman, S. M. 1968. Protein-synthesizing machinery of thermophilic bacteria. *Bacteriol. Reviews 32*:27–38.
Morita, R. Y. 1975. Psychrophilic bacteria. *Bacteriol. Reviews 39*:144–167.
Travassos, L. R. R. G., and A. Cury. 1971. Thermophilic enteric yeasts. *Ann. Rev. Microbiol. 25*:49–74.

Osmotic Effects

Larsen, H. 1967. Biochemical aspects of extreme halophilism. *Advan. Microbial Physiol. 1*:97–132.
Litchfield, C. D. 1976. *Marine Microbiology*. Dowden, Hutchinson & Ross, Inc., Stroudsburg, Pa.
MacLeod, R. A. 1965. The question of the existence of specific marine bacteria. *Bacteriol. Reviews 29*:9–23.
Thayer, D. W. 1975. *Microbial Interaction with the Physical Environment*. Dowden, Hutchinson & Ross, Inc., Stroudsburg, Pa.

Radiation

Ginoza, W. 1967. The effects of ionizing radiation on nucleic acids of bacteriophages and bacterial cells. *Ann. Rev. Microbiol. 21*:325–368.
Moseley, B. E. B. 1968. The repair of damaged DNA in irradiated bacteria. *Advan. Microbial Physiol. 2*:173–194.
Witkin, E. M. 1969. Ultraviolet-induced mutation and DNA repair. *Ann. Rev. Microbiol. 23*:487–514.

Papers

1. *J. Bacteriol. 87*:562 (1964).
2. *J. Bacteriol. 94*:197 (1967).
3. *Biochim. Biophys. Acta 213*:159 (1970).
4. *Amer. Scientist 62*:712 (1974).
5. *Can. J. Microbiol. 19*:1452 (1973).
6. *J. Bacteriol. 121*:907 (1975).
7. *J. Bacteriol. 94*:1970 (1967).
8. *J. Biol. Chem. 245*:6468 (1970).
9. *Nature 233*:494 (1971).

10. *J. Bacteriol.* *107*:303 (1971).
11. *Bacteriol. Reviews* *37*:320 (1973).
12. *J. Bacteriol.* *85*:218 (1962).
13. *J. Bacteriol.* *122*:177 (1975).
14. *Bacteriol. Reviews* *32*:27 (1968).
15. *J. Bacteriol.* *105*:165 (1971).
16. *J. Bacteriol.* *95*:345 (1968).
17. *J. Bacteriol.* *92*:1388 (1966).
18. *J. Bacteriol.* *105*:512 (1971)
19. *J. Bacteriol.* *81*:165, 172 (1961).
20. *J. Bacteriol.* *121*:664 (1975).
21. *J. Bacteriol.* *122*:341 (1975).

12
The Chemical Environment and Antibiotics

The environment closest to the cell is normally water, and it is through materials dissolved in or suspended in water that interaction with the cell is accomplished. We have earlier considered the effects of too little nutrient upon the cell and the effects of too much of an otherwise nontoxic material. We wish now to consider the effects of external substances, particularly those chemical substances that inhibit or kill microorganisms.

First, it is important to note the methods by which the effects are determined and the mechanisms by which cells are injured. The survival of a bacterial population in contact with a toxic agent depends on many factors: on the concentration of the agent and the number and type of bacteria, on the length of the contact period, and on supplementary factors such as pH, temperature, and organic matter in the suspending medium. Survival is an all-or-none phenomenon. We (perhaps arbitrarily) define a living bacterial cell as one that is capable of reproduction, and an organism that survives the ordeal of contact with a toxic agent is, by definition, one that is still capable of reproduction. It may not grow while in the presence of the agent, but may proceed into growth and division when removed from the agent. A compound that acts only to prevent reproduction while in contact with the cell is called *bacteriostatic*, and one that permanently damages the reproductive mechanism is called *bactericidal*. Because much work has been carried out on pathogenic bacteria, the terms *antiseptic* and *disinfectant* have been applied to bacteriostatic and bactericidal agents, respectively, in particular with reference to pathogens. These have become vernacular terms in recent years, and their use among bacteriologists has been decreasing. Furthermore, the distinction between temporary and permanent injury of bacterial reproduction depends to a great extent on the concentration of the agent employed. At low concentrations it may be bacteriostatic, at higher ones, lethal. The actual range from ineffective through inhibitory to lethal concentrations is a function of each particular agent. Many agents will stimulate growth within a fairly narrow zone of concentrations between the ineffective and inhibitory ranges; the causes of this phenomenon have not been explored. The death of a bacterial cell should not be the sole measure of toxicity; as practical a yardstick as it may be, it is, after all, a last resort, from the cell's point of view. Only if the injury is sufficiently severe, and the opportunity for recovery lacking, will reproduction be permanently blocked. Toxicity may be manifested in other ways — by temporary inhibition of growth, as in bacteriostasis, or by growth of abnormal cells. In the presence of sublethal concentrations of many chemical agents, the first (and thus presumably the most sensitive) process to be deranged is that of cell division; although the cells grow, they do not divide, and one observes long spaghetti-like cells, or swollen and bulbous forms.

We also must note that an agent may have more than one activity, and sometimes it is difficult to decide which is the primary effect of the agent and which are sequellae. For example, streptomycin was first noted to precipitate DNA and RNA (and it has proved useful in chemical separations of these substances); subsequently, it was shown to inhibit the metabolism of certain amino acids, then of certain organic acids, then of certain phosphorus compounds. It was then shown to inhibit adaptive enzyme formation, and the metabolic effects noted above seem to be caused mostly by such inhibition. Since enzymes are protein, the effect of streptomycin could be ascribed to inhibition of protein synthesis; but since the mechanism of protein synthesis was not known at the time of these studies, little further progress could be made. Streptomycin was then discovered to alter cell permeability and cause leakage, especially of nucleotides. Finally, it was shown to react with the 30S ribosome, which appears to be its *primary* action. All the various effects noted above are real; streptomycin does indeed do all the things listed. But we believe the actions on the ribosome to be the primary action since it takes place much more rapidly than the other effects, except nucleic acid precipitation. Furthermore, the nature of the 30S ribosomal reaction is sufficiently unique, occurs at proper concentrations, and is specific, so that it can serve to explain the other effects noted, which are therefore considered secondary events resulting from the primary action. But it is neither necessary nor, indeed, likely that every agent will have only *one* effect, or even only a single primary effect, and we may expect some materials to exhibit more than one.

In the following discussion we shall consider various cell locations and various cell processes as the target of action of toxic materials. We shall not attempt a complete survey nor even consider all the major toxic materials. Rather, we shall choose examples that illuminate the nature of the processes affected, that illustrate how such interference damages the cell, and that throw some light upon the problem of why certain agents act with a high degree of specificity.

AGENTS ACTING AT THE CELL SURFACE

Cell Wall

To be toxic, a compound does not necessarily have to penetrate within the cell; it may disorganize, disrupt, or even abort the formation of the outer boundary itself, either wall or membrane. Interference with cell-wall formation and function has been definitely implicated by many studies on the mode of action of penicillin. It was known for some years that penicillin induced aberrant morphological changes in bacteria: the spaghetti-like forms mentioned earlier, large swollen bodies and L forms, and even lysis. Another piece of information (which turned out to be most helpful) was that organisms inhibited by penicillin accumulated unfamiliar uridine nucleotides. The principal nucleotide accumulated by *Staphylococcus aureus* was eventually identified as uridine acetyl muramic tetrapeptide. At the time the first chemical studies on identification of this compound were done, very little was known about the composition of bacterial cell walls, and there was no particular reason to correlate the accumulation of odd uridine nucleotides and the observed effects of penicillin on morphology and amino acid incorporation. But in time the riddle was solved, studies on cell-wall composition, particularly of gram-positive bacteria, clearly showed that the walls contained muramic peptide as a major constituent. It is now known that the action of penicillin inhibits the enzyme, alanine transpeptidase, which knits the murein chains into a complete corset; in addition, penicillin inhibits D-alanine carboxypeptidase, whose physiological function is less clear. As might be expected, most penicillins and the chemically related cephalosporins interfere with cell-wall synthesis in approximately the same manner, but other agents may act at other stages, since the pathway from glucose to the closely knit "bag-shaped molecule" that constitutes the cell wall is indeed rather complicated. Vancomycin and ristocetin seem to act by inhibiting the incorporation of the lipid intermediates into the peptidoglycan. Bacitracin prevents peptidoglycan synthesis indirectly by preventing the dephosphorylation of the lipid carrier pyrophosphate, and thus preventing its regeneration (1,2,3). Bacitracin, in common with the other circular peptide antibiotics, also combines with the membrane.

Cell Membrane

Interaction at the cell membrane may be a little more

complex. One should emphasize that in addition to its osmotic and permeation properties, the bacterial cell membrane is also the site of several other activities (e.g., respiration, energy generation, sometimes protein synthesis, etc.). Chemical agents may interfere with these membrane activities. Agents inhibiting respiration (cyanide, azide) may be regarded as acting at the membrane. Uncouplers of oxidative phosphorylation (dinitrophenol, valinomycin) may also be regarded as acting at this locus. Protein synthesis will be discussed in terms of a cytoplasmic locus, but we should point out that the protein synthesis taking place at the membrane seems to be considerably more efficient and seems to possess somewhat less sensitivity to certain agents (streptomycin, for example).

But the membrane also excludes a variety of materials that, if they were to enter, might be deadly. Protein synthesis, for example, in the cell-free state is inhibited by streptomycin even if the preparations are made from eucaryotic cells. Yet eucaryotic cells themselves are not inhibited by streptomycin because it does not penetrate.

Cyclic peptides. Two types of cyclic peptide antibiotics, produced by various *Bacillus* species, affect the integrity and function of the cytoplasmic membrane of sensitive bacterial cells. One group includes the Circulins, produced by *B. circulans*, and the polymyxins A to E, produced by *B. polymyxa*. The formula for one of these, polymyxin E or colistin, is characteristic for the whole group: a *heptapeptide* in cyclical form and a tail (Figure 12-1). Except for the replacement of D-leucine by D-phenylalanine in polymyxin B, and the replacement of L-leucine by L-isoleucine in circulin A, the rest of the molecule in the three antibiotics is identical. The unusual amino acid, L-α,γ-diaminobutyric acid, functions at the junction of the circle and the tail by peptide linkages through the γ and α amino groups, respectively. The second group of cyclic peptide antibiotics includes the tyrocidins A to C, produced by *B. brevis*, and gramicidin S, produced by an organism closely related to it. Gramicidin S, incidently, differs in structure from the other gramicidins and should be classified in the tyrocidin

The tyrocidins are *decapeptides* (Figure 12-1), which all contain the identical pentapeptide sequence shown in the figure; gramicidin S is a dimer of the pentapeptide, whereas the tyrocidins vary only in the aromatic amino acids between proline and asparagine (L-phenylalanine-D-phenylalanine in tyrocidin A; L-tryptophan-D-phenylalanine in tyrocidin B; L-tryptophan-D-tryptophan in tyrocidin C).

Although the antibacterial spectrum of these various antibiotics of the two groups varies considerably one from another, their effect on sensitive cells produces the same general pattern of leakage of intracellular

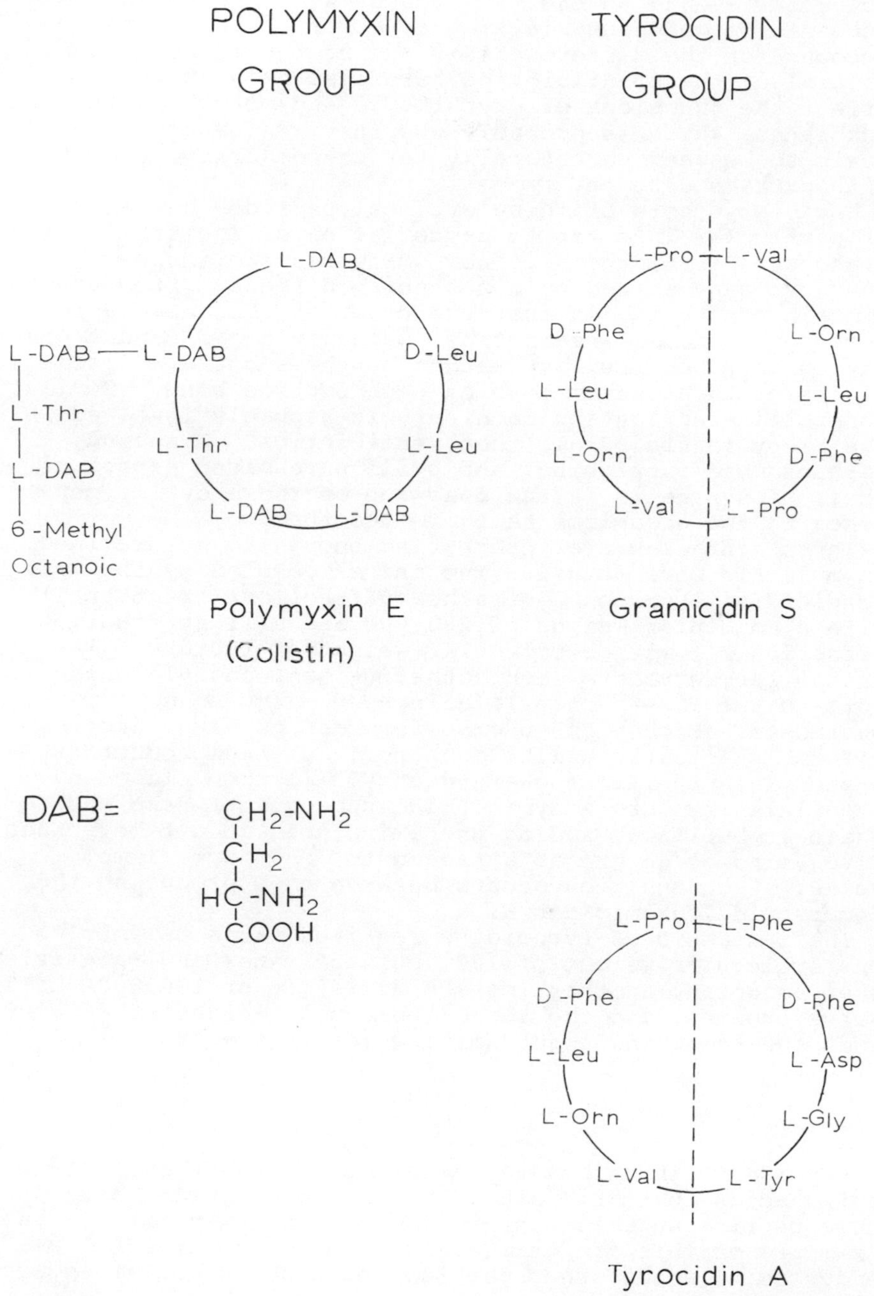

POLYMYXIN GROUP

L-DAB

L-DAB — L-DAB D-Leu

L-Thr

L-Thr L-Leu

L-DAB

L-DAB L-DAB

6-Methyl Octanoic

Polymyxin E (Colistin)

TYROCIDIN GROUP

L-Pro L-Val

D-Phe L-Orn

L-Leu L-Leu

L-Orn D-Phe

L-Val L-Pro

Gramicidin S

DAB = CH_2-NH_2 | CH_2 | HC-NH_2 | COOH

L-Pro L-Phe

D-Phe D-Phe

L-Leu L-Asp

L-Orn L-Gly

L-Val L-Tyr

Tyrocidin A

FIGURE 12-1. *Examples of the circular peptide antibiotics.*

constituents, increased permeability to extracellular substances, and eventually lysis. Still unknown are the molecular events in the membrane damage, and the differences in target bacterial membrane structure, which account for the differences in antibacterial spectrum. Several of these antibiotics have also been shown to affect the functions of erythrocyte and mitochondrial membranes, which is probably why they are usually too toxic to be used parenterally for chemotherapy, except with extreme caution.

The biosyntheis of these cyclical peptides has also been of particular interest to bacterial physiologists for two reasons. First, there is now quite clear evidence that they are synthesized by a mechanism different from that of protein synthesis, in that the order of the amino acid residues is determined not by coding in a mRNA and translation on ribosomes, but rather by the substrate specificity of each enzyme forming each peptide bond. Similar nonprotein-fabricating machinery is probably involved in the biosynthesis of all these antibiotics. A second, perhaps more intriguing, and still unresolved aspect of their biosynthesis is the function of these cyclic peptides to the organisms that produce them.

These syntheses are interesting enough to be examined in a little more detail. The enzyme complex synthesizing gramicidin is composed of a heavy fraction (fraction I) with a molecular weight of 280,000 and a light fraction (fraction II) with a molecular weight of 100,000. The latter enzyme reacts with either D- or L-phenylalanine to form enzyme II — D-phenylalanine—AMP complex and pyrophosphate. Either the D or L form can be used, and the L form is actually a little better, but the product is D-phenylalanine. The charged enzyme II then places phenylalanine onto enzyme I, and subsequently the entire chain is built up, adding one amino acid at a time. When five amino acids are so attached to *two* enzyme I molecules, a condensation occurs between them producing the decapeptide, gramicidin S.

The synthesis of tyrocidine requires three enzymes. One (molecular weight of 100,000) activates and racemizies phenylalanine, another (molecular weight of 230,000) activates proline, and the third (molecular weight of 460,000) adds the remaining eight amino acids in the chain.

BIOSYNTHESIS OF PEPTIDE ANTIBIOTICS[1]

Peptide antibiotics are so-called because they contain amino acids that are joined together by peptide bonds. Some peptide antibiotics are composed of only amino acids. Examples of this are gramicidin S and tyrocidine A. Besides amino acids other peptide antibiotics contain some

[1]Perlman, D., and M. Bodanszky. 1971. Biosynthesis of peptide antibiotics. *Ann. Rev. Biochem.* *40*:449—464. (Reviewed by Anubha R. Mukherjee.)

some fatty acids. Well-known examples in this group are the polymyxins.

Other characteristics of peptide antibiotics are as follows:

1. They are generally smaller than proteins; their molecular weight is in the range of 300-2,000.

2. They have both L and D isomers of the individual amino acids. In some cases they contain unusual amino acids, like *N*-methyl amino acids and diaminobutyric acid.

3. Most of the peptide antibiotics are cyclic in nature and are not attacked by proteases.

The general approaches used in the study of biosynthetic problems are the following:

1. Addition of labeled precursors to growing cultures and isolating labeled intermediates.

2. Addition of labeled amino acids and following their incorporation in antibiotics. It has been shown that D-amino acids in peptide antibiotics are obtained from the corresponding L-amino acids. The bioysnthesis does not follow the same pathway as protein biosynthesis. Addition of chloramphenicol to the actinomycin producing *Streptomyces antibioticus* prevented further protein biosynthesis but did not affect actinomycin biosynthesis. That peptide antibiotic synthesis does not involve ribosomes or mRNA has been proved by the cell-free synthesis of gramicidin S, tyrocidine, edeine, and actinomycin.

The enzyme system that makes gramicidin S has been isolated from *B. brevis* as a multienzyme complex. The enzyme complex, when mixed with ATP, Mg^{2+} and the corresponding amino acids, gives gramicidin S. Thus

$$2(\text{leucine} + \text{ornithine} + \text{valine} + \text{proline} + \text{phenylalanine}) + 10\ \text{ATP} + Mg^{2+} \xrightarrow{\text{enzyme}} \text{gramicidin S}$$

The enzyme system has been separated into two components, factor I of molecular weight 280,000 and factor II of molecular weight 100,000. Of these, factor I catalyzes the activation of phenylalanine, and factor II catalyzes the activation of the other four amino acids. The activation of the amino acids is independent of tRNA. During the activation of phenylalanine by the factor I, the amino acid is bound by the enzyme protein through the sulfhydryl linkage. This enzyme bound phenylalanine initiates on the heavier protein the sequential addition of another amino acid (proline, valine, ornithine, and leucine) to form enzyme-bound polypeptide thioester. Carboxyl activated D--phenylalanine initiates peptide synthesis, by transfer to L-proline, which remains linked to heavy factor as a thioester. The active carboxyl of the dipeptide thioester then reacts with the amino group of the next amino acid, valine, to form H_2N-pehnylalanine-proline-valine-thioenzyme and the process is repeated until we get the pentapeptide H_2N-phenylalanine-proline-valine-ornithine-leucine-thioenzyme. At this stage two enzyme-bound pentapeptides appear to cyclize by interaction between activated luecine carboxyl and the free NH_2 group of phenylalanine. In polymyxin biosynthesis studies it has been shown that chloramphenicol and actinomycin inhibited incorporation of threonine into protein and stimulated incorporation into polymyxin B.

Other important peptide antibiotics are the actinomycins. In the biosynthesis of actinomycins, it has been shown that L-valine was derived from L-valine, L-isoleucine is the source of D-allo-isoleucine, L-proline is the origin of L-proline, hydroxyproline, and 4-oxo-L-proline. Glycine is a direct precursor of sarcosine. L-Methionine is the source of the methyl groups of sarcosine and *N*-methyl valine. Cell-free actinomycin biosynthesis has been achieved with extracts from *S. antibioticus* supplemented with tryptophan, glycine, methionine, valine, threonine, proline, pyridoxal PO_4, and ATP.

Polyenes. Another example of antibiotics acting on the membrane is found in the action of the polyene antibiotics (especially nystatin and amphotericin B), which act upon yeasts and some fungi, while having no action on bacteria. These have been shown to combine with membrane steroids, and thus to affect permeability. Since bacteria have no steroids in their membranes, they are not affected by such materials.

Other agents may act at the membrane in a much less specific fashion. For example, toluene damages the cell membrane, materials begin to leak out, and soon nucleotides derived from the rapid breakdown of ribosomal RNA are in the external medium. Detergents, lipases, solvents, and the like, affect the membrane, and thus set in train a considerable degradation of cell constituents. Why is the chemical integrity of membrane so important? Again we would suggest that much of the cell's metabolic activity is in the membrane, and that here synthetic and degradative enzymes are spatially localized. When the membrane is disorganized, the spatial relations are destroyed and the degradative enzymes begin to attack the cell itself.

Valinomycin in a cyclic peptide containing three molecules each of L-valine and D-valine, three molecules of L-lactic acid, and three molecules of D-α-hydroxyisovaleric acid. It has the curious property of wrapping itself about a potassium ion and holding the ion in the interior of the antibiotic molecule. Although the antibiotic has little clinical significance, it is widely used in the study of membrane transport.

INTERACTION WITH NUCLEAR MATERIAL

The DNA of the cell exists in two forms: as nuclear material and as extranuclear material, an episome or plasmid. The DNA of both types is in the double-stranded form, and certain agents (as mentioned in the discussion of radiation) act by altering the integrity of the strand, the radiomimetic agents (as H_2O_2), for example. The DNA must reproduce, and for this purpose it needs the necessary enzymes (DNA polymerase) plus the deoxynucleotide pool. It must also become, at least for a short while, a single strand, and some substances, such as mitomycin, react with DNA, making permanent cross linking between the strands; such DNA cannot replicate. Apparently, one cross link is sufficient to cause death. DNA in its double-stranded form serves as the template for mRNA, and certain substances inhibit this process. Actinomycin D reacts with DNA to prevent mRNA formation, but DNA replication is slightly less sensitive. The cell is thus deprived of new mRNA, and when it uses up the mRNA formed before the addition of the antibiotic, it can no longer make protein. This action of actinomycin D is used to measure the length of time mRNA survives in the cell.

Although a variety of agents act upon either the synthesis, the activity, or the replication of DNA, there are relatively few of clinical importance. We shall therefore content ourselves with simply listing some of the agents and only detail those with clinical importance.

With respect to DNA, an agent might act in the following ways:

1. Agents that act upon and interfere with nucleotide metabolism. There are a wide variety of agents of this type, some antibiotics, some synthetics, arising largely from cancer research. Among these are the clinically important antitumor drugs, 5-fluorouracil and 5-fluorodeoxyuridine, which, after conversion to 5-F-dUMP *in vivo* inhibit thymidylate synthesis. Some of the agents inhibit purine synthesis (azaserine and DON, are related to glutamine with whose action they interfere), some inhibit pyrimidine synthesis (5-azaorotic acid), some inhibit nucleotide interconversion (hadacin, an analogue of aspartate, and 6-mercaptopurine), while others are actually incorporated into polynucleotides (5-bromodeoxyuridine).

2. Agents that act upon DNA replication. One agent is of clinical importance, nalidixic acid. It selectively inhibits DNA synthesis in bacteria, especially in gram negatives, but just how it acts is not clear. It does not inhibit polymerase I. Phenylethyl alcohol, although not greatly antimicrobial, seems to inhibit DNA synthesis, but the mechanism is not clear.

3. Agents that impair the template function of DNA. The "classical" agents are proflavine, an acridine, and ethidium, a trypanoside. These inhibit DNA polymerase I but seem to do so by interaction with the required DNA template. Evidently they "adsorb" so actively to the double-stranded DNA that they prevent any action on it. They seem to act preferentially on circular DNA, and thus they tend to selectively inhibit plasmids and mitochondria. Other similar intercalating agents are miracil (a

schistosomicidal drug), chloroquine (an antimalarial), and the antibiotics, nogalamycin and daunomycin. The best known and widely used antibiotic is actinomycin D (antimycin D). This has limited use as an antitumor agent, but biochemically it has been very useful since it binds to the DNA and prevents further mRNA synthesis. There are other antibiotics that act by binding to DNA, although the mechanism is not known. These are chromomycin, methramycin, olivomycin, echinomycin, hedamycin, rubiflavin, pluramycin, netropsin, distamycin, phleomycin, luteoskyrin, and kanchanomycin.

4. Agents that cross link the strands of DNA so that the DNA cannot replicate. Such an agent is mitomycin C. Other materials with a similar action are the bifunctional nitrogen and sulfur mustards and anthramycin.

5. Agents that break a DNA strand. Two agents of this type are known, streptonigrin and bleomycin.

6. Agents that inhibit transcription, that is, mRNA formation. One of the most interesting of these are the substances of the rifamycin group (discussed later). Others are streptovaricins, streptolydigins, and amanitins.

7. The ribosomal RNA is also made on the DNA template, and although there seem to be no specific agents acting on ribosomal RNA synthesis, such synthesis is inhibited by chloramphenicol, ethyromycin, and the tetracyclines, although the inhibition is probably indirect. One can also conceive of agents acting on the operator, promotor, and regulator sections of the DNA, but at present no agents are known to act at these loci.

ANTIBIOTICS AND NUCLEIC ACIDS[1]

This article deals with antibiotics that interact with template DNA (and sometimes RNA), those which interact with the enzyme responsible for transcription or replication.

Agents that Complex with DNA.

Inhibitors of template function: noncovalent complex formation. The actinomycin binds to DNA. Gram-negative bacteria may be resistant, owing to permeability properties. Spheroplasts prepared from these cells are sensitive. RNA chain elongation, not initiation, is inhibited. DNA must be helical and have guanine residues. It is the G—C pairs with which the agent complexes and strongly binds by the chromophore portion of the molecule. Actinomycin intercalates into the DNA and causes uncoiling of the double helix.

[1]Goldberg, I. H., and P. A. Friedman. 1971. Antibiotics and nucleic acids. *Ann. Rev. Biochem.* *40*:775—810. (Reviewed by Gail P. Simon.)

Chromomycin A is a potent inhibitor if *in vivo* and *in vitro* RNA synthesis, but it inhibits DNA synthesis to a lesser extent in man; in bacterial cells the inhibition is equal to that of RNA. RNA chain elongation not initiation is inhibited. Like actinomycin, it requires guanine residues and helical DNA, but it also requires the presence of divalent cation (Mg^{2+}) for interaction. Separated strands of DNA show no binding, but the DNA does not uncoil, so it must not intercalate into the DNA.

Anthracyclines (e.g., daunomycin) intercalates between adjacent base pairs of DNA, but binding is stronger than accounted for by simple intercalation. DNA synthesis is more sensitive than that of RNA, and the DNA molecule is uncoiled.

Rubiflavin, hedamycin, pluramycin, in a variety of eucaryotic and procaryotic cells selectively block nucleic acid synthesis. DNA is more sensitive. Rubiflavin and hedamycin bind to DNA by free radicals, and their action is not correlated with G—C content of the nucleic acid. Hedamycin acts by preventing strand separation because inhibitory effects are greater when DNA is double-stranded.

Inhibitors of template function: covalent complex formation. Mitomycin shows a covalent linkage to the DNA template. This bifunctional alkylating agent can form covalent complexes with many biological molecules. Since mitomycin selectively inhibits synthesis of DNA, its lethal action is presumed to be due to its formation of covalent bonds with DNA, resulting in changes in DNA structure that prevent its replication.

There are other antibiotics in which strong binding to DNA is presumed to be covalent, but the nature of the binding in these agents is not known. Among these are anthramycin and bleomycin.

Inhibitors of polymerase function. Rifamycin, streptolydigin, and streptovaricin belong to a group of direct enzyme inhibiting antibiotics that have an enormous specificity in inhibiting procaryotic RNA synthesis.

Inhibitors complexing with template and inhibiting polymerase function. Luteoskyrin is a potent inhibitor of RNA polymerase, and it alters DNA structure. It requires magnesium cation.

Kanchanomycin alters the template function of the DNA and inactivates polymerase.

RIFAMYCINS[1]

These antibiotics act on RNA polymerase and on reverse transcriptase. Rifampin is a semisynthetic derivative of rifamycin (Figure 12-2). It is an orally active antibiotic that has a great promise against tuberculosis. It is active against many gram-positive cocci, *Neisseria*, and *Hemophilus*, less active against other gram-negative species, has demonstrable antileprosy activity *in vivo*, and inhibits certain viruses. It inhibits bacterial DNA-dependent RNA polymerase. Discovered in 1959, it was obtained from *Strep. mediterranei* and was found to be an aggregate of five specific substances, which were designated as rifamycins A, B, C, D, and E, of which B was the most stable. Addition of diethylbarbituric acid to the medium greatly enhanced the production of this substance.

By 1964, the chemical structure of rifamycin was determined, and this opened up the way for the synthesis of a vast number of semisynthetic derivatives. Rifamycin B itself was found to be antibacterial, but was limited to parenteral use since it was not orally active. The rifamycin structure contains a naphthaquinone ring and an ansa bridge. The molecule has certain crucial parts, which if modified structurally will result in the loss of the antibacterial activity. If the three double bonds of the ansa chain at C 16—17, 18—19, and 28—29, were hydrogenated successively, a gradual decrease of bacterial inhibition resulted. Hydroxyl groups on C 21 and 23 are absolutely required for biological activity. The acetyl group of C 25, if reduced, will diminish activity against gram positives but with no effect on activity against *Mycobacterium tuberculosis*.

Rifamycin B can be easily oxidized into rifamycin S, with keto groups on C 1 and 4. Next the latter is reduced to rifamycin SV, with hydroxyl groups on the same positions. Now a variety of side chains are introduced at the C 3 position which has the only aromatic hydrogen. Rifampin will result when a 4-methylpiperazino-iminomethyl group is synthetically introduced at the C 3 position. To date (1972), some 750 derivatives have been synthesized (Figure 12-2).

Rifampin is by far the most important and widely used derivative today. Its chemical name is 3-(4-methylpiperazinyl-iminomethyl rifamycin SV). The WHO nonproprietory name is rifampicin; in the United States it is called rifampin.

On the whole cell level, rifampin inhibits RNA synthe-

[1]Riva, S., and L. G. Silvestri. 1972. Rifamycins: a general view. *Ann. Rev. Microbiol.* *26*:199—224.

Wehrli, W., and M. Staehelin. 1971. Actions of the rifamycins. *Bacteriol. Reviews* *35*:290—309.

Lester, W. 1972. Rifampin: a semisynthetic derivative of rifamycin — a prototype for the future. *Ann. Rev. Microbiol.* *26*:85—102.

(Reviewed by John Chu.)

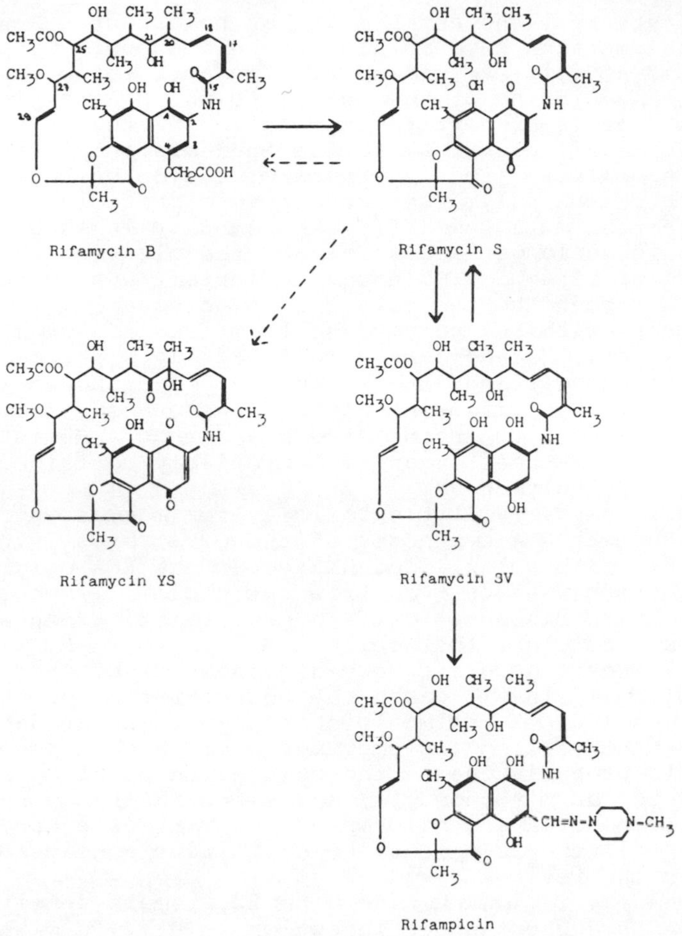

Rifamycin B

Rifamycin S

Rifamycin YS

Rifamycin SV

Rifampicin

FIGURE 12-2. *Formulas of various rifamycin derivatives. Solid arrows represent chemical transformation, dashed arrow are biosynthetic paths. From Wehrli, W., and M. Staehelin, Bacteriol. Reviews 35:291–309 (1971). All formulas should have 23-OH and 24-CH_3. Used with permission.*

sis. Following addition of rifampin, the rate of RNA synthesis decreases exponentially with time. Ribosomal, transfer, and messenger RNA are all equally affected. Protein synthesis stops later, after a time interval corresponding to the life of mRNA. The synthesis of DNA continues until a round of replication has been completed.

On the molecular level, rifampin binds to the free DNA-dependent RNA polymerase and forms a complex. The antibiotic molecule actually binds with the beta subunit of the polymerase. If the chain elongation of the mRNA has already started, the antibiotic has no effect. Resistant bacterial mutants have an altered beta subunit with decreased affinity to rifampin. The antibiotic does not prevent the binding of the enzyme to the DNA template; what is does is prevent the completion of the initiation process. *Bacillus subtilis* rifampin-resistant mutants cannot sporulate, since during sporulation one beta subunit of the RNA polymerase undergoes a proteolytic cleavage; the mutant modifies the subunit in such a way that it is no longer accessible to the cleaving enzyme. One mole of rifampin is enough to inhibit RNA synthesis when it is bound to one mole of RNA polymerase. *In vitro* experiments with RNA polymerase from *Escherichia coli*, *Staph. aureus*, *Micrococcus luteus*, *B. subtilis*, *Azotobacter vinelandii* and *B. sterothermophilus* showed that concentration of rifampin in the neighborhood of 0.02 μg/ml is sufficient to inhibit the polymerase. Resistant strains of all these organisms invariably contained a modified RNA polymerase.

Rifampin has very little toxicity for mammalian organisms. Nuclear RNA synthesis of the eucaryotes is not affected. With solubilized DNA-dependent RNA polymerase, no inhibition was observed from rat nuclei, lymphoid tissue, human placenta, calf thymus, yeast, green algae, protozoa, and plant root cells.

RNA synthesis occurs in eucaryotes in mitochondria and chloroplasts. In all cases the concentration of the antibiotic was 100—1,000 times higher than those needed to inhibit bacterial RNA polymerase. This high concentration effect is probably due to the permeation problem of rifampin into the mitochondria, and since the RNA polymerase from mitochondria and chloroplast of various eucaryotes are not well characterized, no definitive conclusion is given in the reviews.

DNA phages, T4, lambda, and beta 22, remain sensitive to rifampin throughout phage infection. In T7, growth of virus becomes resistant to rifampin inhibition after 5 min of infection, coinciding with the production of T7-specific RNA polymerase. It seems to indicate that rifampin only affects the host bacterial polymerase, but not phage-specific RNA polymerase. Pox virus, which carries its own RNA polymerase inside the virion, is sensitive to rifampin. Intracellular multiplication is inhibited by a rather high concentration (100 μg/ml) of the drug. The virion-associated enzyme was proved to be insensitive to the antibiotic. In rifampin-treated cells, no mature pox virion are formed. Removal of rifampin results in the maturation of viral particles. Herpes and pseudorabies, the larger DNA viruses, are insensitive to rifampin. There are two differences in the mode of action of rifampin against pox viruses as compared to those of bacteria:

1. 100—1,000 times the concentration of bacterial inhibition needed.
2. Inhibition against pox viruses is essentially reversible.

Almost all RNA viruses are unaffected by rifampin except

the oncogenic ones, such as Rous sarcoma virus. In this case, *N*-desmethylrifampin, which has a different side chain than rifampin, was shown to prevent foci formation in chicken fibroblast. The antibiotic here inhibits the action of RNA-dependent DNA polymerase, which is also called reverse transcriptase. As in the case of pox viruses, a large concentration is required to produce the effect.

Rifampin has been very effective against *M. tuberculosis*; the minimum inhibitory concentration is 0.5 μg/ml in the medium. It is well tolerated by human patients with little toxicity and no significant adverse effects. It is recommended that rifampin be used in combination with another antituberculosis drug.

1. Resistance to rifampin occurs regularly and rapidly.
2. There is no evidence of cross resistance of the antibiotic with other antituberculosis drugs.

The other drug of choice is INH (isoniazid). Rifampin is very expensive and is not widely applicable to many initial therapy programs for tuberculosis. Rifampin is also reported to be active against leprosy. One study showed that 4 months of rifampin treatment yielded therapeutic benefits equal to that observed after 3 years of sulfone treatment.

Conclusion

1. The crucial part of the rifampin molecule is in three double bonds in the ansa chain and the 4-hydroxyl group in both the naphthaquinone ring and the ansa bridge.
2. Rifampin inhibits RNA synthesis by binding directly with the beta subunit of the DNA-dependent RNA polymerase.
3. It does not affect in general nuclear RNA synthesis of mammalian cells. Effects on mitochondrial and chloroplastic RNA synthesis require 100 times the concentration.
4. Rifampin is active against certain pox viruses and another derivative of rifampin is active against Rous sarcoma virus, but in both cases requiring a very high concentration. It does not bond with the RNA polymerase of the former, but it inhibits the reverse transcriptase of the latter.
5. Rifampin is an effective antituberculosis agent. Its use is recommended in combination with at least one other antituberculosis drug.
6. Rifampin is active against leprosy.
7. Rifampin itself cannot be used as an antiviral agent. It may serve as a prototype for additional semisynthetic rifamycin derivatives that will have increased activity against viruses.

AGENTS ACTING ON PROTEIN SYNTHESIS

Any of the several steps in protein synthesis could be subject to inhibition. These could be thought of as follows

1. Those steps involving the activation of the amino acid, and the charging of the tRNA. There seem to be no materials known that act on this process.

2. The agent acts on the attachment of mRNA to the 30S (or the smaller portion if one is considering eucaryotic ribosomes) and the proper positioning of the rRNA, indeed, on all those actions associated with the 30S unit. The clinically important streptomycin and tetracycline groups act on the 30S unit, as well as the lesser substances, spectinomycin, kasugamycin, aurintricarboxylic acid, pactamycin, and edeine.

3. The agent acts similarly on the 50S ribosome and inhibits (or even stimulates, as puromycin) the processes associated with the larger unit. Here one finds the clinically important antibiotics, chloramphenicol, the macrolides, erythromycin, lincomycin, as well as the lesser antibiotics, streptogramins, sparsomycin, fusidic acid, thiostrepton, siomycin, bottromycin, cyclohexamide (or actidione), gougerotin, amicetin, blasticidin S, together with other agents with less specific action.

The actions associated with either of the ribosomal units also include the effects on initiation, elongation (translocation of both peptide and mRNA), termination, and breakdown of mRNA, together with the action on the separable factors involved in these processes, and might further include the association or dissociation of the smaller and larger unit. It is not always feasible to attempt to separate these effects, and for our purposes we shall only discuss the streptomycins and tetracyclines in the second group and the chloramphenicol, the macrolides, the erythromycins, lincomycin, and puromycin in the third group.

Agents Acting on the 30S Unit

1. Streptomycin group (Figure 12-3). In general, these substances react with the 30S ribosome to cause "misreading." For example, in an *in vitro* system using poly-U as messenger the presence of streptomycin stimulates the incorporation of isoleucine (code AUU) while inhibiting the normal phenylalanine incorporation (code UUU). The other streptomycinoid antibiotics act similarly but cause a different misreading; for example, using poly-U streptomycin inhibits phenylalanine incorporation, but stimulates serine (UCU), leucine (CUU), and isoleucine (AUU) incorporation. Kanamycin stimulates serine, whereas neomycin stimulates isoleucine, serine, and especially tyrosine (UAU). The site of action is on the 30S ribosome, and if these are prepared from resistant strains, the system is resistant. It was first thought that the

streptomycin, which must get to the 30S ribosome before the messenger, just raised a little lump on it, as it were, which threw the message a bit out of line so that it was misread. But subsequently, when it became possible to separate out the various proteins in the 30S particle, it was found that one of these reacted specifically with streptomycin (and was not present in resistant ribosomes), so the action of the drug is undoubtedly more complex.

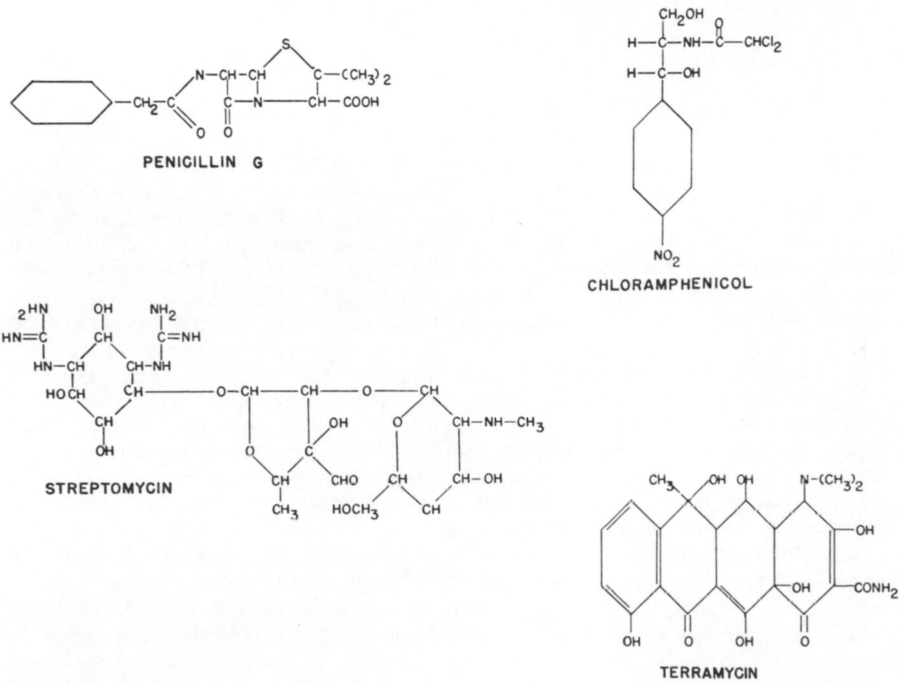

PENICILLIN G

CHLORAMPHENICOL

STREPTOMYCIN

TERRAMYCIN

FIGURE 12-3. *Chemical structures of the well established clinically useful antibiotics.*

Until one knows just what this special protein does in the translation process, we really cannot further specify just how it is that streptomycin causes miscoding. Nevertheless, knowing that miscoding is the basis of its action, one may speculate that streptomycin does not immediately stop protein synthesis but instead promotes the synthesis of proteins with different amino acid substituents. Where they should have had phenylalanine, they have isoleucine, for example. In some proteins this would not greatly matter with respect to their biological activity, but with others it might be critical (4).

Several antibiotics (aminoglycosides) cause miscoding. These are streptomycin, dihydrostreptomycin, bluensomycin, neomycin B and C, paromycin, kanamycin A and B, gentamicin, nebramycin, and hydromycin B.

2. The tetracyclines (Figure 12-3, terramycin) have so many effects that it has been difficult to sort out those which might be related to its antibiotic action. They do not prevent mRNA binding (as earlier supposed) but inhibit the transfer of amino acids from tRNA to the growing peptide and inhibit the binding of charged tRNA.

Agents Acting on the 50S Unit

1. Chloramphenicol (Figure 12-3) inhibits techoic acid synthesis, but this activity is probably unrelated to its antibiotic effect. The degree of inhibition of peptide synthesis varies with the mRNA. Inhibition of poly-U is relatively small; inhibition of natural mRNA is much greater, which would suggest interference by the drug with mRNA binding. But the drug clearly binds to the 50S portion and seems to inhibit the peptidyl transferase enzymes. It does not react with ribosomes from eucaryotic cells.

2. The macrolide antibiotics bind to the 50S portion and prevent or inhibit the binding of certain charged tRNAs but not others. Erythromycin, but not necessarily other members of the group, seems to inhibit translocation. It does not react with mammalian ribosomes.

3. Lincomycin seems to be similar in its action on ribosomes to erythromycin (5).

4. Puromycin causes the release of peptides by being itself incorporated into the carboxy terminus of the peptide. It is thought that this occurs because it structurally resembles aminoacyl adenosine, or charged tRNA.

INHIBITORS OF RIBOSOME FUNCTION[1]

Polysome patterns of antibiotic treated cells provide a way of determining where in the ribosome cycle the antibiotic acts. Inhibitors of initiation cause polysome breakdown, whereas inhibitors of translocation induce stabilization of the polysome pattern. Inhibitors of termination will cause polysomes of abnormally large sizes. Specific effects of an antibiotic often occur at a ratio of one bound antibiotic molecule per ribosome. It is also evident that inhibitors of the 30S subunits can effect the action of 50S subunits, and vice versa.

[1]Pestka, S. 1971. Inhibitors of ribosome functions. *Ann. Rev. Microbiol.* *25*:487–562. (Reviewed by Alexandria Cabrera.)

30S inhibitors. Streptomycin is a 30S inhibitor. Using cell-free extracts from streptomycin-sensitive *E. coli* strains, studies have shown that streptomycin stimulates miscoding. In the presence of poly-U, streptomycin stimulates leucine, isoleucine, serine, and tyrosine incorporation and inhibits the normal phenylalanine. Denatured DNA along with streptomycin in the incubation mixture is required for the miscoding effect. Pyrimidines are misread more frequently than purines.

Neomycin, another aminoglycoside, produces greater coding changes than streptomycin. It can affect two bases of a codon at one time. The degree of miscoding is a function of the concentration of the aminoglycoside used. Polynucleotides, which ordinarily do not serve as templates in protein synthesis, can function as templates in the presence of neomycin. Thus streptomycin and neomycin actually alter the structural requirement for the template in protein synthesis.

Physical effects of bound neomycin and streptomycin are also observed. Streptomycin and neomycin protect ribosomes against thermal denaturation.

The ability of ribosomes to miscode in the presence of streptomycin *in vitro* correlates with the ability of the antibiotic to substitute for an arginine requirement *in vivo*. Among auxotrophic mutants of *E. coli*, an arginine auxotroph can grow in the absence of arginine, if streptomycin is present in the medium.

Binding of streptomycin occurs on a single site on the 30S subunit, which contains a structure including protein P_{10}. Shortly after polypeptide synthesis has ceased, peptidyl tRNA is released from the ribosomes, which themselves are released from mRNA as subunits or 70S monomers. After release the subunits reassociate into 70S monomers, which cannot dissociate. These streptomycin monosomes contain mRNA and are 70S ribosomes, which are irreversibly inactivated by streptomycin. At high concentrations, streptomycin can more directly interfere with translocation and peptide bond formation. The precise causes of lethal effects of streptomycin are still uncertain, but are related to its irreversible binding to ribosomes.

Tetracycline is bound by both the 30S and 50S subunits. Yet, the 30S subunit binds at least twice as much. Tetracycline inhibits aminoacyl tRNA binding to the 30S subunit in response to template.

50S inhibitors. Puromycin is an inhibitor of protein synthesis in intact cells as well as cell-free extracts from diverse organisms. Puromycin has a structural resemblance to the aminoacyl adenosine and to aminoacyl tRNA. Puromycin inhibits protein synthesis by substituting for the incoming coded aminoacyl tRNA, competing for its similar terminal end on the peptide chain. Puromycin does not interfere with aminoacyl tRNA formation or mRNA binding to ribosomes. Several aminoacyl nucleosides are inhibitory to protein synthesis when adenosine comprises the terminal base. Phenylalanyl adenosine was almost as active as puromycin. A second effect caused by the action of puromycin is the breakdown of polysomes. Aminoacyl tRNA is essential in the functioning of the ribosome; replacement by puromycin releases peptidyl tRNA and loosens the junction of the subunits, so that ribosomes fall off mRNA. Thus puromycin stimulates ribosomal subunit exchange. When peptidyl tRNA is released, 50S subunits probably dissociate, and new 50S subunits reassociate with 30S subunits remaining attached to mRNA. The random distribution of *N*-terminal amino acids of puromycin peptides suggests that 30S subunits can remain attached to mRNA. Breakdown of polysomes but not release of peptides by puromycin requires energy and results in 70S ribosomes rather than subunits. Puromycin may also inhibit the normal termination mechanism of the ribosome. Chloramphenicol inhibits incorporation of amino acids into protein in cell-free extracts of *E. coli*, but not in those from eucaryotic cells. In intact bacteria, chloramphenicol inhibits peptidyl—puromycin formation. Chloramphenicol may also be an analogue of the aminoacyl end of aminoacyl tRNA.

Streptogramin A, lincomycin, and most macrolides interfere with radioactive chloramphenicol binding to ribosomes. These antibiotics are presumed to be 50S subunit inhibitors also.

Chloramphenicol prevents polysome breakdown when it inhibits protein synthesis. It can also prevent breakdown caused by puromycin or actinomycin, and can block degradation of mRNA observed in the presence of actinomycin. Chloramphenicol does not inhibit attachment of the ribosomes to mRNA. Experiments suggest that chloramphenicol inhibits ribosome movement along mRNA. Chloroplast ribosomes from tobacco plants show the same inhibition of protein synthesis by chloramphenicol as do bacterial ribosomes, yet, under the same conditions, chloramphenicol does not affect cytoplasmic ribosomes.

All ribosomes of the 70S class containing 16S and 23S RNA are inhibited by chloramphenicol. The resemblance of chloramphenicol to uridylic acid may make it possible for it to compete with uridylic residues on template codons for ribosomal sites. Hence, both mitochondrial protein synthesis as well as bacterial protein synthesis are inhibited by chloramphenicol.

Macrolide antibiotics (erythromycin) are active against gram-positive bacteria, whereas most gram negatives are resistant. Erythromycin inhibits bacterial growth by blocking protein synthesis. All macrolides contain a lactone ring, yet they do not have identical effects on protein synthesis. Macrolides with larger lactone rings tend to inhibit more reactions than those with smaller rings. Erythromycin will inhibit the binding of chloramphenicol to ribosomes. They do not displace each other; the inhibition of both antibiotics on polylysine synthesis is additive. Erythromycin inhibits phenylalanine incorporation into polyphenylalanine directed by the poly-U template. In cell-free extracts of *E. coli*, erythromycin can inhibit polylysine formation directed by poly-A. When cells are exposed to subinhibitory concentrations of erythromycin, along with a complete media for growth, the cells become resistant. If the drug is removed, the cells become sensitive within one or two generations. The mechanism by which this induced resistance occurs is not yet clear.

Erythromycins' most potent inhibitory effect is interference with the proper situation of peptidyl tRNA. Hence the macrolides bind to a common site on 50S subunits, and they are strong inhibitors of transpeptidation, when they inhibit substrate attachment to ribosomes.

The inhibition of ribosomes by antibiotics is generally highly specific. Antibiotics bind to a specific ribosome site. Availability of the site varies with the state of the ribosome.

Other Interactions

Interaction with and denaturation of enzymes does, upon occasion, play a role, but these actions seem to be much less important than originally supposed. Even competitive inhibition (such as the sulfonamides) seems to be a rather rare example of inhibitory action in the whole cell.

ORIGINS OF SPECIFICITY

How can certain materials be used in the animal body to kill off or inhibit the bacteria therein without harm to the host, or how, indeed, can a given substance be toxic to one organism and not to another? These are really two questions, not one, and we shall discuss them separately under *chemotherapeutic activity* and *resistance*. The latter refers to strains of microorganisms unaffected by a substance that is highly toxic to other strains.

Chemotherapeutic Activity

The basis for a chemotherapeutic effect may arise from several causes. In the case of penicillin and sulfonamides, it appears that the reactions inhibited (mucopeptide cross linking and folic acid synthesis, respectively) are not parts of animal metabolism and thus the animal cell is indifferent to whether or not they occur. In principle, then, one may be able to kill off an organism without harming a second organism with an agent inhibiting a reaction that occurs in or is only vital to the first. There is a related but slightly different reason for the specificity of chloramphenicol. It clearly inhibits protein synthesis even *in vitro* when added to ribosomes from sensitive bacteria, but it has no effect on a comparable kind of system from animal cells. That is, the ribosomes of the animal are constituted differently from those of the bacteria, and chloramphenicol reacts only with the bacterial system.

The case of streptomycin, since it reacts with the bacterial ribosome, might be thought to be similar; that is, it does not react with animal ribosomes. And there may well be ribosomes from some animal or plant cells with which it does not react. But it does inhibit the usual ribosomal system prepared from animal cells. There seem to be two factors that protect the animal cell: (1) usually streptomycin penetrates the animal or plant cell very slowly, if at all, and hence its internal concentration at the site of ribosomal protein synthesis is rather low, and (2) streptomycin reacts with the 30S ribosome (and presumably the comparable 40S of the animal) when it is free from mRNA, but when mRNA is present and the system is engaged in peptide synthesis, streptomycin has much less effect.

A reverse example is nystatin, an antifungal agent that reacts with steroids in the membranes of yeasts, fungi, and animal cells (where it causes a delayed damage) but does not affect the bacterial membrane, since the latter does not have steroids as critical parts of its structure. For other chemotherapeutic agents, we simply do not know why they can be used in the animal (or plant) cell without harm to the host.

Resistance

The occurrence and development of resistant mutants is one problem of chemotherapy. In this section we wish to consider the physiological mechanisms by which resistance can occur. There appear to be four physiological bases of resistance. First, if the drug acts internally it may not penetrate; if it acts at any specific site at the surface, this site may be altered. Second, the reaction that is inhibited may be substituted or eliminated. For example, in L-forms derived from *Proteus*, the muramic complex is no longer formed and the cells are resistant to penicillin. Third, the enzyme or material at the locus of action may alter. For example, in strains resistant to streptomycins, a particular protein in the 30S portion has changed and the ribosome no longer reacts with streptomycin. Fourth, the organism may possess or develop an enzyme for destroying or inactivating the substance. An example is the formation of penicillinase.

The drug-destroying system does not have to be absolutely specific, however. Some few years ago epidemics of enteric disease in Japan stimulated the isolation of enteric bacteria resistant to a wide variety of agents (tetracyclines, streptomycin, chloramphenicol, and sulfonamides) whose mode of action was clearly different. The resistance could be transferred to sensitive strains via an episomic *R factor*. Initially it was thought that the toxic substances were not destroyed, but more detailed studies showed that the toxic materials were in fact inactivated by acetylation, phosphorylation, and the like. More details will be found in the next section.

MECHANISMS OF ANTIBIOTIC RESISTANCE[1]

There are two broad classes of mechanisms of resistance to antibiotics: (1) mutation of a component in the cell can result in the antibiotic never interacting normally with its target site in the cell, and (2) destruction of the antibiotic itself; resistance is normally associated with an extrachromosomal element (R factor).

Mutants

Streptomycin-resistant mutants have a mutation in the S12 protein of their 30S ribosome that prevents the ribosome from binding the drug. Spectinomycin-resistant bacteria have a mutation in the S5 protein of the 30S subunit, eliminating the drug binding site. Kasugamycin resistance results from an alteration in the ribosomal RNA. The binding of kasugamycin depends on the methylation of the 16S rRNA, and mutation eliminates the activity of the methylating enzyme. Erythromycin-resistant mutants are an example of 50S alteration in the ribosomal protein in *E. coli* and *B. subtilis* and in the 23S rRNA in *Staph. aureus*.

Mutants have also been found that are concerned with RNA synthesis. Rifamycin and streptolydigins inhibit bacterial transcription by interacting with RNA polymerase. Resistant mutants have an RNA polymerase altered at its catalytic site. Rifamycin inhibits the initiation of the RNA chain synthesis; streptolydigins block chain elongation. Little study has been done concerning resistant mutants involved with DNA. One drug, nalidixic acid, is a specific inhibitor of DNA synthesis, yet studies have failed to reveal the mechanism of action of resistant mutants.

Antibiotic Destruction

The second class of mechanisms are those in which the antibiotic itself is inactivated. This is the type normally associated with clinical bacterial isolates, and these resistant bacteria usually possess an extrachromosomal element. Penicillin prevents the completion of cell wall synthesis in bacteria. Resistance to penicillin is due to the presence of a beta lactamase, which hydrolyzes the beta lactam ring present in all penicillin and

[1]Benveniste, R., and J. Davies. 1973. Mechanisms of antibiotic resistance in bacteria. *Ann. Rev. Biochem.* 42: 471—506. (Reviewed by Elaine M. Ehlers.)

cephalosporin drugs. In gram positives, the genes for penicillinase synthesis are located on a plasmid. In gram negatives, resistance is also mediated by extrachromosomal elements (R factors), and the resistance is also the result of a beta lactamase.

Chloramphenicol resistance results from inactivation owing to the acetylation of the drug to yield 3-acetyl and 1,3-diacetyl esters, neither of which has any antibacterial activity. R factors determining chloramphenicol resistance code for the acetylating enzyme.

The mechanism of resistance to tetracycline is not really understood but seems to involve a decreased uptake of the drug by strains carrying an R factor. There is no evidence for destruction of target sites or enzymatic inactivation. The majority of clinical isolates of antibiotic-resistant bacteria harbor extrachromosomal elements, such as penicillinase. R factors can be transferred between a wide variety of bacteria. They can be transferred between pathogenic and nonpathogenic bacteria. And as R factors often mediate resistance to more than one antibiotic at once, drugs to which bacteria are still sensitive will have the effect of selecting for resistant bacteria. This could lead to R factors mediating resistance to increasing numbers of drugs.

ANTIBIOTIC DEPENDENCE

Certain mutants have been found that require the presence of the toxic antibiotic for growth. An early example are strains of bacteria that require sulfonamide. It turned out that these produced so much PABA that it itself inhibited, and small amounts of sulfonamide permitted growth. Another example is found with streptomycin. Since it acted by causing miscoding, one might assume that dependence would mean a mutation in which the amino acid sequence was wrong (and that the protein was inactive), but that streptomycin could correct the error by miscoding and thus permit growth. There are probably some of these, but a single mutation would have to cause an error in *all* the proteins of the cell, a highly unlikely circumstance. Hence, if one had the situation where only one or two vital proteins were made wrong (and corrected by streptomycin), one would expect only limited growth, since while streptomycin was correcting the one wrong protein, it would make the others incorrect. Most *conditionally dependent mutants* are probably of this type, and their growth is usually limited. On the other hand, there are dependent strains that grow very well and in which nothing will replace the streptomycin requirement for growth. One might assume that these result from a mutation in the streptomycin-reacting protein of the 30S ribosome such that it is nonfunctional except when streptomycin is present.

ANTIBIOTIC PRODUCTION

An interesting question in comparative physiology is why a substance (the antibiotic) produced by one organism should kill another organism. The name "antibiotic" itself suggests teleological importance in the regulation of populations, but, as the name is of man's ascribing, so probably also is the teleology. We have no notion yet of the function of the antibiotics in the organisms that produce them, but the very fact that they act on specific metabolic loci in many other organisms implies that they may be analogues with a structure rather similar to that of essential metabolites.

However, there are some slight indications of why an organism may produce an antibiotic. It may, for example, play a role in its metabolism. A mutant of *B. licheniformis* does not produce bacitracin and is not inhibited by Mg^{2+} concentrations that inhibit the parent. It has therefore been supposed that the antibiotic is involved in ion transport (6).

Another hypothesis is that the antibiotic properties, by which we recognize antibiotics, are not the important point from the microorganism's point of view. Possibly, all organisms produce *secondary metabolites*, sometimes in large quantity, some of which "just happen" to be antibiotics, and others that have no such detectable activity. Some evidence suggests that such secondary metabolites represent the maintenance of mechanisms involved in specific cell functions, such as cell division or cell wall expansion. When these functions are prevented, the mechanisms keep on running, producing intermediates that accumulate (and possibly interact or alter the products via feedback inhibition), because they are not used up in the function that they were designed to serve.

A further problem is why the organism producing the antibiotic is not itself inhibited by it. In the case of penicillin, the fungi producing it do not have peptidoglycan cell walls. In the case of other antibiotics, very little is known, but with streptomycin-producing strains there appears to be no streptomycin in the cytoplasm of the cells; but cell extracts develop such activity on standing, which suggests that the streptomycin is present in an inactive form (7).

BIOCHEMISTRY OF STREPTOMYCIN SYNTHESIS[1]

Streptomycin is a trisaccharide composed of streptidine, L-streptose, and *N*-methyl-L-glucosamine. All three glycosidic bonds are in the alpha configuration. To date, no information is available on the enzymatic mechanisms by

[1]Demain, A. L., and E. Inamine. 1970. Biochemistry and regulation of streptomycin and mannosidostreptomycinase (α-D-mannosidase) formation. *Bacteriol. Reviews 34*:1—19. (Reviewed by JoAnn Sciacchitano.)

which the three moieties are joined.

Most of the biosynthesis of the streptidine portion was derived from radioactive labeling. The pathway given is simply a suggestion since the true position of the phosphate group is not known, nor which group of the streptamin is first to be transamidinated.

D-glucose is converted to myoinositol. After phosphorylation of the glucose, the ring is expanded from five to six members by a cyclase. Next carbon 2 is subject to dehydrogenation, transamination, phosphorylation, transamidination, and dephosphorylation. This whole process is repeated on the carbon 4, and the result is streptidine (Figure 12-4).

D-GLUCOSE
ATP ADP
1 hexokinase
D-GLUCOSE 6-PHOSPHATE
2 glucose-6-phosphate cyclase (NAD^+, NH_4^+)
D-myo-INOSITOL-1-PHOSPHATE
3 inositol-1-phosphatase (Mg^{++})
myo-INOSITOL
4 dehydrogenase (NAD^+)
2-KETO-myo-INOSITOL (INOSOSE-2)
5a
scyllo-INOSITOL
αKG-N glu-N
5 transaminase (pyridoxal ph.)
scyllo-INOSAMINE
ADP ATP
6 kinase (Mg^{++})
O-PHOSPHORYL-scyllo-INOSAMINE
orn arg
7 amidino-transferase
O-PHOSPHORYL-N-AMIDINO-scyllo-INOSAMINE
8 phosphatase
N-AMIDINO-scyllo-INOSAMINE
9 dehydrogenase
N-AMIDINO-3-KETO-+ scyllo-INOSAMINE
ala pyr
10 transaminase (pyridoxal ph.)
N-AMIDINO-STREPTAMINE
ATP ADP
11 kinase (Mg^{++})
O-PHOSPHORYL-N-AMIDINO-STREPTAMINE
amidino-transferase
12 arg orn
O-PHOSPHORYL-STREPTIDINE
phosphatase
13
STREPTIDINE

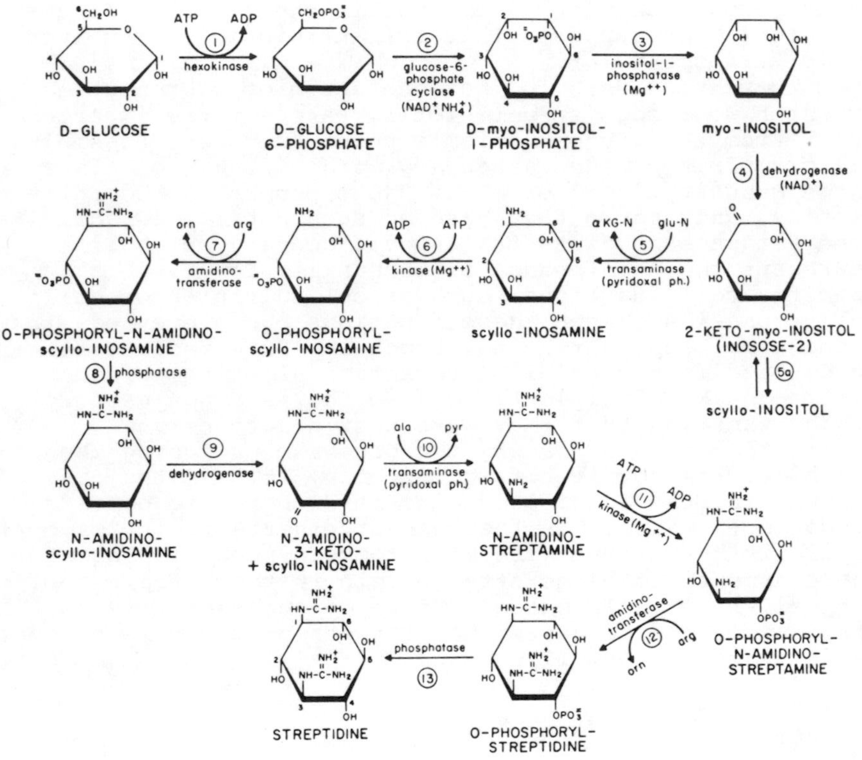

FIGURE 12-4. *Probable route of synthesis of the streptidine portion of streptomycin. From Demain, A. L., and E. Inamine, Bacteriol. Reviews 34:1–19 (1970). Used with permission.*

For streptose, glucose is attached to thiamine diphosphate. It is believed that streptose formation is from an intermediate of TDP-rhamnose formation. The entire glucose molecule is the streptose; only a rearrangement is made. The carbon 3 of glucose becomes a branch carbon atom.

N-methyl-L-glucosamine also arises from glucose, and the source of the methyl group is L-methionine. Little is known about the conversion of glucose to *N*-methyl-L-glucosamine. The pathway probably involves D-glucose, D-glucose-6-phosphate, D-fructose-6-phosphate, D-glucosamine-6-phosphate, as in other organisms, and the amino group is derived from glutamate. What follows is an epimerization of the four asymmetric carbon atoms, removal of the phosphate, and *N*-methylation.

There are some regulatory mechanisms. It has been found that the enzyme amidinotransferase, which acts at two steps of the streptidine synthesis, is only found in those *Streptomyces* strains which produce one or another type of streptomycin. A dramatic increase in activity of this enzyme occurs at the start of streptomycin synthesis. Also, in media that do not support synthesis of streptomycin, this increase of activity does not occur. The sudden appearance of this enzyme is due to *de novo* synthesis, since it can be inhibited by chloramphenicol. High phosphate interferes with release of streptomycin into the medium; excess phosphate interferes with the formation or action of one or more of the phosphatases involved in streptomycin synthesis, and also represses synthesis of amidinotransferase. Feedback inhibition may also play a role in streptomycin formation, since streptomycin inhibits streptidine kinase.

REFERENCES

Anderson, E. S. 1968. The ecology of transferable drug resistance in the enterobacteria. *Ann. Rev. Microbiol.* *22*:131—180.

Borick, P. M. 1973. *Chemical Sterilization.* Dowden, Hutchinson & Ross, Inc., Stroudsburg, Pa.

Gale, E. F., E. Cundliffe, P. E. Reynolds, M. H. Richmond, and M. J. Waring. 1972. *The Molecular Basis of Antibiotic Action.* Wiley, New York.

Hamilton-Miller, J. M. T. 1973. Chemistry and biology of the polyene macrolide antibiotics. *Bacteriol. Reviews 37*:166—196.

Harold, F. M. 1970. Antimicrobial agents and membrane function. *Advan. Microbial Physiol. 4*:46—105.

Helinski, D. R. 1973. Plasmid determined resistance to antibiotics: molecular properties of R factors. *Ann. Rev. Microbiol. 27*:437—470.

Kleinschmidt, W. J., and E. B. Murphy. 1967. Interferon induction with statolon in the intact animal. *Bacteriol. Reviews 31*:132—137.

Meynell, E., G. G. Meynell, and N. Datta. 1968. Phylogenetic relationships of drug-resistance factors and other transmissible bacterial plasmids. *Bacteriol. Reviews 32*:55—84.

Newton, B. A. 1965. Mechanisms of antibiotic action. *Ann. Rev. Microbiol. 19*:209—240.

Pestka, S. 1971. Inhibitors of ribosome functions. *Ann. Rev. Biochem. 40*:697—710.

Richmond, M. H., and R. B. Sykes. 1972. The β-lactamases of gram-negative bacteria and their possible physiological role. *Advan. Microbial. Physiol. 9*:31—88.

Scaife, J. 1967. Episomes. *Ann. Rev. Microbiol. 21*:601—638.

Schaeffer, P. 1969. Sporulation and the production of antibiotics, exoenzymes, and exotoxins. *Bacteriol. Reviews 33*:48—71.

Weinberg, E. D. 1970. Biosynthesis of secondary metabolites: roles of trace metals. *Advan. Microbial Physiol. 4*:1—45.

Weisblum, B., and J. Davies. 1968. Antibiotic inhibitors of the bacterial ribosome. *Bacteriol. Reviews 32*:493—528.

Papers

1. *J. Biol. Chem. 243*:783 (1968).
2. *Proc. Natl. Acad. Sci. 57*:767 (1967).
3. *Proc. Natl. Acad. Sci. 68*:3223 (1971).
4. *Nature 254*:161 (1975).
5. *Fed. Proc. 33*:2303 (1974).
6. *Nature 254*:879 (1975).
7. *Can. J. Microbiol. 21*:463 (1975).

13
Metabolism

A classic in the early literature of bacterial physiology was R. E. Buchanen and E. I. Fulmer's *Physiology and Biochemistry of Bacteria* (3 volumes, William and Wilkins, Baltimore, Md., 1930); the title recognized two aspects of the study of bacteria, their physiology and their biochemistry. We have been concentrating on physiology, using aspects of biochemistry as necessary to clarify the physiological problem under discussion. But we have neglected a discussion of metabolic pathways and biochemistry of bacteria in part because this has been covered by J. R. Sokatch in *Bacterial Physiology and Metabolism* (Academic Press, Inc., New York, 1969), and in part because it is in fact a separate subject. Some knowledge of bacterial biochemistry is necessary for bacterial physiology, but it is the overall concepts of the ebb and flow of materials through the channels of the cell that we wish to emphasize, rather than details of the mechanisms or pathways.

THE FLOW OF CARBON

As the flow of streams and rivers is controlled and utilized by man for a variety of purposes, so the flow of carbon compounds taken up from the environment is controlled and utilized by the bacterial cell. One of the most important purposes of this carbon flow is the generation of energy-rich compounds; we might liken this to the use of rivers in the generation of electricity. Another major purpose is the diversion of some of the carbon flow, at strategic points, for biosynthetic reactions; we might liken this to the diversion of water for irrigation, and thus, to put it simply, for growth. The bacterial cell may utilize the flow to get rid of substances for which it has no further use; we might liken this to the use of streams for carrying off human and industrial wastes. So what have been recognized for many years as the products of bacterial carbon metabolism (lactic acid, carbon dioxide, ethanol, butanol, etc.) represent what is left over after the flow has passed the power plants and the irrigation ditches. Since for a long time such end products were the only available index of the patterns of carbon flow preceding them, the spectrum of products produced by any one organism was, and still is, a major factor employed by bacteriologists in its physiological classification. It is only in the past 20 years that bacterial physiologists have come to comprehend in any depth what happened upstream: the intricacy of the carbon flow, the nature of the power plants, the sites of the irrigation shunts to biosynthesis. We shall concern ourselves in this chapter with these aspects of the metabolism of carbohydrates. However, one will find some mention of other materials, since a compound made, as it were, in one pathway may be used as an intermediate in another.

Let us begin with D-glucose, a logical starting point, since it is metabolized by more different kinds of bacteria than is any other carbohydrate, and since the pathways through which its carbons flow are joined and entered by the metabolic pathways for other hexoses and for pentoses.

There are three very useful ways of studying the pathways of metabolism of carbon compounds, not only of glucose, as in the present case, but also many others. One involves the use of radioactive carbon, a carbon isotope that has an atomic weight of 14 instead of the normal 12, and decays very slowly (half-life of about 6,000 years) to ^{14}N with the emission of β rays (electrons). If the glucose is labeled so that one or more of the ^{12}C atoms is replaced by ^{14}C, the products of glucose metabolism by intact cells or cell extracts can be analyzed for radioactivity, the position of the label in the products determined by degradation, and predictions made concerning the derivation of the labeled products. For example, if one finds that carbon dioxide is much more radioactive when glucose is labeled with ^{14}C at carbon 1 than at any of the other five carbons, one can predict as a reasonable possibility that glucose is split by the organism to carbon dioxide and a pentose, from which the rest of the products are derived. Although there are problems inherent in the use of ^{14}C tracing methods, such as the recycling of initial products, it has proved to be an extremely valuable tool in the analysis of pathways.

Another method is the study of the individual steps of a pathway with enzymes purified from bacterial extracts for proof that a specific substrate is metabolized to a particular product. Purified preparations free of the enzymes for preceding, subsequent, and competing reactions are preferable or obligatory for determination of the kinetics, cofactor requirements, and other parameters of an individual enzyme. A third method is to use mutants blocked at one or more steps in a pathway, which has provided information on the physiological importance of a pathway or a single step by the alterations in behavior of the mutant.

The first step in glucose metabolism is the conversion of glucose to a phosphate ester, glucose-6-phosphate, by transfer of the terminal phosphate group from adenosine triphosphate or ATP. With very few exceptions, it might be said that this is the only reaction glucose undergoes. The enzyme carrying out this transfer is termed hexokinase, if it can phosphorylate hexoses in general, or glucokinase if it acts specifically on glucose. From glucose-6-phosphate, three alternative paths are possible, two of which are of very fundamental importance (Figure 13-1).

The first of the alternative paths for glucose-6-phosphate consists of a shift of phosphate from position 6 to position 1, carried out by the enzyme phosphoglucomutase and yielding glucose-1-phosphate. This reaction does not thereafter lead to the breakdown or splitting of the glucose molecule, but rather is concerned with the formation of polysaccharides and the interconversion of hexoses.

The second and more important transformation consists of an alteration in the ring structure that yields fructose-6-phosphate. This compound enters a series of reactions yielding two C_3 compounds, both of which are eventually convertible to pyruvate ($CH_3COCOOH$), whose further metabolism may follow several alternative pathways. This pathway from glucose-6-phosphate through fructose-6-phosphate to pyruvate is frequently called the Meyerhof—Embden system (Figure 9-1). It is the major pathway of glucose breakdown in animal muscle and yeast, and is widely distributed among the bacteria.

The third transformation consists of oxidation of glucose-6-phosphate at the carbon 1 position to yield 6-phosphogluconic acid. This product may enter two alternative pathways by (1) dehydration and tautomerization to 2-keto-3-deoxy-6-phosphogluconic acid, or (2) oxidation at the 3 position and decarboxylation to CO_2 and ribulose-5-phosphate, which itself stands at another crossroad of metabolism.

All the degradative pathways, whether through fructose-6-phosphate or through 6-phosphogluconic acid, result in the formation of a triose-phosphate, glyceraldehyde-3-phosphate, from the bottom half of the glucose molecule (i.e., from carbons 4, 5, and 6). Where they differ is in what happens to the top half before cleavage. Furthermore, once the glyceraldehyde-3-phosphate is formed, it is metabolized to pyruvic acid by the same sequence of enzymatic steps in all bacteria, so far as is known, no matter what pathway was taken from glucose to triose phosphate. The fermentation types of microorganisms differ not only in what they do to the top half of glucose, but even more strikingly in what they do to pyruvic acid. We assume that you are familiar with the Meyerhof—Embden system, and we shall therefore discuss the phosphogluconic acid pathways.

The very weight of the evidence in favor of the Meyerhof—Embden system, and the fact that the pathway

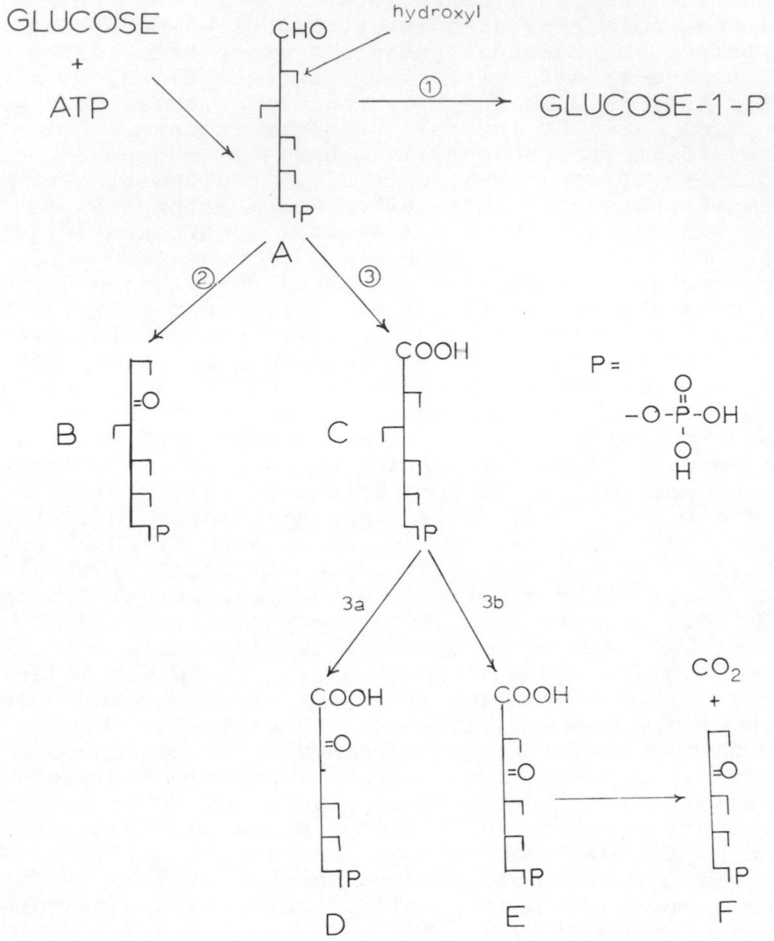

FIGURE 13-1. *Schematic diagram illustrating the varied pathways followed by glucose-6-phosphate. Substance A is glucose-6-phosphate, B is fructose-6-phosphate (see Figure 9-1 for its further metabolism), C is phosphogluconic acid, D is 2-keto-3-deoxy phosphogluconic acid (see Figure 13-2), E is 3-keto-phosphogluconic acid which is decarboxylated to F, ribulose-5-phosphate (see Figure 13-3).*

through hexose diphosphate was capable of explaining so much about carbohydrate metabolism, inhibited for some time search for other routes of metabolism. However, when ^{14}C tracer techniques were introduced, it became apparent that in some organisms the labeling patterns of the products could not be explained by, or reconciled with, the anaerobic glycolytic pathway. Several alternative pathways have since been found, and it is not yet certain that all have been discovered. Those which we shall describe all arise from the shunt of glucose-6-phosphate to 6-phosphogluconic acid, instead of to fructose-6-phosphate as in the Meyerhof—Embden pathway.

THE ENTNER—DOUDOROFF PATHWAY

This pathway was first described in *Pseudomonas saccharophila*, and has since been shown to occur in other species of *Pseudomonas*; it can also function in other bacterial genera. The route of metabolism of 6-phosphogluconic acid by this sequence of reactions is shown in Figure 13-2. The first step involves the dehydration of 6-phosphogluconic acid to an enol compound, which then undergoes tautomerization to 2-keto-3-deoxy-6-phosphogluconic acid. The latter compound is then cleaved by a specific aldolase, different from that which cleaves fructose-1,6-diphosphate, to result in pyruvate directly from carbons 1, 2, and 3, and glyceraldehyde-3-phosphate, again from carbons 4, 5, and 6 of glucose. The evidence for such an alternative pathway was first indicated when it was found that the carbon 1 of phosphogluconate (or of glucose) appeared as the carboxyl carbon of pyruvate, a labeling pattern inconsistent with the Meyerhof—Embden pathway. The "case of the upside-down pyruvate" was solved when the enzymatic steps were demonstrated in cell-free systems. However, since the other product, glyceraldehyde-3-phosphate, is metabolized exactly the same way as it is in the Meyerhof—Embden pathway, the pyruvate derived from carbons 4, 5, and 6 is labeled identically: carbon 4 becomes the carboxyl group of pyruvate, carbon 5 the keto group, and carbon 6 the methyl group.

The net yield of ATP by this pathway of hexose to pyruvate is only one ATP/hexose, rather than two: one ATP is utilized in the phosphorylation to hexose monophosphate, and two ATP are produced in the process of pyruvate formation from triose phosphate. Why then should an organism utilize this less profitable pathway of forming pyruvate from glucose? The answer is presumably the proverbial "half a loaf is better than none," since pseudomonads generally lack phosphofructokinase and are thus incapable of carrying out the Meyerhof—Embden pathway. The Entner—Doudoroff pathway can therefore be considered as a bypass channel for carbon flow around an insuperable obstacle.

PATHWAYS THROUGH RIBULOSE-5-PHOSPHATE

The other alternative route of 6-phosphogluconate metabolism involves its decarboxylation of CO_2 and ribulose-5-phosphate. The CO_2 is derived from carbon 1 by an oxidative decarboxylation carried out by 6-phosphogluconate dehydrogenase, involving $NADP^+$ reduction to NADPH (or occasionally NAD^+ to NADH), with 3-keto-6-phosphogluconate

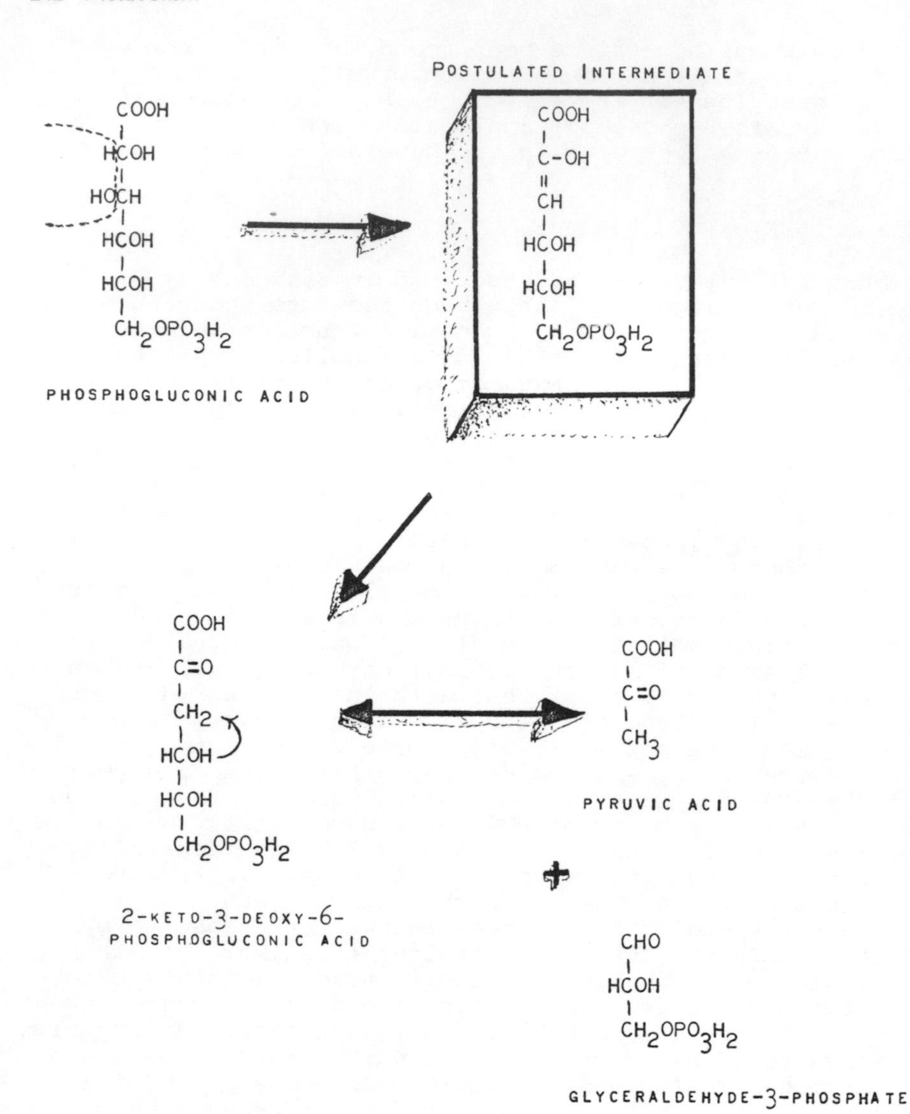

FIGURE 13-2. *The metabolism of phosphogluconic acid via the 2-keto derivative; the Enter—Doudoroff pathway. From Umbreit, W., Metabolic Maps II, Burgess Publishing Co., Minneapolis, 1960.*

as intermediate (Figure 13-1). The ribulose-5-phosphate produced stands at the branch point of a meshwork of metabolic streams (Figure 13-3). It is one of the most important such branch points in bacterial metabolism, and in the metabolism of higher organisms as well. From it originate the following pathways:

1. The formation of phosphoribosylpyrophosphate for synthesis of nucleotides and other compounds (Figure 13-3, path 1).

2. The pentose phosphate cycle for the conversion of pentose to hexose and triose, by way of heptose and tetrose (path 2).

3. The phosphoketolase, or Gibbs—DeMoss pathway, for cleavage of pentose into 2-carbon and 3-carbon fragments (path 3).

4. The formation of ribulose diphosphate, the CO_2 acceptor in both photosynthetic and chemosynthetic autotrophic CO_2 fixation (path 4).

The first two of these pathways have very broad and general distribution in bacteria, whereas the last two occur within a narrower range of physiological types. We shall discuss them in the same order as listed above.

Phosphoribosylpyrophosphate Formation — (Figure 13-3, path 1)

The phosphates of the pentoses ribose and deoxyribose make up the backbone of the nucleic acids. In addition to these functions, nucleotides containing ribose phosphate also serve as part of a good many coenzymes. The pentose moiety of the nucleic acids and coenzymes is derived in the following manner. The first step is the isomerization of ribulose-5-phosphate (substance A) to ribose-5-phosphate (B). The next step involves the transfer of a pyrophosphate group from ATP to carbon 1 of the pentose. The product of this reaction, commonly abbreviated as PRPP (C) is then utilized in the biosynthesis of pyrimidine ribonucleotides and of purine ribonucleotides, with the ribose-5-phosphate appearing as such in the final product (Figure 13-5). It is also utilized in the biosynthesis of the amino acids histidine and tryptophan, and of the coenzymes NAD and NADP (and perhaps of FMN and FAD as well). A partially purified PRPP synthetase from *Escherichia coli* has been shown to be markedly sensitive to inhibition by ADP, which as a purine ribonucleoside diphosphate, is an end product of one of the reaction sequences utilizing PRPP. The ADP inhibition can be reversed by increasing the ATP concentration. Interestingly, this preparation exhibited very little inhibition by either GDP or CDP, two other nucleotide products, or by tryptophan. Considering the number and importance of the biosynthetic reactions in which PRPP participates, it seems to us quite possible that bacterial cells (or other cells) may contain more than one PRPP synthetase, each regulated by different product(s).

PRPP is also a precursor of the deoxyribonucleoside phosphates, since these are derived, in *E. coli* and mammalian tissue, by reduction of a ribonucleoside diphosphate to the corresponding deoxyribonucleoside diphosphate. The reaction in both bacterial systems has been

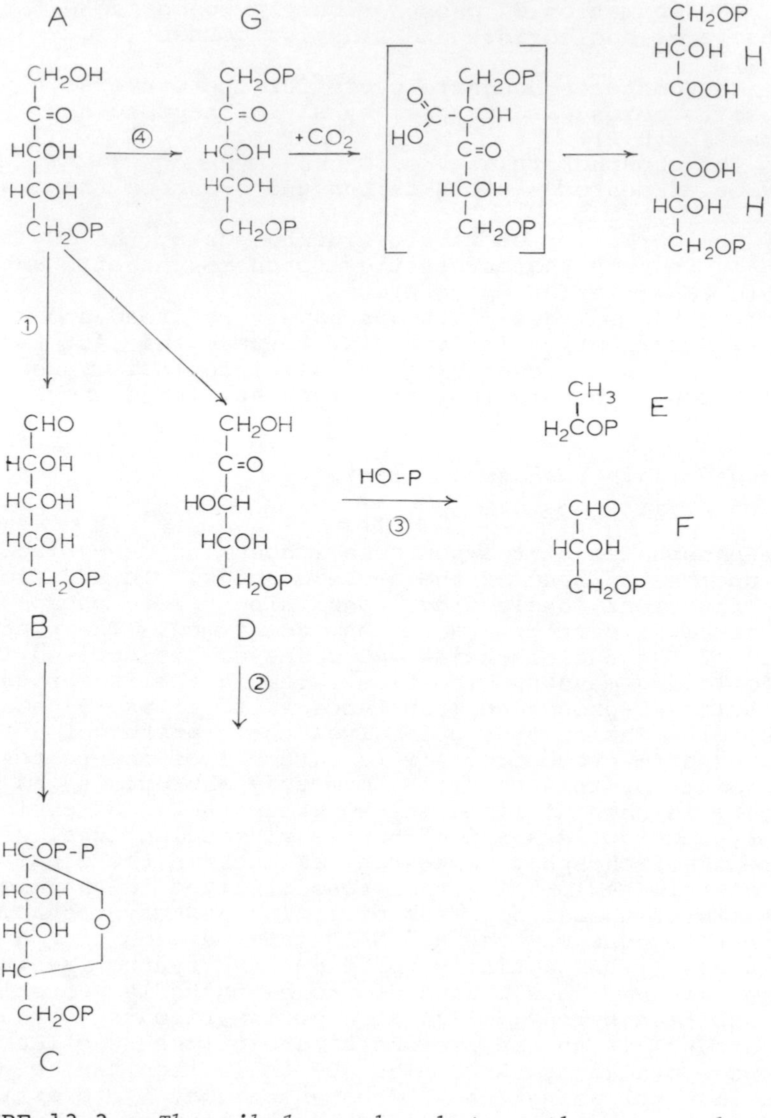

FIGURE 13-3. *The ribulose-phosphate pathways. Substance A is ribulose-5-phosphate; B is ribose-5-phosphate, C is phosphoribosylpyrophosphate commonly called PRPP and is the substance upon which the purine nucleotides are synthesized. Substance D is xylulose-5-phosphate (note inversion of hydroxyl at position 3) which is either metabolized further via pathway 2 (Figure 13-4) or split (pathway 3) into E, acetyl phosphate and F, glyceraldehyde-3-phosphate. Substance A is further transformed by reaction 4 to G, ribulose-diphosphate, to which CO_2 may be added to form an intermediate which rapidly breaks down to two molecules of H, phosphoglyceric acid.*

shown to require not only ribonucleotide reductase enzyme, but also a sulfur-containing protein, reduced thioredoxin, and a coenzyme form of vitamin B_{12}. The active form of thioredoxin is generated by NADPH reduction of a disulfide bridge. *Escherichia coli* appears to have only a single ribonucleotide reductase, which is active on the derivatives of all four of the commonly occurring bases, purine, and pyrimidine. However, the activity with each substrate is allosterically controlled by the concentration of one or more deoxyribonucleoside triphosphates other than the ultimate product of the reaction sequence.

The Pentose Phosphate Cycle — (Figure 13-3, path 2, and Figure 13-4).

This series of reactions has been named in a variety of ways other than the pentose phosphate cycle: hexose monophosphate (HMP) shunt, sedoheptulose pathway, phosphogluconate pathway, and others. The sequence of interconversions results in the return of pentose carbon to the "main" stream of carbon flow, by way of fructose-6-phosphate and glyceraldehyde-3-phosphate, or conversely in the formation of pentose phosphate from the hexose and triose phosphates, without loss of organic carbon through decarboxylation. Basically, the sequence consists of three enzymatic steps mediated in order by transketolase (TK, Figure 13-4), transaldolase (TA, Figure 13-4), and then transketolase again, no matter whether the direction of flow is to or from pentose phosphate. The transketolase reaction involves the transfer of a 2-carbon fragment, glycolaldehyde, bound to the thiamin pyrophosphate coenzyme moiety of *transketolase*. You will note that the illustration shows the H and OH on the carbon below the keto group of the donor ketose to be in the L configuration (i.e., as drawn with the OH group on the left side, since this configuration is required for transketolase activity). There is, on the other hand, very little specificity for the nature of the acceptor aldose, and a very general formula for aldose is therefore presented. The product ketose has the same configuration as the substrate ketose.

The second step again involves transfer from ketose to aldose, this time of a 3-carbon fragment, dihydroxyacetone, bound to the enzyme *transaldolase*. Although transaldolase is relatively specific for the ketoses fructose-6-phosphate and heptulose-7-phosphate, and for the aldoses glyceraldehyde-3-phosphate and erythrose-4-phosphate, it can also utilize as substrates other aldoses and other ketoses with the correct configuration.

To begin the pathway from either direction, one must have then both aldose and appropriate ketose. When it is begun with pentose phosphates, both such compounds are derived from ribulose-5-phosphate: ribose-5-phosphate by an isomerization and xylulose-5-phosphate by an epimerization (note that ribulose-5-phosphate itself cannot be a

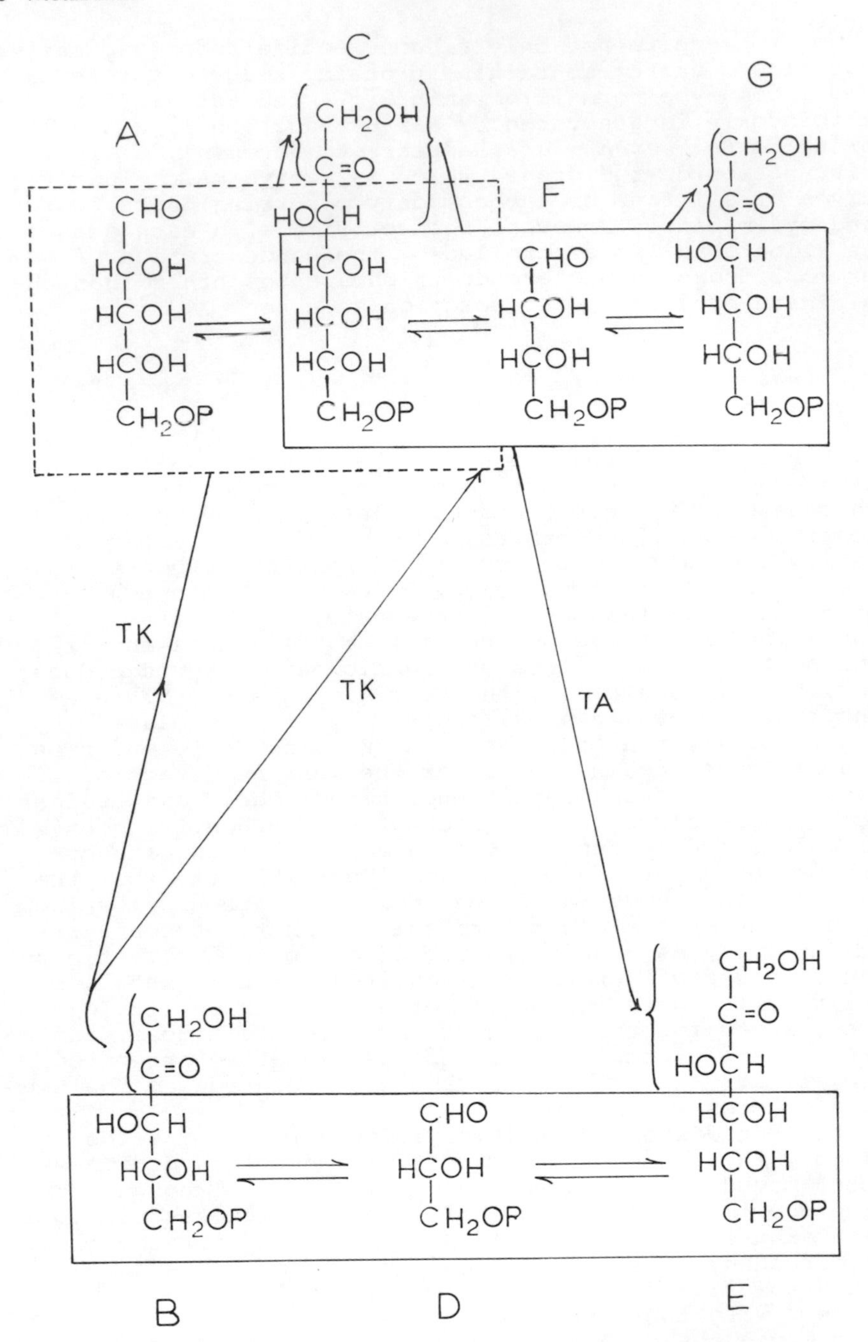

FIGURE 13-4. *The pentose cycle. Substance A is ribose-5-phosphate, B is xylulose-5-phosphate, C is heptulose-7-phosphate, D is glyceraldehyde phosphate, F is erythrose-4-phosphate and E and G are fructose-6-phosphate, TK represents transketolase, TA represents transaldolase.*

substrate, because of incorrect configuration). To begin (Figure 13-4), xylulose phosphate (B) transfers (via transketolase) its top two carbons to ribose phosphate (A) to form heptulose or sedoheptulose phosphate (C) and glyceraldehyde phosphate (D). The heptulose phosphate (C) transfers (via transaldolase) its top three carbons to glyceraldehyde phosphate (D) to form fructose-6-phosphate (E), leaving its four remaining carbons as erythrose phosphate (F). This substance now accepts the 2-carbon moiety from xylulose phosphate (B) (via transketolase) to form another fructose-6-phosphate (G; same as E) and glyceraldehyde phosphate (D). Thus one ribose phosphate and two xylulose phosphates form two fructose-6-phosphates and one glyceraldehyde phosphate.

The pentose cycle or pentose shuttle has several functions in bacterial cells. From it is derived erythrose-4-phosphate, which on condensation with phosphoenolpyruvate provides the starting material for the biosynthesis of aromatic rings. A second function is the formation of ribose-5-phosphate, required for nucleotide synthesis, from hexose and triose in an organism lacking glucose-6-phosphate dehydrogenase and/or 6-phosphogluconate dehydrogenase. Third, the pathway permits those pentoses convertible to xylulose-5-phosphate or ribose-5-phosphate to be used by a bacterium as carbon source. In addition, the function of this cycle in the regeneration of pentose phosphate for autotrophic CO_2 fixation is of major importance. These functions refer specifically to the pentose cycle itself. Coupled with the two $NADP^+$-reducing reactions leading from glucose-6-phosphate through 6-phosphogluconate to ribulose-5-phosphate, the pentose phosphate cycle permits NADPH generation with return of the pentose carbons to the main hexose—triose axis.

The reactions from glyceraldehyde-3-phosphate to glucose-6-phosphate involve mainly the reversal of steps already considered in the "downward" flow through the Meyerhof—Embden pathway: triose phosphate isomerase, aldolase, and glucose-phosphate isomerase, plus the hydrolytic enzyme fructose diphosphatase. The NADPH produced through this scheme is then available for use in a variety of biosynthetic reactions (e.g., those for fatty acids and amino acids). These various and multiple functions of the pentose phosphate cycle are correlated with its ubiquity in bacteria, and we are inclined to the opinion that no normal, wild-type bacterium will be found to be totally devoid of a functional pentose phosphate cycle under any and all conditions. Indeed, in some organisms, such as *Acetobacter suboxydans*, the pentose phosphate pathway serves as the main and primary route for glucose metabolism, whereas in bacteria with functional Meyerhof—Embden systems, such as *E. coli*, only a fraction of the glucose (10—20 percent) may be metabolized through pentose phosphates. One way of determining the proportion of flow through the Meyerhof—Embden and hexose monophosphate pathways involves the incubation of bacterial cells in replicate flasks with glucose labeled with ^{14}C in various positions, under experimental conditions such that the $^{14}CO_2$ produced is flushed into a KOH trap and assayed at intervals. Operation of the Meyerhof—Embden pathway would result in early release of $^{14}CO_2$ from positions 3 and 4 of glucose, whereas metabolism through pentose phosphate would result in early release from position 1; labeled CO_2 from other positions would be derived only through subsequent reactions, and would be released later. From such data, one can calculate by means of appropriate equations, flow rates through alternative pathways, and, by using cells grown under different cultural conditions, determine physiological shifts in pathway utilization.

The Phosphoketolase (Gibbs—DeMoss) Pathway — (Figure 13-3, path 3)

The group of organisms known as the *heterolactic* bacteria includes the genus *Leuconostoc* and several species of the genus *Lactobacillus*. These organisms differ from the *homolactics* (*Streptococcus*, other species of *Lactobacillus*) in that they are incapable of fermenting glucose through the Meyerhof—Embden pathway because of their lack of the enzyme aldolase. The pathway of glucose metabolism in the heterolactic organisms involves instead the cleavage of xylulose-5-phosphate (D) to glyceraldehyde-3-phosphate (F) and acetyl phosphate (E), with the esterification of inorganic phosphate (Figure 13-3, reaction 3). The cleavage enzyme is termed phosphoketolase. The acetyl phosphate can thereafter be reduced to acetaldehyde, and the latter to ethanol. The glyceraldehyde-3-phosphate (again from carbons 4, 5, and 6 of glucose!) is metabolized to lactic acid by the same sequence of enzymes functional in the Meyerhof—Embden pathway. The dehydrogenations are mediated by NADH-requiring enzymes, and, interestingly, the glucose-6-phosphate and 6-phosphogluconate dehydrogenases of *Leuconostoc mesenteroides* can reduce either NAD^+ or $NADP^+$; NADH generated in these steps can then be used for the reductions to ethanol or lactic acid. This may well be true for the heterolactic organisms in general. The phosphoketolase pathway also functions in the fermentation of pentoses by the homolactic *Lactobacillus* species, which use the Meyerhof—Embden pathway for glucose metabolism. The products of such pentose fermentations are acetic acid and lactic acid.

The phosphoketolase enzyme, like transketolase, requires thiamin pyrophosphate as coenzyme. The phosphorolytic cleavage of xylulose-5-phosphate and the subsequent reactions could at least theoretically result in a net gain of two ATP per mole: one via triose phosphate dehydrogenase, and one via transfer from acetyl phosphate. However, since the actual yields are considerably lower, probably most of the phosphate of acetyl phosphate is not transferred on further metabolism of this compound.

Ribulose Diphosphate and Autotrophic CO_2 Fixation — (Figure 13-3, path 4)

The photosynthetic and chemosynthetic autotrophic bacteria share with green plants the ability to live on CO_2 as the sole source of carbon. This ability, in the simplest terms, requires a mechanism for the biochemical entrapment of CO_2 and the conversion of the entrapment product into compounds on the mainstream of metabolic flow, with concomitant or subsequent regeneration of the trapping agent in order to permit continuous operation. This mechanism, as it is now considered to function in both plants and bacteria, is termed the reductive pentose phosphate cycle or the Calvin cycle (or the Calvin—Benson-Bassham cycle). It is dependent on the function of two key enzymes acting in sequence, phosphoribulokinase and ribulosediphosphate carboxylase. Ribulose diphosphate is derived by phosphorylation from ATP of ribulose-5-phosphate by phosphoribulokinase. The next step, mediated by the carboxylase, results in the fixation of CO_2 to ribulose diphosphate, with the formation of an unstable intermediate, the structure of which is presently presumed to be that illustrated in Figure 13-3, reaction 4. The intermediate is cleaved into two molecules of 3-phosphoglycerate, an intermediate in the Meyerhof—Embden pathway. It is the compound most rapidly labeled in short-term experiments, of a few seconds or so, on incubation of $^{14}CO_2$ and extracts of photosynthetic cells.

The enzyme, ribulose diphosphate carboxylase, turns out to be of very great interest. First, it has a dual function and can either fix or release CO_2 (1). It catalyzes the crucial reaction in photosynthesis, the fixation of CO_2, but it also catalyzes the crucial reaction of photorespiration, in that it can split ribulose diphosphate in the presence of oxygen to form one molecule of phosphoglyceric acid and one molecule of phosphoglycollic acid which seems to be the principle substrate for photorespiration (2). Furthermore, the enzyme comprises more than 50 percent of the soluble leaf protein of green plants and thus is probably the most abundant protein in nature. It has a molecular weight of 560,000 and has eight large subunits (molecular weight 55,000) which contain the catalytic sites and eight small subunits (molecular weight 12,500) involved in regulation. In tobacco, chloroplast genes control the large subunit made on chloroplast ribosomes (70S), while nuclear genes code for the small subunit which is made on the cytoplasmic eucaryotic ribosomes (80S) (3,4). This enzyme has also been found in crystals in *Thiobacillus thiooxidans*, an organism capable of obtaining all of its cell substance from carbon dioxide.

Although we shall not explore metabolism further, the reaction series shown in Figure 13-5, the synthesis of the pyrimidine, UMP, illustrates two characters. First, one may note that intermediates are drawn from what at first seems to be quite unrelated pathways. In this case the pyrimidine ring starts from aspartic and glutamine and CO_2 (Figure 10-5) and this is combined with a product of carbohydrate metabolism PRPP (Figure 13-3). Second, this reaction series is subject to considerable feedback control, as discussed in the next section.

PYRIMIDINE METABOLISM IN MICROORGANISMS[1]

Carbamyl phosphate is required for arginine and pyrimidine biosynthesis. Therefore, its activities are modulated by intermediates of both. Glutamine is the preferred amino donor.

The activity of carbamyl phosphate synthetase is subject to feedback inhibition of UMP (a negative allosteric effector) and is activated by ornithine (a positive allosteric effector). Ornithine is an intermediate in the biosynthesis of arginine. The mechanism of this allosteric effect is unknown. The affinity of the enzyme for ATP appears to be altered by the various allosteric effectors. The addition of ornithine along with ATP enhanced the inhibition. The addition of UMP with ATP decreased the inhibition. The data suggest that in the presence of ornithine the enzyme exists in a conformational state that is different from that which exists in the presence of UMP. The binding of ATP to the enzyme is probably greatly altered by the presence of ornithine or UMP.

In some organisms there are two distinct CPSases for both arginine and pyrimidine biosynthesis. In these organisms (e.g., *Neurospora* and *Saccharomyces cerevisiae*) the carbamyl phosphate produced by one reaction is not freely available for the other pathway.

Aspartate transcarbamylase (ATCase) catalyzes the first reaction unique to pyrimidine biosynthesis and consists of two protein subunits. The larger of these possesses all the catalytic activity and bears the active sites for the substrate sensitive to feedback inhibition by CTP or activation by ATP. The smaller subunit is the regulatory subunit that binds CTP and is responsible for inhibition. The enzyme is regulated by feedback inhibition by (in decreasing order) CTP, CDP, CMP, and cytidine. Uridine derivatives do not inhibit ATCase either. dCTP is the best inhibitor. ATP activates the enzyme, and both ATP and CTP appear to compete for a single site on the regulatory subunit. Since ATCase is the first enzyme unique for pyrimidine biosynthesis, it does differ from organism

[1]O'Donovan, G. A., and J. Neuhard. 1970. Pyrimidine metabolism in microorganisms. *Bacteriol. Reviews 34*:278–343. (Reviewed by John O'Keefe.)

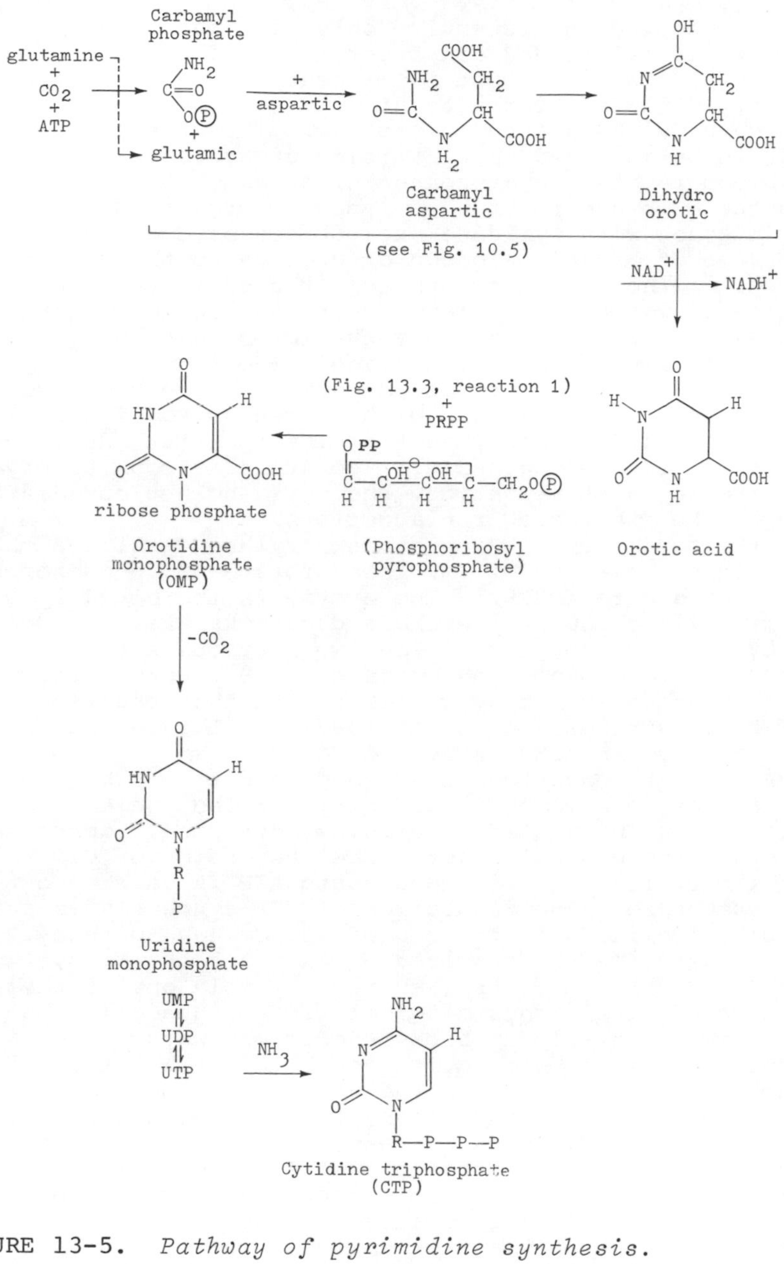

FIGURE 13-5. *Pathway of pyrimidine synthesis.*

to organism with different inhibitors and activators.

Dihydroorotase catalyzes the reversible cyclodehydration of carbamyl aspartate to dihydroorotate, which is then converted to orotate by dihydroorotate dehydrogenase. Dihydroorotate is the first ultraviolet-absorbing pyrimidine. Dihydroorotase is a very unstable enzyme. It is highly specific and is active only with the L-isomers of carbamyl aspartate and dihydroorotate. The enzyme is not inducible in *E. coli*, but is present regardless of carbon source. Cell-free extracts of *E. coli* converted aspartate plus carbamyl aspartate to orotate. Thus the enzyme was shown to be involved in pyrimidine biosynthesis.

Dihydroorotate dehydrogenase of *E. coli* is associated with the membrane portion of lysed spheroplasts and does not interact with pyridine nucleotides as does the dehydrogenase of organisms grown on orotate as the sole carbon source. Although it appears that the catabolic dihydroorotate dehydrogenases are pyridine nucleotide linked, this is not true of the biosynthetic enzymes that appear to link to oxygen or ferricyanide. Two functionally different dihydroorotate dehydrogenases have been isolated from *Pseudomonas* strains. With oxygen as the final electron acceptor, the biosynthetic dihydroorotate dehydrogenase is particle bound and will donate electrons to oxygen. With the degradative enzyme, the pyridine nucleotides are used as the final electron acceptors.

In the next step, OMP pyrophosphorylase catalyzes the reversible formation of OMP from orotate and phosphoribosylpyrophosphate (PRPP). The enzyme is inhibited by OMP.

Next follows the irreversible decarboxylation of OMP to UMP by OMP decarboxylase. The two previous steps (dihydroorotate to orotate and orotate to OMP) were unfavorable for biosynthesis, but by coupling with this reaction (OMP to UMP) biosynthesis proceeds readily. The enzyme is inhibited by UMP, CMP, AMP, and GMP.

UMP is next converted to UDP and then to UTP by kinases that appear to have no specificity for the nucleoside moiety of the nucleotides; uridine, cytidine, adenosine, and guanosine can all react. The conversion of UTP to CTP with the utilization of ammonia and ATP is carried out by CTP synthetase. The regulation of CTP synthesis is rather unique since it is at the end of the *de novo* pathway. Both UTP and CTP are required equally for the biosynthesis of RNA and DNA. A control exerted by only one of these would be of little physiological use to the cell. The most common controls in biosynthetic pathways are end-product inhibition and activation by a metabolite, which accumulates in response to the end product becoming limited. At high UTP concentration, CTP acts as a competitive inhibitor, with UTP thus preventing its own synthesis. At subsaturating concentrations of GTP and glutamine, CTP activates the reaction. Activation control is also exerted by UTP and ATP. The isolated CTP synthetase is a dimer of two identical polypeptide chains. The enzyme exhibits positive homotropic effects for ATP and UTP, which indicates that the enzyme shows strong site-to-site interaction.

Thymidine nucleotides are unique in that their ribose-containing counterpart, ribosylthymine nucleotides, are not normal metabolites of the cell. Thymidylate synthetase has been extracted and catalyzes a tetrahydrafolatedependent methylation of dUMP to dTMP. Except for a rather weak product inhibition of the reaction by dTMP, competitive with dUMP, no naturally occurring nucleotide has been reported to influence the activity of the enzyme. dUMP seems to be the immediate substrate for thymidine nucleotide biosynthesis. UDP is reduced by ribonucleoside diphosphate reductase of *E. coli* to dUDP. The further conversion of dUDP to dUMP seems to occur via dUTP by dUTP pyrophosphatase.

MICROBIAL COMETABOLISM AND THE DEGRADATION OF ORGANIC COMPOUNDS IN NATURE[1]

Cooxidation was first reported by Leadbetter and Foster in 1959 when they saw the oxidation of ethane to acetic acid, propane to propionic acid and acetone, and butane to butanoic acid and methyl ethyl ketone during growth of *Ps. methanica* on methane. (Methane is the only hydrocarbon capable of supporting the growth of this organism.)

The term cooxidation, now used interchangeably with cometabolism, describes a process in which a microorganism oxidizes a substance without being able to utilize the energy derived to support growth. In other words, the substance will be metabolically altered but no increase in growth occurs. Resting cells may also cometabolize a substance.

Other evidence of cometabolism was found in enzyme induction experiments in which one compound, not utilizable as an energy source, stimulated the synthesis of enzymes necessary for growth on a second compound.

If the cell derives no energy from the cometabolic process, why does it occur? The biochemical mechanism of cometabolism is yet inadequately explained. The inability of the organism to degrade the substrate to a point where the carbon can be assimilated, the accumulation of toxic oxidation products, and the specificity of degradative enzymes have been offered as reasons for the lack of substrate utilization.

Cometabolism may account for a large percentage of the degradation of many environmental pollutants (i.e., pesticides and other halogenated hydrocarbons) that do not sustain microbial growth. This has been demonstrated in pure, mixed, and natural (simulated) cultures. It has been shown that application of both pesticide and a biodegradable simulant of the pesticide may allow man to have both the benefit of action of the pest control agent and a rapid oxidation of the pest compound, thereby eliminating a potential environmental hazard.

Besides being environmentally important, cometabolism can be utilized as a useful biochemical technique. It has been used to isolate and identify products resulting

[1]Horvath, R. S. 1972. Microbial co-metabolism and the degradation of organic compounds in nature. *Bacteriol. Reviews 36*:146–155. (Reviewed by Lynda A. Kiefer.)

from ordinarily rapid reactions in which the utilizable substrate products would not accumulate in sufficient quantity for analysis. It also can be used to trace various metabolic pathways through the interpretation of the accumulated oxidative products. Cometabolism has also been used to accumulate,in a profitable way, at least one important pharmaceutical product, equilin.

REFERENCES

Anderson, R. L., and W. A. Wood. 1969. Carbohydrate metabolism in microorganisms. *Ann. Rev. Microbiol.* *23*:539—578.

Dagley, S. 1971. Catabolism of aromatic compounds by microorganisms. *Advan. Microbial Physiol.* *6*:1—46.

Doelle, H. W. 1969. *Bacterial Metabolism.* Academic Press, Inc., New York.

Doelle, H. W. 1974. *Microbial Metabolism.* Dowden, Hutchinson & Ross, Inc., Stroudsburg, Pa.

Fraenkel, D. G., and R. T. Vinopal. 1973. Carbohydrate metabolism in bacteria. *Ann. Rev. Microbiol.* *27*:69—100.

Gatt, S., and Y. Barenholz. 1973. Metabolic alterations of fatty acids. *Ann. Rev. Biochem.* *42*:61—90.

Goren, M. B. 1972. Mycobacterial lipids: selected topics. *Bacteriol. Reviews* *36*:33—64.

Kates, M. 1966. Biosynthesis of lipids in microorganisms. *Ann. Rev. Microbiol.* *20*:13—44.

Kurahashi, K. 1974. Biosynthesis of small peptides. *Ann. Rev. Biochem.* *43*:445—459.

Ljungdahl, L. G., and H. G. Wood. 1969. Total synthesis of acetate from CO_2 by heterotrophic bacteria. *Ann. Rev. Microbiol.* *23*:515—538.

Morris, D. R., and R. H. Fillingame. 1974. Regulation of amino acid decarboxylation. *Ann. Rev. Biochem.* *43*: 303—325.

Quayle, J. R. 1972. The metabolism of one-carbon compounds by microorganisms. *Advan. Microbial Physiol.* *7*:119—204,

Ramakrishna, T., P. S. Murthy, and K. P. Gopinathan. 1972. Intermediary metabolism of Mycobacteria. *Bacteriol. Reviews* *36*:65—108.

Reeves, H. C., R. Rabin, W. S. Wegener, and S. J. Ajl. 1967. Fatty acid synthesis and metabolism in microorganisms. *Ann. Rev. Microbiol.* *21*:225—256.

Ribbons, D. W., J. E. Harrison, and A. M. Wadzinski. 1970. Metabolism of single carbon compounds. *Ann. Rev. Microbiol.* *24*:135—158.

Schlegel, H. G., and O. Eberhardt. 1972. Regulatory phenomena in the metabolism of Knall gas bacteria. *Advan. Microbial Physiol.* *7*:205—242.

Shapiro, B. M., and E. R. Stadtman. 1970. The regulation of glutamine synthesis in microorganisms. *Ann. Rev. Microbiol.* *24*:501—524.

Stadtman, T. C. 1967. Methane fermentation. *Ann. Rev. Microbiol. 21*:121—142.
Stanier, R. Y., and L. N. Ornston. 1972. The β-ketoadipic pathway. *Advan. Microial Physiol. 9*:89—152.
Stebbing, N. 1974. Precursor pools and endogenous control of enzyme synthesis and activity in biosynthetic pathways. *Bacteriol. Reviews 38*:1—28.
Truffa-Bachi, P., and G. N. Cohen. 1973. Amino acid metabolism. *Ann. Rev. Biochem. 42*:113—134.
Wegener, W. S., H. C. Reeves, R. Rabin, and S. J. Ajl. 1968. Alternate pathways of metabolism of short-chain fatty acids. *Bacteriol. Reviews 32*:1—26.
Wolfe, R. A. 1971. Microbial formation of methane. *Advan. Microbial Physiol. 6*:107—146.

Papers

1. *Ann. Rev. Biochem. 44*:123 (1975).
2. *Science 188*:626 (1975).
3. *Science 191*:429 (1976).
4. *Nature 259*:325 (1976).

14
Variations on a Theme: Diversity in Organisms

Inasmuch as one cannot completely cover a field of knowledge as large as bacterial physiology in a single volume, we shall discuss certain interesting aspects. These are of two types. One has to do with organisms differing from bacteria, about which relatively little is known, at least of their physiology. These offer an opportunity for future work and suggest how physiological studies might be applied. The other (Chapter 15) are unusual kinds of processes possessed by some bacteria, but not by all. To this is added a consideration of pathogenicity (and possibly virulence) considered as a physiological problem.

BDELLOVIBRIO[1]

In 1962, in Berlin, Heinz Stoep discovered a unique bacterium, which he named *Bdellovibrio bacteriovorus*. The name describes some of the outstanding characteristics of the organism; *Bdello* meaning leech, refers to the ability of the organism to firmly attach to a second host cell, *vibrio* describes its vibroid shape, and the species name, *bacteriovorus*, meaning bacteria-eater, describes its parasitic mode of existence.

These parasitic bacteria differ from most genera of bacteria in that they are extremely small, measuring about 1—2 μm in length and about 0.35 μm in width (they thus are capable of passage through bacterial membrane filters), they possess a thick polar flagellum, which, being three to four times the thickness of a typical pseudomonad flagella, affords them a motility of up to 100 cell lengths per second. The end opposite the flagellum has a characteristic structure referred to as a

[1]Starr, M. P., and R. J. Seidler. 1971. The Bdellovibrios. *Ann. Rev. Microbiol.* *25*:649—678. (Reviewed by Gail Dargenzio) See also (1), (2).

"holdfast" or an "infection cushion," because it is in this place that the *Bdellovibrio* and host are attached. Associated with this area are mesosomes, which are thought to have some involvement with the enzymes functioning during the parasitic stages, and fine filaments, which may also play a role in the host—parasite attachment.

The cell wall contains muramic acid, glucosamine, and 13 amino acids, and stains as a typical gram-negative bacterium. The cytoplasm contains a centrally located nucleoplasm, distinctive in that it occupies two thirds of the volume of the cytoplasm and seems a large volume of DNA for an organism of such a small size. Dense particles surround this, which are assumed to be ribosomes.

The most outstanding characteristic of *Bdellovibrio* is its parasitic life cycle. Host cells are restricted primarily to gram-negative bacteria, although parasitic interaction with a few gram-positive species has been reported. The first stage of the parasitic interaction appears to be a recognition of susceptible host cells, presumably by chemotaxis. The *Bdellovibrio* then swims rapidly toward it at an astonishing speed and collides with it. Some reports say *Bdellovibrio* are able to carry a host cell 10 times its mass across a 1,000 times magnified oil immersion field. This aids the organisms' attachment to and eventual penetration of its host cell.

Once attached, *Bdellovibrio* begins a rapid rotation of up to 100 revolutions per second, which is described as an "arm in socket" movement, for the firm attachment of host and parasite would otherwise cause the host cell to rotate also and it does not. The host cell wall becomes damaged and a pore of small diameter is formed through which the *Bdellovibrio* squeezes. The pore formation is probably due to a combination of mechanical (the rapid collision and drilling rotations) and chemical means. Production of muramidase, proteases, and lipases by *Bdellovibrio* and possible host autolysis may very well be involved. Studies have shown that antibiotics that inhibit protein synthesis inhibit penetration but not attachment.

During the early stages of host—parasite interaction, even before penetration, the host cell rounds up and appears as a spheroplast, but it is not osmotically sensitive. Of the total number of *Bdellovibrios* capable of attachment only two thirds appear capable of actual penetration into the host. All stages of host cell growth and even ultraviolet-killed and heat-killed cells are satisfactory hosts.

The actual penetration involves the squeezing of the *Bdellovibrio* through a pore in the host cell wall. The parasite situates itself between the cell wall and cell membrane and begins to grow and elongate into a U- or C-shaped cell as it feeds off its host. The host cell cytoplasm becomes progressively more disorganized as the parasite robs it of its nutrients. At no time does the parasite actually enter the host's cytoplasm but remains between the wall and membrane and draws its nutrients out from there. The host cell contents are the sole source of nutrient for the parasite.

Eventually, the elongated parasite constricts and segments into between five and seven actively motile daughter cells, which emerge from an empty host cell to begin again their complex life cycle.

Isolation procedures involve differential filtration of these tiny organisms through membrane filters and plating of the filtrates from 0.45, 0.65, and 0.8 μm pore sized filters. These filtrates are mixed with a 0.5-ml portion of the prospective host cells in 3 ml of soft agar and plated on top of a 15-ml bottom layer of a solid agar, most frequently yeast extract—peptone agar. This double-layer agar technique is that usually employed in the isolation of bacteriophage. Those plaques which form in 24 h are marked and assumed to be due to phage lysis; plaques appearing in two to six or more days of incubation are possible *Bdellovibrio*.

Host-independent (HI) forms of *Bdellovibrio* have been isolated. These appear as small yellow colonies of yeast extract—peptone media when plated with concentrated lysates of *Bdellovibrio* cells. A procedure used in isolation of these HI strains involves the streptomycin-resistant mutants of HD (host dependent) cells, which are then grown on streptomycin-sensitive hosts which upon transfer to streptomycin-containing media are prevented from growing. Using this technique, HI forms have been isolated from all 16 HD cultures tested. HI strains liquefy gelatin, produce ammonia from peptone, and are oxidase positive. They do not produce indole from tryptophan nor do they reduce nitrate, and they are unable to use carbohydrates or organic acids as carbon or energy sources. During HD growth they can use only host cell protein, and evidence seems to indicate that their metabolic system restricts them to utilization of only proteins, peptides, and amino acids.

Some studies have been done on the DNA base compositions of several isolated strains. The GC ratio of the majority ranges from 50—51 percent, but two strains had a GC ratio of 43 percent.

VIROIDS, THE SMALLEST KNOWN AGENTS OF INFECTIOUS DISEASE[1]

A viroid is a fragment of viral nucleic acid. To date, all known viroids are short strands of RNA, with a molecular weight of 75—100,000 daltons, and are able to replicate in a susceptible host and cause disease. Viroids were first discovered in 1967, during an attempt to purify and isolate the suspected viral agent causing PSTV (potato spindle tuber disease). The infectious agent was free RNA, and virus particles were not present in infected

[1]Diener, T. O. 1974. Viroids: the smallest known agents of infectious disease. *Ann. Rev. Microbiol.* *28*:23-39. (Reviewed by Arlene Potts.)

tissue. Since this discovery, it has been established that two other plant diseases are caused by viroids. They are chrysanthemum stunt and citrous exocortis.

Early work with crude extracts from potato leaves infected with PSTV showed that it sedimented out at very low rates (ca. 10S), making it unlikely that the extracted infectious agent is a conventional virus particle. Since the agent was found to be insensitive to various organic solvents, lipid-containing virus particles of low density were also ruled out. Treatment of the crude extracts with phenol did not affect the infectivity nor the sedimentation properties. In view of these results, it was proposed that the agent was free nucleic acid. Incubation of the crude extract with ribonuclease in media of high ionic strength showed that the agent partially survived. This property, along with the elution pattern from columns of methylated serum albumin, suggested that the agent may be a double-stranded RNA.

When the extracted infectious nucleic acid from PSTV was compared with PSTV in crude extracts, it was found that, compared with the agent in a crude extract, the RNA had a greater infectivity dilution end point and higher thermal inactivation point and was more sensitive to ribonuclease. Also, when infectivity of a crude extract was removed by heating or incubation with ribonuclease, it could be restored by treatment with phenol. From comparative centrifugation studies it was concluded that no free RNA was present in crude extracts, but did not specify what the RNA was bound to.

T. O. Diener showed that the sedimentation properties of the faster-sedimenting infectious material which is present in crude extracts are not significantly altered by treatment with phenol or with phenol in the presence of sodium dodecyl sulfate, making it unlikely that the RNA is bound to protein or is present in the form of complete or partially degraded virions. Similarly, purified RNA preparations from PSTV-infected tissue also contain infectious material that sediments out faster than the free RNA. If the faster sedimenting infectious material were composed of viral nucleoprotein particles, they would have most unusual properties. Their protein coat would have to be loose enough to allow access to ribonuclease (since all infectivity is sensitive to treatment with ribonuclease), and yet be resistant to treatment with phenol and SDS. Since there are no known nucleoproteins with these properties, Diener concluded that the faster-sedimenting infectious material is not composed of viral nucleoprotein.

If the RNA in the faster-sedimenting infectious material is not contained in virions, what is it bound to? Because of its strong dissociation properties with phenol, belief that it was not bound to protein was strengthened. More likely, the RNA is bound to cellular constituents in complexes that are not degraded by phenol, or it occurs in aggregates of varying size.

Bioassays of subcellular fractions from PSTV-infected

tissue disclosed that appreciable infectivity is present only in the original tissue debris and in the fraction containing nuclei. When chromatin was isolated from the infected tissue, most infectivity was associated with it and could be extracted as free RNA with phosphate buffer. Chloroplast, mitochondria, ribosomes, and the soluble fraction contained only traces of infectivity. This suggests that, *in situ*, PSTV is associated with the nuclei and particularly with the chromatin of infected cells.

Although the low molecular weight of viroids has been conclusively demonstrated, their exact structures cannot be definitely stated. Two models are compatible with the properties determined: (1) the RNAs may be single-stranded molecules with some sort of hairpin structure, involving extensive base pairing, or (2) the RNAs may be double-stranded but incompletely base-paired molecules.

How can such RNAs contain sufficient genetic information to induce replication in a susceptible host? The molecular weight of PSTV is only adequate to code for 70 or 80 amino acids, which is barely sufficient to code for a very small protein. It is conceivable, however, that PSTV is not a singular molecular species, but rather a population of several RNA molecules of similar length with different nucleotide sequences which together may comprise a viral genome of conventional size. Because of their small size and the fact that very few viroids need be present to elicit disease, it does not seem likely that the host's synthesis of nucleotides is hampered. Interference to the host seems to occur by the viroid acting as an abnormal transfer RNA, causing synthesis of faulty proteins, or they might interfere with the host genome transcription.

Although viroids so far identified cause diseases of higher plants, it is now reasonable to search for viroids in animals, where a viral etiology of an infectious disease has been assumed, but where no causative agent has ever been found. Scientists are now working on "slow viruses" or "stalled diseases," which some feel are being caused by incomplete viruses that can remain inactive in the body for many years until triggered by some unknown mechanism. Various bits of evidence link measles virus to the development in later life of multiple sclerosis and a rare, fatal nerve disease called SSPE. In most cases, the SSPE victim contracted measles in the first year of life, before the immune mechanisms were fully developed. Other infectious diseases where viroids are suspect are Kuru and Creitzfeldt—Jacob disease, which causes premature senility and mental deterioration. It has been proposed that hepatitis may be caused by viroids.

FLEXIBACTERIA[1]

Flexibacteria is a common name for bacteria with a flexible body, that exhibit a strange and unexplained mechanism of motility by gliding on solid surfaces without flagella. Morphologically, the flexibacteria are similar to the *Cytophaga*; both have no rigid cell wall, a peculiar method of locomotion, and exhibit flexing movements. They were therefore included in the order Myxobacterales. However, the myxobacteria have fruiting bodies. In 1965, the DNA base composition of the myxobacteria was determined to have GC values between 68 and 71 percent, whereas the cytophaga/flexibacteria cultures (without fruiting bodies) had GC ratios in the range of 30—40 percent. Over 100 cultures were isolated from soil and water samples that possessed the basic flexibacteria characteristics. Some were found to be cellulolytic and some not. Some were carbohydrate dependent and used inorganic nitrogen; others utilized a wide range of carbon sources and were able to grow in the absence of carbohydrates, but could not grow with inorganic nitrogen.

PROSTHECATE BACTERIA[2]

The prostheca is a semirigid appendage extending from a procaryotic cell, with a diameter which is always smaller than that of the mature cell, and which is bounded by the cell wall, containing murein. Examples of the prosthecate bacteria are *Caulobacter*, which have cellular stalks, and the budding bacteria, *Hyphomicrobium* and *Rhodomicrobium*, which have hyphae or filaments. *Caulobacter* and *Asticcacaulis* have cellular stalks, which play no role in the generation of daughter or swarmer cells, and contain no DNA. The differences between these two genera are the inequality in size of the daughter cells of *Asticcacaulis*, the swarmer cell being much smaller than the stalked cell, the absence of adhesive properties in the *Asticcacaulis* prostheca, the stalk originating from an excentral position rather than from the center of the pole as in *Caulobacter*, and the difference in GC ratios. Buds are not considered to be prosthecae, but some of the budding bacteria have prosthecae. In *Hyphomicrobium* and *Rhodomicrobium* the buds are produced at the tips of hyphae or filaments, with cell walls continuous between the hyphae and bacterial cells. Lecithin has been found to be present in the membrane systems of the *Hyphomicrobium* and *Rhodomicrobium*, and it is rarely present in any bacterial membranes. A new group of prosthecate bacteria has been found that oxidize iron and manganese, are similar morphologically to the *Rhodomicrobium*,

[1]Soriano, S. 1973. Flexibacteria. *Ann. Rev. Microbiol.* *27*:155—170. (Reviewed by Ramona M. Slepetis.)
[2]Schmidt, J. M. 1971. Prosthecate bacteria. *Ann. Rev. Microbiol.* *25*:93—110. (Reviewed by Ramona M. Slepetis.)

and develop hyphae from several sites on the cell. This organism was named *Pedomicrobium*. Other rare prosthecates are the *Planctomyces/Blastocaulis* with long slender stalks, and the *Pasteuria*, which are pear-shaped motile cells with very short stalks. Several prosthecate bacteria have been recently found in open water; they have many prosthecae emanating from diverse locations on the cell surface, giving a spiked outline. *Prosthecomicrobium* has species that are motile, and the *Ancalomicrobium* are nonmotile and budding. *Prosthecochloris* has approximately 20 appendages per cell, and is a strictly anaerobic green sulfur bacterium. Other prosthecate bacteria have been identified, their prosthecae being of minute size. Among these are the *Helicoids, Tuberoidobacter, Kuznetsovia*, and *Caulococcus*. There has been some controversy over the nature of the iron-containing ribbons of *Gallionella*. One view is that they are noncellular secretions, but recent investigations show that they possess an organic matrix and exhibit some reproductive capacity. Bacterial flagella and pili are nonprosthecate appendanges; however, some bacteria, such as *Bacillus brevis, Vibrio metchnikovii, Bd. bacteriovorus*, and *Pseudomonas stizolobii*, have sheathed flagella.

The function of the prosthecae in the budding bacteria is known, but prosthecae of nonbudding bacteria are not involved in reproduction. Several theories for their function have been postulated. Since they greatly increase cell surface, they might enhance membrane-associated activities such as respiration and nutrient uptake. However, the only correlation has been with oxygen uptake. Since the caulobacters are obligate aerobes, J. S. Poindexter believes that the stalk, as a flotation organelle, maintains the cells close to the surface of the water. The idea of the prosthecae as an attachment organelle is no longer very popular, as holdfasts are not present in all species. It has been suggested that the stalk is used to parasitize other microorganisms, but there has been no proof of this. Morphological observations have led to the assumption that the rosette formation represents a conjugation phenomenon, although there has been no genetic evidence concerning this. Since there is no one function of the prosthecae, it is possible that they may have multiple functions or none at all.

BUDDING BACTERIA[1]

Bacterial division by fission is preceded by growth of the cell wall that increases the length or diameter of the cell. Bud formation prior to fission is common among microorganisms. In this process a small area of the parental cell wall is weakened, which allows localized

[1] Hirsch, P. 1974. Budding bacteria. *Ann. Rev. Microbiol. 28*:391–444. (Reviewed by Kathleen G. Clodius.)

membrane and wall growth. The resulting structure, the bud, is commonly thought of as a small and spherical daughter cell attached to a larger mature mother cell. Although budding is a common process, only certain microorganisms bud regularly before cell division. These are considered "truly budding" microorganisms.

If one defines a bud as "a local protuberance that is smaller than the cell forming it," then there are several types of bacterial budding processes. First, there is budding for cell multiplication. A second type of budding is for cell or hyphal branching. Bacterial cells, such as *Pelodictyon clathratiforme*, or hyphal extensions, such as in *Hyphomicrobium*, *Rhodomicrobium*, or *Pedomicrobium*, have been reported to form branches through a true budding process. Similar branching is found in actinomycetes and in cyanobacteria.

The third type of budding is for sporulation and spore germination. The germination of various types of actinomycete spores has been termed "budding" and is probably similar in mechanism to hypha formation by *Hyphomicrobium* and other related gram-negative bacteria; a local swelling leading to annular wall growth results in formation of a tubular cell extension of constant diameter. In addition to these types of budding, there are bacterial forms that resemble budding stages. One example is bacteria that undergo constrictive cell division and another is bacteria that undergo uneven cell division owing to an asymmetric site of septum formation, which may result from unipolar cell wall growth.

For the purpose of this review, a process of new cell wall formation in a bacterium will be considered to be a truly budding process if the following criteria apply: (1) *Morphological*: the new cell must be initially smaller in size (i.e., narrower and shorter) than the mother cell; even after separation; this usually holds true. (2) *Developmental*: all or most of the bud wall must be newly synthesized, although the age distribution pattern in the bud wall can vary in different organisms. Usually, buds are preformed before they become nucleated. Also, spherical and small buds increase in cell diameter during cell wall growth contrary to elongation of rod-shaped nonbudding bacteria. (3) *Functional*: budding must be the only mode of new cell formation. A bud is the creation, by the mother cell, of a new space outside, into which the new cell constituents either have to migrate or where they have to be formed *de novo*. Thus, budding does not necessarily describe a mode of cell separation (multiplication), but rather one of new wall formation.

Budding appears to be characteristic of bacteria which live in habitats that undergo frequent and sudden changes. They inhabit many extreme locations and show great resistance to such stressful conditions. An example of this is their tendency to attach to surfaces and frequently to form rosettes. This is advantageous in habitats poor in nutrients or with a changing nutrient supply. The formation of buds is necessary for survival of some organisms in which the main body of the cell becomes encrusted with mucilage, humus substances, or metal oxides. The bud represents the mobile propagative stage of the organism.

In investigating the natural life cycles of the budding bacteria, several problems arise. Laboratory cultures may contain forms or cell sizes that are not detected in nature. Conversely, the appearance of pure cultures when transferred into the natural habitat is often not known. To solve these problems, growth chambers sealed by membrane filters or dialysis membranes were inoculated with pure cultures and brought into the natural habitat. Another method consisted of direct, water-immersion microscopy of organisms growing in the undisturbed habitat. Observation of bacteria attached to glass slides or use of glass capillaries were further useful techniques. Seminatural conditions could be obtained by using two-dimensional, agar diffusion gradient techniques.

We shall consider five of the more than 20 budding bacteria described.

1. *Hyphomicrobium*: The type species *Hyphomicrobium vulgare* has small rods with pointed ends and polar filaments that carry terminal knobs. A pellicle is formed on liquid media, and a preference for some unusual carbon or nitrogen sources has been pointed out. *Hyphomicrobium* appeared in soil suspensions or enrichments. Methane or natural gas favors their development. Freshwater types are found in lakes, brooks, springs, acid bog water, forest marl ponds, water distribution pipes, acid mine drainage water, or borewell water. They also occur in aquarium water, laboratory water baths, and in seawater incubated in the laboratory. Although capable of growing in very dilute media, hyphomicrobia are also found in sewage sludge or polluted streams, and they live in algal jellies. Fine structural studies of *Hyphomicrobium* and *Rhodomicrobium* hyphae demonstrated the presence of cytoplasmic membranes, a dense peptidoglycan layer, and gram-negative cell walls. A "stalk" originating by excretion had been postulated previously. *Hyphomicrobium* strains from soil formed a turbidity in liquid media, and the cells had tubular, membranous invaginations. These tubules could be interconnected and distributed throughout the whole cell. They contained a fine granular matrix similar to that in the periplasmic space. Older cells had one or more large storage granules of poly-β-hydroxybutyrate bounded by one single-track, stainable layer.

2. *Hyphomonas*: This organism was isolated from nasal mucoid of a sinus-infected patient, which was surprising since the hyphomicrobia known so far are nonpathogenic. Although they resemble *Hyphomicrobium* cells, they are physiologically quite different. They grow best at 37°C on coagulated serum, blood agar. Pathogenicity against mice cannot be established. There is no growth on mineral salts media with a variety of carbon sources, which normally support growth of hyphomicrobia.

3. *Pedomicrobium*: These budding hyphal bacteria were observed to decompose organomineral humus complexes. They

resemble *H. vulgare*; their distinguishing characteristics are (1) formation under some conditions of one to four branching hyphae, which could arise laterally and give the cell a pleomorphic appearance, (2) the hyphal outgrowth from still small, attached and immature buds, (3) the often highly branched mycelium, and (4) the lack of motile cells. The species *Pedomicrobium ferrugineum* deposits iron, and *P. manganicum* deposits manganese. Pedomicrobia occur worldwide in podzolic soils, freshwater, or brackish water.

4. *Rhodomicrobium: Rhodomicrobium vannielii* was described as a photosynthetic, budding, hyphal bacterium with a morphology almost identical to that of hyphomicrobia. Cell suspensions, cell-free extracts, or fractions exhibit a peculiar, light-stimulated oxygen uptake. *Rhodomicrobium* buds are motile, and both peritrichous and monotrichous flagellation have been observed. One strain isolated produced heat-resistant, angular spores. Hyphal outgrowth occurred at one to four sites. The spores were sometimes arranged in chains and resembled conidia. Since heat resistance of angular, nonmotile spores could not be shown conclusively, it was suggested that the term spore be replaced by cyst.

5. *Anclomicrobium*: These budding bacteria have long, tapering cellular appendages (prosthecae). The cell body consists mainly of arms and is irregularly shaped. Slide cultures were used to study its peculiar mechanism of multiplication. Buds grew from either the central part of the cell or from the tip of a prosthecae. One of the two strains so far isolated is *A. adetum*, which is gram-negative, catalase +, and facultatively anaerobic. It has two to eight tapering appendages with lengths up to 3 μm, and contains gas vesicles, some of which are in the prosthecae.

Other budding bacteria described in the review are: *Ancalochloria*, *Prosthecomicrobium*, *Prosthecochloria*, star-shaped bacteria, *Planctomyces*, *Pasteuria*, *Blastobacter*, *Methylosinus*, *Naumanniella*, budding *cocci*, *Siderococcus*, *Seliberia*, *Myococcus* and nocardioform organisms, *Geodermatophilus*, *Blastococcus*, bacterial protoplasts and L-forms, *Mycoplasma*, *Acholeplasma*, *Thermoplasma*, *Metallogenium*, *Caulococcus*, and *Kusnezovia*.

MYCOPLASMA AND THEIR MEMBRANES[1]

Mycoplasma are organisms that have no cell wall and are bounded only by a plasma membrane. This membrane is of the single membrane type, a trilaminar structure consisting of a unit membrane with a protein coat on both sides. The chemical composition of the membrane has been reported to be 50—59 percent protein, 32—40 percent lipid, and

[1]Maniloff, J., and H. J. Morowitz. 1972. Cell biology of the mycoplasmas. *Bacteriol. Reviews 36*:263—290. (Reviewed by Virginia Glaser.) See also (3,4,5,6).

0.5—2 percent carbohydrate.

Electrophoresis of the mycoplasma membrane proteins in polyacrylamide gel shows 15 to 30 different protein bands. These probably do not represent all the proteins in the membrane, as several proteins may migrate together and more than one may be present on each band. These banding patterns are species specific, reproducible, and can be used for identification and classification. Proteins do not completely cover the surface of the membrane, as cells exposed to a high titer of antibody specific for the membrane glycolipids agglutinate.

Mycoplasma contain little or no lipid except that in the membrane. The lipid consists of glycolipids, phospholipids, and cholesterol (in the sterol requiring mycoplasma), carotenoids in acholeplasma, and in *Acholeplasma axanthum* a sphingolipid. The glycolipids along with the proteins are responsible for most of the antigenic determinants of the organism. There are more cross reactions indicating antigenic similarities between various mycoplasma strains and bacteria than between species of mycoplasma. One especially useful one has been the cross reaction between *Mycoplasma pneumoniae* and *Streptococcus* MG antigens. This was thought to reside in the diglucosal diglyceride moiety, but has been shown to be in the trigalactosyl component. More recently, this same antigen has been extracted from the membrane of spinach cells, making available a convenient and inexpensive source of glycolipid for use in serological identification of mycoplasmal pneumonia. It has been postulated that glycolipids may also perform a function as structural components.

The phospholipids are mainly glycerophospholipids. The T-strains have as well phosphotydyl ethanolamine and an *O*-acyl-*N*-acyl-diamino lipid, and *A. axanthum* contains a ceramide phosphoryl glycerol.

The fatty acids in the medium are usually incorporated into the polar lipids, although some are found as free fatty acids in T-strains and *M. hominis*. The few mycoplasma strains tested did not seem to be able to synthesize, elongate, unsaturate, or interconvert any of the long chain fatty acids. *Acholeplasma laidlawii*, however, is capable of synthesizing long chain fatty acids from acetate. There is an acyl carrier protein similar to that of *Escherichia coli* but more sensitive to heat. The need for malonyl CoA is absolute possibly because the acetyl-CoA carboxylase activity of the organism is very low. If long chain fatty acids are present in the medium, they are incorporated into the membrane, and synthesis is cut off, apparently by end-product inhibition.

It is thought that carotenoids, sterols and the sphingolipids carry out the same function, despite the dissimilarity of their chemical composition (i.e., to produce an orientation of the lipid bilayer structure so that the long axes of the lipids are perpendicular to the lamellar plane). *Acholeplasma laidlawii* seems to be able to regulate the chain length and degree of saturation of the fatty acids in the first and second positions in the phospholipid in order to maintain a certain degree of fluidity. Mycoplasma selectively incorporate fatty acids. The longer the chain of the saturated fatty acid on the first position, the shorter and more saturated the fatty acids on position number 2. The amount of cholesterol taken up also seems to be controlled in living cells. Being able to control membrane fluidity in this way is crucial to such functions as permeability, membrane-bound enzyme activity, transport of certain nutrients, and osmotic stability.

It was thought at one time that cholesterol played a part in the transport of nutrients across the membrane. However, a number of transport systems have been reported. *Mycoplasma gallisepticum*, *M. mycoides* var. *mycoides* and var. *capri*, as well as goat strain Y have a PEP-dependent phosphotransferase system; *M. fermentens* and *M. hominis* transport systems for amino acids behave like typical permease systems. They show typical transport kinetics, accumulate metabolites against a gradient, require energy, and are inhibited by sulfhydryl blocking agents. K^+ is also actively transported, being energy dependent and sulfhydryl mediated.

RICKETTSIA[1]

Rickettsia are obligate intracellular parasites that inhabit such anthropods as fleas, ticks, lice, and mites, in which they cause injury. They are often pathogenic to man. They range from 0.3—0.5 μm in width and from 0.3—2 μm in length. In most, the cell wall has three layers, but with higher magnification it is sometimes possible to distinguish a five-layer structure. The cell wall contains, in addition to sugars, amino sugars, and amino acids, muramic acid and diaminopimelic acid. No techoic acid has been found. There is evidence of endotoxic activity, which suggests the presence of a lipopolysaccharide layer, typical of gram-negative bacteria. There is also evidence of capsules and a cytoplasmic membrane. The internal structure consists of electron dense granules and five strands, indicating the presence of ribosomes and DNA, and both DNA and RNA have been isolated. Rickettsiae have not been grown in the absence of host cells, so they must be cultivated in eggs, tissue cultures or small lab animals, eggs usually giving the greatest return of viable organisms.

Rickettsia prowazeki is strictly confined to cytoplasm. It does not invade vacuoles,and it does not change the cell morphology of the host. In contrast, *Coxiella burnetii* multiplies in vacuoles and, in doing so, converts the cell into one huge vacuole while compressing the

[1]Weiss, E. 1973. Growth and physiology of Rickettsiae. *Bacteriol. Reviews 37*:259—283. (Reviewed by Gary Jones.) See also (7).

nucleus and remaining cytoplasm toward the periphery of the host cell. *Spotted fever* and *scrub typhus* rickettsiae multiply in the cytoplasm, and spotted fever rickettsiae have also been seen occasionally in the nucleus. Scrub typhus aggregate in the region adjacent to the nucleus. Most rickettsial species form dense masses in the cytoplasm of mouse lymphosarcoma cells, but spotted fever rickettsia remain scattered throughout the cells, even at the peak of infection. Plaques can be seen if grown in heart—brain infusion broth at 32°C. The size and amount of time to form a plaque can be used to show differences between rickettsiae. The times for plaque formation run from 5—6 days for spotted fever rickettsia to 17 days for *R. tsutsugamushi*.

The penetration into host cells seems to be related to the absorption time in contact with the host. In general, only viable rickettsiae penetrate host cells, are toxic for mice, or lyse red blood cells. For example, *R. tsutsugamushi* exposed to ultraviolet light cannot penetrate the host. It has been shown with typhus rickettsia that viability, toxicity for mice, hemolytic activity, and respiration stimulated by glutamate are all associated with penetration. Some factors affecting penetration influence metabolism directly; others alter the rate of inactivation of extracellular rickettsiae. The stability of rickettsiae plays a role in penetration because there is an appreciable decline in infectivity of extracellular scrub typhus rickettsiae within 2 h at 37°C in complete medium. (Complete medium-balanced salt solution, BSS + beef embryo extract + horse serum.) When the medium is supplied with BSS alone, penetration of *R. tsutsugamushi* is reduced to one half.

With many of the rickettsiae, it was shown that only glutamic acid and glutamine, and combinations of α-KG + asp (which yield glutamic acid with the proper transaminase), are effective in increasing penetration rate. Survival of rickettsiae in the absence of host cells was increased by glutamate. Chloramphenicol affects growth of rickettsia, but it has no effect on host penetration.

Transverse binary fission appears to be the only means of replication. The view that the infected cell fills and bursts, releasing many organisms, is oversimplified. Rickettsia are sometimes trapped by microfibrillar structures protruding from the edge of the cell. When these microfibrils contract, they either take the rickettsia back into the cytoplasm or release them into the extracellular fluid. *Coxiella burnetii* grows in the vacuoles, filling these organelles with organisms. As the cultures age, the numbers of organisms decrease, suggesting that the rickettsia are released slowly over time instead of all at once. There is also some evidence that rickettsia are released directly into neighboring cells.

Rickettsiae are unaffected by sulfonamides, but all except *C. burnetii* are affected by *p*-amino benzoic acid which reacts with NAD to inhibit malate dehydrogenase activity. Penicillin and streptomycin have no therapeutic effects, but these two antibiotics do affect the growth of cells in culture. The tetracyclines work best in controlling rickettsiae, which are inactivated when exposed directly or when in tissue culture during absorption. Inhibition does not occur if the drug is added after the infection is already established. Antibiotic-resistant strains have been isolated in the laboratory, but there have been none isolated in nature.

In regard to the following discussion on metabolism, mainly typhus and spotted fever are referred to. Respiration is stimulated most vigorously by glutamate. Glutamine is the second most stimulating. Pyruvate and TCA-cycle intermediates also stimulate respiration to a lesser extent. Glucose, G-6-P, lactate, sucrose, and the other naturally occurring amino acids do not stimulate respiration.

There are three end products of glutamate utilization: (1) ammonia, (2) carbon dioxide, and (3) aspartate. Some of the amino group of glutamate is released as ammonia, but most of it is transferred to oxaloacetate, from which aspartate is formed under the action of the enzyme glutamate-oxaloacetate transaminase, one of the most stable of rickettsial enzymes. This enzyme can also catalyze the reverse reaction and is extracellular, but may be associated with a surface structure. Carbon dioxide is produced from all of the carbon atoms of glutamic, which may not be completely oxidized. Intermediates of this pathway are α-KG, succinate, fumarate, oxaloacetic acid, and pyruvate, none of which accumulates without specific inhibitors. The three carbons of pyruvate are all converted into CO_2. TCA-cycle intermediates have not been isolated, but this may be due to technical difficulties rather than to the absence of corresponding enzymes. *Rickettsia typhi* has a flavin—enzyme—iron—cytochrome system, which probably includes cytochromes a and b, but the activity of the cytochromes is low.

Direct measurement of ATP content in starved rickettsia (which are grown for 3 h at 36°C with no substrate) showed no measurable ATP. When glutamate was added to the media, the ATP level rose to 1.5—2.0 μole/mg of rickettsial protein. If adenylic acid was also added, the levels of ATP were somewhat higher. As compared to viable bacteria, ATP levels in rickettsia are generally lower.

Rickettsia are capable of synthesizing small amounts of protein and lipid. One amino acid, glutamine, must be available in far larger concentrations since it serves a dual function of providing energy as well as a source of glutamate for protein synthesis. Glutamate and glutamine can be substituted interchangeably. Two sources of ATP are essential to rickettsiae. One is generated endogenously from glutamine and the other must be provided exogenously by the host. The amount of lipid synthesis that takes place is not really known. Requirements are similar to those of protein synthesis, but a mixture of amino acids, except glutamine, are not required.

Q fever rickettsia are somewhat different from typhus and spotted fever. Q fever is caused by *C. burnetii*, one of the sturdiest of nonsporogenic organisms. It can remain viable when suspended in distilled water or sterile milk at 4°C or room temperature. Neither flash pasteurization nor chemical sterilization is usually effective in destroying it.

Disrupted cell preparations contain some enzymes of glucose metabolism: hexokinase, glucose-6-phosphate dehydogenase, 6-phosphogluconate dehydrogenase aldolase. Q fever rickettsia may have all the TCA-cycle enzymes. Synthesis of citric acid from oxaloacetic acid and to a smaller extent from acetate and acetyl phosphate in the presence of NAD, CoA, and ATP has been demonstrated. Oxidation of isocitrate, glutamate, and malate with reduction of NAD and NADP has also been shown.

BLUE-GREEN ALGAE[1]

Blue-green algae have been considered as a class of algae by botanists because they are photoautotrophs that use water as an electron donor, and also because they contain chlorophyll a and b carotene, the two photopigments responsible for plant photosynthesis. But it has been long recognized that in other respects they resemble the bacteria. Electron microscopic investigation, together with analytical data on cell wall composition and ribosomal structure, have now revealed that blue-green algae are procaryotes. Blue-green algae can now be recognized as a major group of bacteria, distinguished from other photosynthetic bacteria by the nature of their pigment system and by the production of oxygen during photosynthesis.

Purification

The study of any microbial group is largely dependent on the solution of a technical problem: the isolation and maintenance of its members in pure culture. Although the nutrient requirements of these organisms are simple, the primary obstacle in obtaining pure cultures lies in the synthesis of a copious extracellular slime layer that harbors bacteria. In principle, the purification of unicellular blue-green algae should be simpler than that of the filamentous forms, since most of the unicellular

[1]Stanier, R. Y., R. Kunisawa, M. Mandel, and G. Cohen-Bazire. 1971. Purification and properties of unicellular blue-green algae (order Chroococcales). *Bacteriol. Reviews 35*:171—205. (Reviewed by Carolyn M. Hirst.) See also (8-13).

members of the group are immotile and therefore will form small compact colonies on solid media. However, these organisms typically occur in nature as minor components of mixed algal populations. Since their nutritional requirements do not differ significantly from those of many other algae, they cannot be specifically enriched from such populations by nutritional selection.

Many mesophilic blue-green algae, both filamentous and unicellular, can be enriched successfully from mixed algal populations by temperature selection since their maximum temperatures are significantly higher than those of nearly all eucaryotic algae from the same environment. When samples of soil or freshwater are placed in a suitable mineral medium and incubated in the light at 35°C, the population that develops consists almost exclusively of blue-green algae. This enrichment technique is particularly valuable for the isolation of unicellular blue-green algae, because, if present in the inoculum, they eventually far outnumber filamentous members of the group as the enrichment culture develops.

Pure cultures are essential for a study of the utilization of organic compounds by blue-green algae. Consequently, information about this aspect of their physiology is still very limited. Reviewing the information then available, O. Holm-Hansen concluded that most blue-green algae are obligate phototrophs, although there are well-authenticated reports that a few strains can grow in the dark at the expense of organic compounds. In analyzing this question, a clear distinction must be made between the utilization of organic compounds in the light and in the dark. Provided that it is permeable to organic compounds, an obligate phototroph may well be able to assimilate them in the light, using photochemical reactions for the generation of ATP and reducing power. The photosynthetic rate may be increased by the provision of organic substrates if they can serve as general sources of cellular carbon and if the rate of photosynthetic growth is normally limited by the rate of CO_2 assimilation. However, to grow in the dark with organic compounds, a photosynthetic organism must be able to use them as sources of ATP and reducing power generated through respiration or fermentation, and not merely as sources of assimible carbon.

A study was done by A. J. Smith, J. London, and R. Y. Stanier on the utilization of organic compounds in the light by three unicellular strains. All three could assimilate organic acids, amino acids, and glucose, although none of the compounds tested increased the rate of photosynthetic growth. The uptake of these compounds was light dependent. Even acetate, the most rapidly assimilated compound that was examined, contributed only some 10 percent of the carbon incorporated into newly synthesized cell material. An investigation of the patterns of incorporation of ^{14}C labeled acetate and other substrates into the cellular amino acids suggested that the tricarboxylic acid cycle was not operative. This was supported by the failure to detect α-ketoglutarate dehydrogenase in cell-free extracts, which is a key enzyme in this cycle. Similar results in other experiments have confirmed that no blue-green algae examined so far contains a functional tricarboxylic acid cycle. From this it is evident that a normal respiratory metabolism of organic compounds in the dark cannot occur. Growth in the dark therefore probably takes place at the expense of either glycolysis or of the pentose phosphate cycle. It has also been discovered that carbohydrates are the only organic substrates that have been able to support the growth of blue-green algae in the dark. The strict dependence on carbohydrates as energy-yielding organic substrates suggests a very limited respiratory capacity.

Until recently, the capacity for nitrogen fixation among blue-green algae appeared to be invariably associated with the ability to form heterocysts. Heterocysts are not produced by unicellular blue-green algae, with the exception of *Clorogloea fritschii*, which can fix nitrogen. Two non-cyst-forming strains can also fix nitrogen.

The photopigments characteristic of blue-green algae are chlorophyll a, phycobiliproteins (C-phycocyanin, "allo" phycocyanin, and in some strains phycoerythrin), and a variable array of carotenoids. In intact cells, the main visible absorption maxima are attributable to chlorophyll a, C-phycocyanin, which always predominates over "allo" phycocyanin, and phycoerythrin, if this phycobiliprotein is present.

Nearly all bacteria, including the photosynthetic purple and green bacteria, contain only saturated and monounsaturated fatty acids. The majority of filamentous blue-green algae contain large amounts of polyunsaturated fatty acid. Unicellular blue-green algae fall into two general categories. Some have a high content of polyunsaturated fatty acids, which represent between 12 and 50 percent by weight of the total fatty acids of the cells. Such strains always contain significant amounts of triple unsaturated fatty acids. Other strains either do not contain detectable polyunsaturated fatty acids or have a very low content ranging from 1—4 percent of the total fatty acid content of the cells.

Fluorescent light intensities of 2,000—8,000 lux are not deleterious to most unicellular strains. However, some are unusually light sensitive, and they become partly bleached owing to the destruction of phycobiliproteins if exposed to light intensities in this range. The highest fluorescent light intensity at which these strains show normal growth and pigmentation is approximately 500 lux.

Most unicellular blue-green algae are extremely sensitive to penicillin G. There is one comparatively resistant strain that can grow in the presence of at least 50 units/ml. Five strains can grow in the presence of 1 unit/ml, but not of 10 units/ml. The remaining 34 strains are inhibited by penicillin concentrations of 1 unit/ml or less. Therefore, the inhibitory concentrations for the majority of unicellular blue-green algae are similar to those for gram-positive eubacteria and far below the level required to inhibit growth of most gram-negative eubacteria.

Excluding three thermophilic isolates from hot springs for which the temperature maximum is 53°C or higher, 35 of 37 strains examined had temperature maxima between 35 and 43°C. Most strains isolated at Berkeley had been preselected in the course of their isolation for the ability to grow at 35°C. However, 11 or 12 nonthermophilic strains received from other laboratories had temperature maxima in the range of 35—43°C, even though they had been isolated at temperatures below 30°C. A relatively high temperature maximum therefore appears to be characteristic of many unicellular blue-green algae isolated from nonthermal habitats.

There seems to be a correlation between DNA base composition and temperature maximum among strains in group IA. Of the 11 strains of high GC content, 10 have relatively low temperature maxima (25—37°C). Of the 11 strains of low GC content, 7 have temperature maxima in excess of 40°C.

REFERENCES

Other Unusual Organisms

Ahmadjian, V. 1965. Lichens. *Ann. Rev. Microbiol. 19*: 1—20.

Bonner, J. T. 1971. Aggregation and differentiation in the cellular slime molds. *Ann. Rev. Microbiol. 25*:75—92.

Dworkin, M. 1966. Biology of the myxobacteria. *Ann. Rev. Microbiol. 20*:75—106.

Hanks, J. H. 1966. Host-dependent microbes. *Bacteriol. Reviews 30*:114—135.

Kingsley, V. V., and J. F. M. Hoeniger. 1973. Growth, structure, and classification of *Selenomonas*. *Bacteriol. Reviews 37*:479—521.

Lechevalier, H. A., and M. P. Lechevalier. 1967. Biology of actinomycetes. *Ann. Rev. Microbiol. 21*:71—100.

Poupard, J. A., I. Husain, and R. F. Norris. 1973. Biology of the Bifidobacteria. *Bacteriol. Reviews 37*: 136—165.

Starr, M. P., and V. B. D. Skerman. 1965. Bacterial diversity: the natural history of selected morphologically unusual bacteria. *Ann. Rev. Microbiol. 19*:407—454.

Papers

1. *Advan. Microbial Physiol. 8*:215 (1972).
2. *J. Bacteriol. 121*:1131, 1137, 1145, 1158 (1975).
3. *Ann. Rev. Microbiol. 19*:379 (1965).
4. *Ann. Rev. Microbiol. 23*:317 (1969).
5. *Advan. Microbial Physiol. 10*:2 (1973).
6. *Bacteriol. Reviews 35*:206 (1971).
7. *Ann. Rev. Microbiol. 23*:275 (1969).
8. *Bacteriol. Reviews 31*:315 (1967).
9. *Ann. Rev. Microbiol. 22*:47 (1968).
10. *Ann. Rev. Microbiol. 22*:15 (1968).
11. *Bacteriol. Reviews 37*:343 (1973).
12. *Bacteriol. Reviews 31*:180 (1967).
13. *Bacteriol. Reviews 37*:32 (1973).

15 Variations on a Theme: Unique or Unusual Processes

AUTOTROPHIC BACTERIA

We intend to discuss a group of problems associated with the obligate autotrophic bacteria. These are also called the *strictly* autotrophic bacteria or the obligate or strict *chemolithotrophic* bacteria. These organisms have three distinctive characteristics:

1. They derive their energy from the oxidation of an inorganic material.
2. They can use carbon dioxide as their sole source of carbon.
3. They do not grow on heterotrophic media. Indeed, as their discoverer, S. Winogradsky, pointed out, organic substances are frequently toxic to them.

Some people consider photosynthetic bacteria as "autotrophic," since the word means "self-sufficient," but we shall confine our attention to the chemosynthetic autotrophic bacteria, that is, those organisms which can grow in the dark using an inorganic material as their energy source and which are capable of obtaining all of their carbon from CO_2. Of these chemosynthetic autotrophs, there are two main groups: one, the facultative autotrophs (e.g., the hydrogen bacteria) are capable of using a specific inorganic energy source but can also use glucose or other materials if they are available; the second group comprises the strict or obligate autotrophs, which, within most limits, are incapable of obtaining their energy for growth from any substance other than their specific inorganic energy source.

The obligate autotrophs, therefore, operate under somewhat restricted conditions. They are confined to a simple, usually highly specific inorganic energy source. For example, *Nitrosomonas* uses ammonia, oxidizing it to nitrite and is restricted to this substance. Some strains of *Thiobacillus thiooxidans* use sulfur only; some others can use sulfur, thiosulfate, or hydrogen sulfide (but not with equal facility). A closely related group can utilize either sulfur or iron as energy sources. But the range of materials utilized is highly restricted.

The relatively few obligate autotrophic bacteria are divisible into four groups:

1. *Nitrosomonas* sp. ($NH_4^+ \rightarrow NO_2$).
2. *Nitrobacter* sp. ($NO_2 \rightarrow NO_3$).
3. Acid sulfur ($S \rightarrow SO_4$), *Thiobacillus thiooxidans; Ferrobacillus* (also $Fe^{2+} \rightarrow Fe^{3+}$). The ferrobacillus could actually be grown on electricity (1).
4. Neutral thiobacilli (thiosulfate $\rightarrow SO_4$).

Research in the past decade has clarified much about the nature of these organisms. They possess essentially unique systems for the oxidation of their inorganic substrates. These systems, however, consist of removing one or more electrons from their substrates and passing these through cytochrome systems to generate phosphate bond energy or its equivalent. They fix the major part of their CO_2 in essentially the same fashion as photosynthetic organisms, that is, via the ribulose diphosphate system. Internally, the enzymes appear to be similar both in principle and in detail to their counterparts from heterotrophic organisms. Although there is much to be said about autotrophic bacteria, we shall confine our attention to three questions:

1. How do sulfur-oxidizing bacteria manage to attack insoluble sulfur?
2. How do autotrophic bacteria obtain the ability to reduce the carbon dioxide fixed?
3. Why are obligate autotrophic bacteria unable to grow on preformed organic matter for energy or carbon?

How Do the Sulfur Bacteria Manage to Oxidize Insoluble Sulfur?

Most living cells obtain their energy from soluble materials moving into or through the membrane by extracting the energy from them by processes located at or within the cell. In bacteria, a large part of the respiratory system appears to be in the membrane. Insoluble materials are rendered soluble, usually by some hydrolytic process, and the soluble fragments brought within the cell. But sulfur, which can be attacked by several facultative and obligate autotrophic bacteria, is insoluble. How then do sulfur-oxidizing bacteria manage to attack it and to derive from it adequate amounts of energy for growth?

To approach this problem we have used the obligate autotrophic bacterium *T. thiooxidans*, an organism growing on elementary sulfur from a pH of 5 downward, actually ending at 5—10 percent sulfuric acid. It was isolated by S. A. Waksman and J. S. Joffe in 1922, and the Waksman strain is still available. It has the curious characteristic that it will grow only on sulfur and not on thiosulfate (except as the thiosulfate is converted into sulfur), but in this respect it appears to differ from other strains, clearly *T. thiooxidans*, which do grow on thiosulfate. We shall concentrate on studies using the Waksman strain and refer to others where pertinent.

It was first established by H. G. Vogler and W. W. Umbreit in 1941 (2) that there was a need for direct contact between the bacteria and the sulfur before oxidation could occur. Subsequent evidence from the electron microscope (3) shows that the organism lies on the surface of the sulfur and is tightly bound to it. On the sulfur crystal one may see small hollows have been evidently dissolved away just where the organisms were located. If a high inoculum is used, and the crystal completely covered with the bacteria, it erodes the crystal according to the molecular cleavage planes, and one obtains a replica looking much like a contour map. There is, therefore, a firm attachment between the sulfur and the bacteria. When so attached, the organism "dissolves," if that is the word, that portion of the sulfur crystal in its immediate vicinity. The net result is the erosion of the sulfur crystal in a layered fashion.

The first theory to explain these observations was based upon the relatively high fat content of the cells plus some morphological observations (4), the general idea was that the organism contained a globule of highly unsaturated fat that it placed in contact with the sulfur, dissolved the sulfur in the fat, and thus brought it into the cell. This theory, of course, had the difficulties associated with any theory. A theory has to be reasonable; a fact does not. Two unreasonable facts developed in the course of time, which required modification of the theory. The first developed very early when G. Knaysi reported (5) that *T. thiooxidans* was capsulated; how indeed could the organisms get at the sulfur to dissolve it if there was a large capsule interposed? The second developed when shaking machines came into use, and it was demonstrated that *T. thiooxidans* could grow under the violently agitated conditions of a shaking machine. However, it developed that, using low inoculum, a quiet period was necessary before the organism could oxidize sulfur and that if cultures were shaken immediately, growth was markedly inhibited (6). Jones and Starkey (7) proposed that the organism secretes a wetting agent, which coats the sulfur particle to a suitable thickness. The bacterium is embedded in this material and is so firmly attached to the sulfur particle that it cannot be dislodged on shaking. The material comprising this wetting agent was first identified, primarily on its chromatographic properties, as phosphotidylinositol. Its physical effect can be reproduced by phosphotidylinositol, but further chemical identification showed the substance to be phosphotidylglycerol (8,9). With this information at hand, we reexamined Knaysi's capsule and found that it could be demonstrated, if at all, not by the normal capsule stains, but only by certain cell-wall stains that were high in detergents. These same detergents precipitate phospholipid; in short, the "capsule" was and is the external phospholipid layer about the bacterium which far from being an insurmountable barrier to the penetration of sulfur is the very material that facilitates it. Of course, a cytologist in calling a structure a "capsule" is referring to its morphology, not to its chemical composition, but we tend to become blinded by conceptions and frequently do not see the significance of the unreasonable facts. It is perhaps appropriate that sulfur should be brought to the cell in a material surrounding the cell's surface, for it is becoming clear that it is here that much of the activity of the bacterial cell takes place.

Granting that sulfur arrives in the membrane area, we do not so far know the real mechanism of its oxidation. It is true that in the whole cell the respiration on sulfur is inhibited by cyanide, azide, and carbon monoxide. Furthermore, cytochrome c and coenzyme Q have been isolated from the organism (10), and cytochrome oxidase has been demonstrated in cell-free extracts.

But information available from cell-free systems actually demonstrating the path of sulfur oxidation is confusing. For example, it was early shown that sulfur could be oxidized by cell-free extracts of *T. thiooxidans* only in the presence of reduced glutathione and, from the data presented, evidently "tween" was used to "solublize" the sulfur (11). The products appeared to be thiosulfates and polythionates, and sulfate was not formed; in fact, the system yields several confusing results. We, ourselves, have tended to discount this system as representing the enzymatic path of sulfur oxidation in the whole cell for the major reason that, in our hands, the system can be boiled and one can still obtain comparable oxidation rates. But even worse, if glutathione is boiled and added to sulfur, there is an actual oxygen uptake, even in the absence of enzyme, boiled or fresh, so the system seems replete with artifacts; it is difficult to know whether this system bears any relation at all to that which the living cell employs. The end product of this system is thiosulfate; that of the cell is sulfate.

Adair (12) isolated a membrane fraction, closely resembling the vesicles used in permeability studies, that oxidized sulfur to sulfate at a neutral pH (the cell itself does so under acid conditions); but aside from the fact that CoQ was involved in the oxidation, little further was learned about the mechanism of oxidation.

The general picture that one obtains from studies on various autotrophic bacteria is that there is a relatively specific enzyme in each which is capable of removing electrons from the specific inorganic substance involved. These electrons are passed through the cytochrome system and produce energy-rich phosphate, which the cell can use to fix and reduce CO_2.

Reduction of Carbon Dioxide

Our second problem is, how do autotrophic bacteria obtain the ability to reduce the carbon dioxide fixed?

Since the carbon dioxide is found in fats, proteins, and in fact in the entire cell material, such reduction is necessary. Yet, so far as we can tell, the process of sulfur oxidation consists of the removal of electrons and their passage through the cytochrome system to generate energy-rich phosphate. Is it possible for the autotrophic bacteria to generate reducing activity in these circumstances (13,14)? Again we turned to *T. thiooxidans*, since we regard it as one of the most extreme examples of living forms and thus feel that it makes a better test case. We also know something about working with it. The first clue came from unpublished studies by Adair in which carbon dioxide fixation was followed under anaerobic conditions. If sulfur is supplied together with CO_2 to cells of *T. thiooxidans*, it enhances the fixation of CO_2, by being itself oxidized to provide energy under these completely anaerobic conditions, and to provide for the reduction of the CO_2.

For example, under absolutely anaerobic conditions in which only CO_2 is supplied, at the end of 3 h, no detectable CO_2 was found in the cells, whereas when sulfur was added with the CO_2, a considerable count was found, approximately 3 percent of that available to the organism. It has long been supposed that the oxygen which appears in the sulfate formed originates not from that in the air, but from the water, and it should be possible to demonstrate the formation of sulfate under anaerobic conditions. This can also be done. A suspension of *T. thiooxidans* was separated into four aliquots and placed in a nitrogen atmosphere freed from oxygen by continued passage through alkaline pyrogallol. After 1 h in the nitrogen, radioactive CO_2 was added to one aliquot, and after 8 h, only 800 cpm were found in the cells; that is, there was some CO_2 fixation, but it was rather small. To a second aliquot, the same quantity of radioactive CO_2 was supplied together with sulfur, the sulfur being unlabeled. The carbon found in the cells was now 3,461 cpm; the addition of sulfur increased the fixation of CO_2 four-fold. To the third aliquot, a suspension of radioactive sulfur was supplied, and after incubation the unchanged sulfur and cells were removed, the supernatant extracted with carbon disulfide to remove any traces of unoxidized sulfur, and the sulfate was precipitated as barium sulfate. The latter was dried under red heat on a planchet and 4,480 cpm were found; elemental sulfur had been converted to sulfate under completely anaerobic conditions. To the fourth aliquot, the S^{35} was added together with unlabeled CO_2, and 6,360 cpm were found in the barium sulfate; the presence of CO_2 permitted a greater sulfur oxidation. This appeared to be a reasonable confirmation of Vogler's contention (15) made over 30 years ago that CO_2 could act as an oxidizing agent for sulfur.

As we have pointed out in the earlier discussion of photosynthesis, electrons (from whatever source) and hydrogen ions, with the appropriate enzymes, can yield $NADH_2$ and $NADPH_2$; thus the removal of electrons from the inorganic substrate can generate reducing power, that is, the reduced coenzymes.

The Utilization of Organic Matter

Tracer studies show that during autotrophic growth many organic materials can be utilized and metabolized by obligate autotrophic bacteria. Enzymes capable of metabolizing a wide variety of organic material have been prepared from autotrophic bacteria. From what we know of their metabolism, they ought to be able to grow heterotrophically. Why, then, do they not grow on heterotrophic media?

A variety of theories have attempted to explain why heterotrophic growth does not occur, but all such theories fail if it is possible to obtain heterotrophic growth. We therefore concentrated on attempts to grow obligate autotrophs on heterotrophic media, and we eventually succeeded by using a dialysis flow-through system, which consists of two chambers separated by a dialysis membrane. One chamber is inoculated. To the other, sterile fresh medium is continually added and pumped out. Both chambers are aerated and stirred. Using this system, we have been able to grow the following obligate autotrophs on glucose in the absence of their specific inorganic energy source: *N. agilis*, *Nitrosomonas europaea*, *T. thiooxidans*, *T. denitrificans*, *T. thioparus*, and, rather poorly, *T. neopolitanus*. These are, in fact, the total number of obligate autotrophs we have tried. Growth on glucose is at least comparable in amount to growth on the inorganic source (16,17). For some reason that we do not understand, we are never able to get very good growth with *T. neopolitanus*. In addition, we have been able to obtain substantial growth on glucose in the dark of the photoautotroph, *Plectonema boryanum*, obtained from M. Shilo. The dark-grown cells lack chlorophyll, which is rapidly synthesized when the cells are exposed to light (18). Ribbons has pointed out that in many respects the methane-utilizing bacteria resemble the obligate autotrophs. Using strains supplied by Ribbons, we were unable to grow *Methomonas methanooxidans* on glucose using the dialysis system, and in this case we found that the organism lacks a glucokinase. It could be grown, without dialysis, on glucose-6-phosphate (19).

We are not the only laboratory to claim growth of obligate autotrophs on heterotrophic media. A. J. Smith and D. S. Hoare (20) claimed to have grown *N. agilis* on a supplemented acetate medium. F. Shafia and R. F. Wilkinson, Jr. (21) reported that *F. ferrooxidans* could grow on glucose after adaptation, and R. Tabit and D. G. Lundgren (22) reported growth of the same organism, now called *T. ferrooxidans*, on glucose. Clearly, then, obligate autotrophic bacteria can grow on proper heterotrophic media under proper conditions. Whatever enzymes they may lack (which seems to vary with strain and culture condition (23,24)) is not a sufficient cause to prevent heterotrophic growth since they do not grow heterotrophically when given the proper chance.

Some have seen these facts as eliminating the concept of obligate autotrophic bacteria entirely. We do not see this difficulty. All we need do is to slightly change the third character in the definition. Instead of saying that they do not grow on heterotrophic media, we can say that they do not grow on the usual heterotrophic media under the usual conditions, and we will still have delineated an interesting group of microorganisms.

Knowing that they will grow on glucose, for example, does not tell us why they usually do not grow on it, so we are still left with the problem of the cause of obligate autotrophy. Still, we are farther along in that we know that they can obtain sufficient energy and intermediates from glucose to grow. Perhaps even more pertinent, protein-synthesizing systems have finally been obtained after considerable struggle. Using log phase cells of *T. thiooxidans* with high metabolic activity, K. Amemiya has been able to prepare a cell-free preparation capable of synthesizing polyphenylalanine using poly-U as messenger (25). The system differs in no striking aspect from those obtained from heterotrophic bacteria, and indeed autotrophic and heterotrophic ribosomes and supernates may be exchanged. The system has a rather broad pH optimum of 6—7.5, a surprising temperature optimum of 37—45°C (the latter with spermidine) for an organism that is almost a psychrophile, a marked sensitivity to the buffer composition, and sensitivity to ribonuclease, streptomycin, and chloramphenicol. To the best of our knowledge, this is the first cell-free protein-synthesizing preparation successfully prepared from an obligate autotrophic bacteria.

But this still leaves us with the problem of why we have obligate autotrophic bacteria. If they can grow on glucose, why don't they? Why is dialysis or an adaptation necessary? The fact that they will grow on glucose with dialysis suggests that dialysis removes a toxic material being produced from glucose. With *T. thiooxidans* based upon growth on sulfur, we have supposed this toxic material to be pyruvate (26) although we do not know that growth on glucose produces pyruvate. What we do know is that when cultures enter the stationary phase the usual content of pyruvate is of the order of 10^{-4} *M*. It may decrease after this point and end up at less than 10^{-6} *M*, but evidently its damage has been done. If pyruvate is added externally at 10^{-4} *M*, it usually prevents growth on sulfur but is gradually utilized, and after a lag period of perhaps as long as 20 days the culture may grow. If 10^{-4} *M* pyruvate is added to the dialyzing fluid of a culture growing on glucose, growth stops immediately, even in some cultures, such as *T. denitrificans*, where pyruvate at these levels has no effect upon growth on thiosulfate (17). Not all obligate autotrophs are inhibited by pyruvate when given their specific inorganic nutrient, but we thought that if we could find out how pyruvate acted on *T. thiooxidans* it would at least give us a clue as to what such toxic materials might be doing. We first found that sulfur oxidation was not greatly inhibited by pyruvate at the 10^{-4} *M* level or below. Indeed, it frequently enhances oxidation. Such enhancement is not evident when sulfur is absent. We then found that, if cells were grown on sulfur with dialysis so that such toxic materials were removed, the rate of respiration on sulfur was not increased, but their ability to fix CO_2 was enhanced, although pyruvate added externally inhibited to the same extent.

Taking these and other facts into consideration, we devised the following hypothesis, which is capable of explaining most of the phenomena observed. We think that the obligate autotrophic bacteria have a permease for CO_2 coupled to sulfur oxidation, probably through phosphate bond energy. This permease is not entirely specific and can also mediate the transport of certain organic materials, including glucose and organic acids. In the presence of sulfur, these are readily transported into the cell; but they compete with CO_2 for such entry, and inasmuch energy, in most circumstances is not the limiting factor in the cell, one does not obtain enhanced growth from such transported organic materials.

Such a permease can also transport glucose, in the absence of sulfur, although at a considerably lower rate. Such glucose, after entry, can be metabolized to yield phosphate bond energy and thus permit growth and further glucose transport. But in so doing it produces pyruvate (or similar materials), which inhibit further transport.

We can demonstrate that this permease is inhibited by pyruvate. If one measures the uptake of glucose (over a 3 h period) by log phase resting cells of *T. thiooxidans* on a minimal salts medium at pH 5.2 (and 28°C), in the absence of pyruvate, the presence of sulfur markedly enhances the glucose uptake. If pyruvate is added at 10^{-4} *M*, there is little effect upon the glucose uptake with sulfur, but essentially complete inhibition of glucose uptake alone. It requires a ten-fold increase in pyruvate to stop glucose uptake when sulfur is present. At levels of 10^{-4} *M* or below, pyruvate itself is transported into the cell when sulfur is oxidized. The evidence suggests that CO_2, glucose, and pyruvate compete for the same site, which we postulate to be the permease. When phosphate bond energy is high, as it is during sulfur oxidation, the materials are readily transported. When phosphate bond energy is low, as when sulfur is absent, the transport of glucose becomes critical. When organisms mutate to glucose utilization, their surface properties change, and we assume that the sensitivity of the glucose permease to inhibition by pyruvate also decreases. This permease hypothesis at least has the virtue of suggesting many new kinds of experiments, most of which remain to be done.

ENDOGENOUS METABOLISM

To the best of our present knowledge, all cells carry out an endogenous metabolism, which may be defined as the metabolism of the cells in the absence of external substances that may be used by them. Endogenous metabolism employs substances "stored" within the cell and to some extent the more vital elements of the cell itself. The usual "storage materials" are polysaccharides (especially glycogen), lipids (especially poly-β-hydroxybutyrate), and polyphosphate; protein, RNA, free amino acids, and peptides in the cell may be used in some circumstances. There is at present no evidence that DNA, cell wall polymers, or cell membrane materials are used in endogenous metabolism.

There are two physiological problems as yet not solved, which involve endogenous metabolism. The first is the question as to whether endogenous metabolism continues on at its normal rate (or is enhanced or depressed) by the presence of external utilizable materials. This may be studied experimentally by growing the cells with isotopically labeled material and comparing the release of isotope when the cells metabolize endogenously and when they metabolize in the presence of external materials. However, all possible cases are found; that is, in certain organisms and with certain external materials, endogenous metabolism (or at least the release of isotope) is completely repressed, in others it is partly repressed, and in still others the endogenous metabolism seems to proceed to the same extent in the presence of external substrate (27). Ad hoc explanations of such phenomena can readily be devised; for example, if the major endogenous substrate were to use a metabolic pathway (e.g., the cytochrome system), which is completely taken over by the metabolism of the external substrate, one would expect endogenous metabolism to be markedly inhibited in the presence of external substrate. But it has proved to be very difficult *to predict* whether or not endogenous metabolism will be suppressed by external substrate, and at the moment no general rules can be given.

The second problem involves the question of a need for endogenous respiration. It has been proposed that there is an *energy of maintenance*, even in nongrowing cells, that is required primarily for the resynthesis of labile compounds which are continually being broken down within the cell (28). Endogenous metabolism supplies this energy. However, if there exists a required energy of maintenance, it ought be demonstrable.

This has proved difficult to do. Therefore, an alternative viewpoint has been developing, which is that the endogenous metabolism, although it does exist and it may provide energy to the cell, is merely the consequence of the complement of enzymes in the cell, which continue to act until the pertinent substrates are used up. If there were a way to stop their action, the cell could still survive the lack of nutrient, and presumably this is what happens in lyophilized cells.

There are many subtle matters in question here, and we can do no more than point out the main lines of differences in the two viewpoints, recognizing that a wide variety of opinion is prevalent. Just what would constitute critical experiments to distinguish among the viewpoints is also a matter of present debate.

NITROGEN FIXATION[1]

Nitrogen fixation is an enzymatic reduction of atmospheric nitrogen to ammonia. Prior to the isolation of a cell-free nitrogen-fixing system in 1960 by Carnahan, studies on nitrogen fixation had been carried out entirely with whole cells. Findings from these studies established the following: (1) nitrogen-fixing organisms require molybdenum as well as iron and calcium; (2) in certain organisms vanadium may replace molybdenum, but molybdenum is preferred; (3) ammonia appears to be the first stable product of nitrogen fixation; (4) possible intermediates such as nitramide, diimide, hydrazine, hyponitrate, and hydroxylamine, are not reduced or utilized as substrates in place of nitrogen; (5) there are relatively large numbers of compounds that inhibit nitrogen fixation. Included among them are oxygen, nitrate, nitrite, ammonia (feedback inhibition), hydrogen, nitrous oxide, cyanide, and carbon monoxide.

In 1960, Carnahan achieved reproducible nitrogen fixation in *Clostridium pasteurianum*. Using cell-free extracts, he established the following: (1) nitrogen fixation requires anaerobic conditions regardless of whether or not the organism is aerobic or anaerobic; (2) a good electron donor must be present; (3) it needs a sufficient supply of ATP; (4) substances are required to act as intermediary electron carriers; (5) an enzyme system (nitrogenase) is required to catalyze the reduction of atmospheric nitrogen to ammonia; (6) an enzyme system is needed to simultaneously hydrolyze ATP to ADP and inorganic phosphate; (7) such a system requires sodium pyruvate, the role of which was found to be two-fold. Oxidation of pyruvate led to the reduction of ferredoxin, which acted as a source for nitrogen reduction and to production of acetyl coenzyme A, which coupled with phosphotransacetylase and acetokinase supplied ATP, an absolute and specific requirement for nitrogen fixation.

As stated previously, nitrogen fixation is an anaerobic process. Aerobic nitrogen fixers, such as *Azotobacter*, some blue-green algae, and the *Rhizobium*—legume association, fix nitrogen in the presence of air, but the ability of aerobes to catalyze this oxygen-sensitive process is reconcilable only if the process occurs in an

[1]Dalton, H., and L. E. Mortenson. 1972. Dinitrogen (N_2) fixation (with a biochemical emphasis). *Bacteriol. Reviews 36*:231–262. (Reviewed by Bridget A. Walsh.)

internal anaerobic environment. This has been evidenced in nitrogen fixation by the *Rhizobium*—legume association in which leghemoglobin serves to bind the oxygen, thus maintaining anaerobic conditions in the legume root nodule.

It is now generally agreed that the basic energy requirements for enzymatic nitrogen fixation include ATP and a powerful reductant, such as reduced ferredoxin or flavodoxin. Low-potential electron transport chains couple cellular reducing power to the nitrogenase system. Other supporting reactions include a "low NH_4^+ pathway" of ammonia assimilation, so far described in bacteria only, as well as a poorly defined mechanism of oxygen protection of the nitrogenase system and ammonium repression and genetic regulation of nitrogenase biosynthesis.

The origin and nature of electron donors utilized in nitrogen fixation vary among the different physiological groups of nitrogen-fixing organisms. The supply of electron donors comes from relatively few cellular metabolites. For example, anaerobic bacteria such as *Cl. pasteurianum* produce the electron donor pyruvate from fermentation of sugars; strictly aerobic bacteria such as *Azotobacter* develop reducing power for nitrogen fixation by catabolism during the Kreb's cycle, utilizing organic acids, alcohol, and sugars. Photosynthetic bacteria and blue-green algae exhibit photochemical production of powerful reductants, which may be used for nitrogen reduction. The symbiotic root nodule bacteria (*Rhizobium*) present a special case in which sugars produced during photosynthesis in the leaf are transported to the nodule site, where they are broken down to supply reductant as well as ATP and carbon skeletons needed for nitrogen reduction and amino acid biosynthesis. That is, metabolites for electron donors are supplied by the host plant.

Nitrogen fixation also requires a good energy source. In photosynthetic organisms, light energy is utilized. In chemotrophs, carbohydrates serve as an energy source.

The enzyme that catalyzes nitrogen fixation has been termed nitrogenase. Nitrogenase binds molecular nitrogen and converts it to ammonia in a series of oxidation—reduction reactions, which require a total of six electrons and fifteen ATP. Nitrogenase has very similar properties in all organisms in which it is found. It is the only protein known to be unique to nitrogen fixation.

Nitrogenase consists of two parts: one, the smaller part, is the azoferredoxin or iron protein; the other part is the molybdenum—iron protein or the molbdoferredoxin.

Nitrogenase reduces N_2, CN^-, N_3^-, N_2O, and isocyanides as well as hydrogen. For these reductions magnesium—adenosine triphosphate, reduced ferredoxin, and the Mo—Fe and Fe proteins are required. The products of ATP utilization are ADP and Pi, which if allowed to accumulate inhibit further nitrogenase activity.

Ferredoxin and flavodoxin are not unique to nitrogen fixation, and possess no catalytic function; they serve merely as electron carriers. Ferredoxins that have been isolated from plants, blue-green algae, and bacteria are iron- and sulfur-containing proteins. The iron and sulfur are essential for the ferredoxin's activity as an electron carrier. Ferredoxins appear to be present in all nitrogen-fixing organisms, and in certain species they seem to be the only electron carrier.

Flavodoxins are riboflavin proteins. They can replace ferredoxins as the sole electron carrier in *Cl. pasteurianum*. In *Acetobacter vinelandii*, both ferredoxins and flavodoxins are active and required.

As for the general scheme of nitrogen fixation, the following mechanisms have been proposed. Cellular energy is supplied by light in phototrophs and by carbohydrates in heterotrophs. The electron donors, supplied by various cellular metabolites, are dehydrogenated, and then their electrons are trapped by very specialized electron carriers, ferredoxin and flavodoxin. They in turn are capable of reducing the nitrogenase. The nitrogenase then catalyzes the reduction of ATP → ADP + Pi, and the conversion of atmospheric nitrogen to $2NH_3$.

Recently, it has become possible to study the nitrogen fixation genes in some detail in *Klebsiella pneumoniae*, a nitrogen-fixing relative of *Escherichia coli*. Most nitrogen-fixation genes (Nif) were found to be clustered on the *Klebsiella* molybdenum—iron protein near the histidine operon. At least three genes, corresponding to the molybdenum—iron protein and the iron protein, code for nitrogenase. Mutations in these genes might completely destroy nitrogenase activity, since neither component can fix nitrogen without the other.

UNIQUE BIOCHEMICAL EVENTS IN BACTERIAL SPORULATION[1]

The purpose of this review was to evaluate the evidence for the existence of sporulation-specific genetic determinants. There are seven stages of development of the spore. Cytological changes are divided into seven stages based on observations by electron microscopy (Figure 15-1).

Stage 0. End of logarithmic growth phase.

Stage 1. The formation of an axially disposed filament of condensed chromatin.

Stage 2. Completion of forespore septum results in the segregation of the nuclear material into two compartments, referred to as the mother cell and the forespore cytoplasmic units.

Stage 3. The forespore protoplast is engulfed as a result of unidirectional growth of the cytoplasmic membrane of the sporangium.

Stage 4. Coat and exosporium formation.

Stage 5. Coat and exosporium formation, spore coat

[1]Hanson, R. S., J. A. Peterson, and A. A. Yousten. 1970. Unique biochemical events in bacterial sporulation. *Ann. Rev. Microbiol.* *24*:53—90. (Reviewed by Janice R. Hulbert.)

thickens.

Stage 6. Maturation or ripening process occurs. The refractibility of the endospore increases, and heat resistance develops.

Stage 7. Liberation of the mature endospore via autolysis of the mother cell.

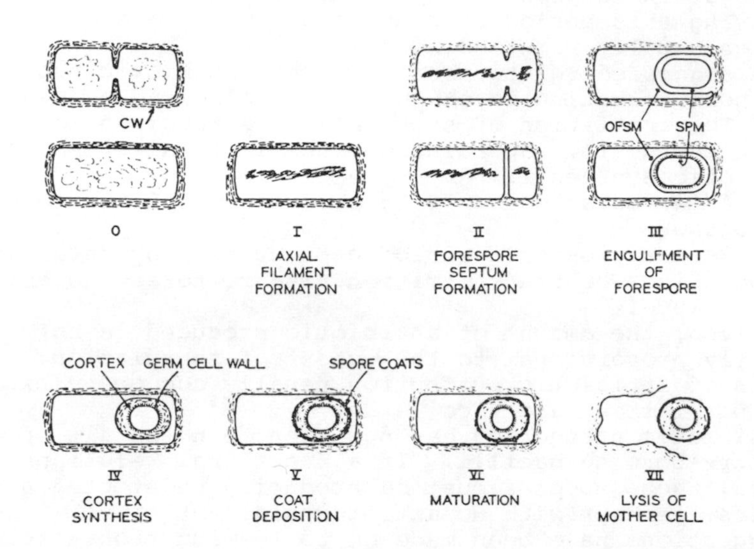

FIGURE 15-1. *Cytological changes during spore formation in Bacillus. From Hanson, R. S., J. A. Peterson, and A. A. Yousten. Ann. Rev. Microbiol. 24:53–90 (1970). Reproduced with permission from Annual Reviews of Microbiology, Volume 24.*

Evidence suggests that several enzymes and a few peptide antibiotics are either sporulation specific or are one of a series of sequentially related events necessary to sporulation.

Some species produce several extracellular enzymes simultaneously. Work done with *Bacillus subtilis* showed that 20 percent of its proteolytic activity produced at the end of logarithmic growth was due to a metal-requiring neutral proteinase, and 80 percent was due to alkaline proteinase sensitive to diisopropyl fluorophosphate (DFP). Two other DFP-sensitive enzymes have been described using the same strain of *B. subtilis*. It has been suggested that as many as six extracellular enzymes are produced by *B. subtilis*, and all are essential to sporulation. The neutral proteinase is not sporulation specific since a mutant devoid of it sporulated well. The function of the individual proteinases has not been described at present, nor can the role of any one proteinase in protein turnover and sporulation be specified.

Some species of *Bacillus* produce structurally related compounds that differ in one or a few amino acids and are considered members of the same family of peptides, which include the bacitracins, gramicidins, and tyrocidines.

Antibiotic production is one of the earliest biosynthetic events associated with sporulation. It begins in the postlogarithmic phase and precedes visible sporulation. Following this period of synthesis, the levels of antibiotics usually remain constant.

The suggested relationship between antibiotic production and spore formation is shown by the following evidence:

1. The initiation of sporulation is required for antibiotic synthesis. Sporulation inhibitors also inhibit antibiotic synthesis.
2. A few mutants unable to produce antibiotics are asporogenous.
3. Restoration of antibiotic production by reversion, transduction, or transformation also restores the ability to sporulate.

However, the amount of antibiotic produced is not directly proportional to the extent of sporulation. Low levels of antibiotic production usually accompany excellent sporulation, and vice versa.

Antibiotic production has not been found in all species of spore-forming bacilli. If a functional relationship exists, these species must be producing undetected antibiotics, perhaps with no antimicrobial activity.

Suggestions have been made as to the functional roles of antibiotics. One is in the regulation of cellular activities. Edeine, produced by *B. brevis*, inhibited DNA synthesis *in vivo* in *E. coli* and affected the activity of DNA polymerase *in vitro*. Bacitracin inhibited the induction of β-galactosidase in *Staphylococcus aureus*. This is a plausible hypothesis, but there are no data for its support.

Some peptides selectively transport ions across mitochondrial membranes, and ion transport could be essential for sporulation. Antibiotic molecules, being lipophilic, can easily pass through bacterial membranes. Molecules that cannot pass out of the cell could either attach to antibiotic molecules or pass through pores created by them.

Peptide antibiotics could be inducers of certain proteinases synthesized during sporulation. The fact that antibiotics are resistant to proteolysis and antibiotic negative strains produce less proteinase than parent strains is evidence for this possible function.

Even though antibiotic synthesis may not have a direct role in sporulation, it does reduce internal amino acid pools, which derepresses enzymes. The synthesis and excretion of antibiotic is one way of eliminating toxic materials from the cell.

PATHOGENICITY AS A PHYSIOLOGICAL PROBLEM: THE CASE OF DIPHTHERIA

Interest in bacteria as a distinct form of life arose during the latter half of the nineteenth century. For almost 50 years, roughly from 1870 to 1920, the attention of bacteriologists was focused, with only rare exceptions, on one striking capacity of certain bacterial cells: their ability to produce disease in animals, man, and plants. Yet only few species among the great variety of known bacteria have this capacity for disease production. Several other species of bacteria live in association with hosts of higher biological organization without damaging the tissues of their hosts, and in many cases they actually further the welfare of the hosts. We speak, indeed, of the *normal* bacterial flora of the throat or the intestinal tract of man or of the rumen of cattle. Many of these organisms synthesize B vitamins and thus provide the host with nutrients that it cannot make for itself. Many other examples of commensalism and symbiosis can be found in the literature, and they furnish an interesting chapter on the integration of the processes of nature.

Nevertheless, man is logically most interested in the forces of nature that affect his own well-being. The progress made in the control of infectious diseases of man during the past 50 years is enough to stagger the imagination, even on sober reflection. It is no longer commonplace to hear of fatal cases of pneumonia, typhoid fever, or diphtheria. The mechanisms of this control over infection are three-fold: (1) disinfection of carrier media, (2) immunization of susceptible hosts, and (3) treatment of infected cases with antiserum or antibiotics. The remarkable fact about these mechanisms of control is that they were established as practical methods before there was very much knowledge about how and why bacteria cause disease, except for the bacteria producing extracellular toxins, or poisons. Somewhat more is known today, but there are still great gaps in our understanding of the means by which bacteria are capable of invading tissue, of counteracting the defensive forces of the host, and of producing the lesions of the disease. It is our purpose to examine some of the pertinent information now available in one case where the mechanism of pathogenicity is partially known. Pathogenicity, in the broadest sense, is the capacity to produce disease in the living host. The virulence of any particular strain of bacteria depends on several factors: (1) the susceptibility of the host, (2) the route of inoculation, and (3) the physiological properties of the bacteria. When all three factors are strongly in favor of setting up the disease process, the bacterium is said to be highly virulent, but note that it is not solely on the properties of the bacterium that the judgement is made. A particular culture may cause disease when injected intraperitoneally but not intravenously, may infect mice, but not rats, and is therefore pathogenic, as qualified, for mice by intraperitoneal infection.

Given the same route of infection in the same host, a condition that is experimentally possible, we are able to measure the third factor with some degree of accuracy. The virulence of different strains of the same bacterial species has been found to vary considerably. Generally, strains freshly isolated from a susceptible host are more virulent than those that have been cultivated in the laboratory for some time. Passage of the laboratory culture through serial transfers in a susceptible host will, in many cases, sharply enhance the virulence. It is considered that such animal passage operates as a selection mechanism for the few more virulent organisms in the original laboratory culture, rather than as a medium of adaptation of the entire culture to greater pathogenicity. In the same fashion, the cultivation on laboratory media would operate in the reverse direction by selection for the relatively avirulent organisms more capable of free-living existence.

While in most cases there is no physiological criterion by which pathogenicity can be measured except by inoculation into the susceptible host, some few bacteria are pathogenic primarily because they produce potent exotoxins. In one case, at least, we know something of the process — the case of diphtheria. With it as an example one can approach some of the problems associated with pathogenicity. Granted the original premise, that an organism is pathogenic because it produces one factor or many factors that cause the clinical symptoms of its specific disease, what peculiarity of the physiology of the organism results in the production of that factor? How do virulent strains differ from avirulent strains of the same species? What is the distinction in genetic pattern between them?

For one organism, *Corynebacterium diphtheriae*, we know at least part of the answer to all these questions. The information is the result of one of the few explorations into the physiological and biochemical nature of virulence, and it deserves more than cursory mention. The pathogenicity of *C. diphtheriae* strains is due to their production of a lethal exotoxin. Nonpathogenic strains do not produce this toxin. For this organism, therefore, we can equate the terms "pathogenic" or "virulent" and "toxigenic," contrasted to "avirulent," and "nontoxigenic."

Furthermore, the pathogenic strains produce toxin on *in vitro* cultivation. This has provided an approach to the study of the mechanism of toxin formation, and the results obtained were curious indeed. The amount of toxin produced is dependent on the amount of iron in the medium. At minimal concentrations of iron, both the growth of the organism and the production of toxin are quite meager. With increasing concentrations of iron the growth is improved, and the amount of toxin formed rises. At about 100-μg of iron/liter of medium, maximum toxin production is achieved. If the concentration of iron is raised above

100 μg/liter, the growth of the organisms improves, but the amount of toxin produced decreases sharply, until at 500–600 μg of iron/liter almost none is formed.

Paralleling the dependence of *C. diphtheriae* toxin production on iron concentration is the production of porphyrin, also present in the culture filtrate. The data show that, for each four molecules of iron supplied, four molecules of porphyrin and one molecule of toxin fail to appear in the medium. The added iron can be recovered quantitatively from the bacterial cells. These results strongly indicate that the cells form a substance composed of porphyrin, iron, and toxin protein. When the iron concentration is low (100 μg or less), the cells form porphyrin and toxin protein, but are unable to link them into a complete molecule of useful function in the absence of the third component, iron. These substances, therefore, are excreted into the medium. Addition of more iron allows for conjugation of the three components, the porphyrin and toxin are bound within the cell, and their excretion into the medium is decreased.

The ability to form toxin is known to be a consequence of infection with certain bacteriophage strains, that is, a consequence of lysogenicity. Exposure of avirulent, nontoxigenic cultures of *C. diphtheriae* to certain bacteriophage strains (particularly one designated as β) results in the conversion of the cultures to lysogenicity and toxigenicity simultaneously. Infection of an avirulent culture with other phage strains (e.g., strain γ) may result in lysogenicity without concomitant toxin production. The phenomenon appears to be different from transduction, since toxigenicity is an inevitable consequence of infection with the effective phage strain; if the phage had merely carried over a gene for toxin production from the donor strain on which it had been propagated, one would expect only a very small fraction of the receptor organisms to have received any particular gene. In the conversion of β phage, *all* the organisms that become infected by the phage are toxigenic. The mechanism by which the phage alters the metabolism of the nontoxigenic *C. diphtheriae* cell to toxigenicity is still obscure. Comparative studies of a toxigenic β-lysogenic strain of *C. diphtheriae* and the parent nontoxigenic, non-β-lysogenic strain from which it was derived show that both strains liberate protein and porphyrin at the same rate. However, the protein material released from the nontoxigenic strain is markedly different from that of the toxigenic strain. The former is nontoxic, relatively nonantigenic, and electrophoretically inhomogenous, with a much more diffuse band at the mobility where diphtheria toxin is present as a sharply defined peak. Infection with β phage thus alters not the response to iron deficiency, but rather the type of protein released.

Toxin can be synthesized in cell-free ribosomal systems from *E. coli* programmed with DNA from the β phage carrying the tox^+ gene. The toxin itself is an acid globular protein with a molecular weight of 62,000–63,000, which contains no unusual amino acids and really nothing to chemically distinguish it from other proteins.

The diphtheria toxin inhibits the growth of cells in cell culture, and the sensitivity of the cell culture is closely the same as that from the animal and the specific organ that provided the cells for such culture. For example, the guinea pig is quite sensitive to diphtheria toxin, whereas the rat is resistant. So are the tissues in cell culture (29,30). The heart seems to be especially susceptible to toxin; so it is in cell culture.

Mild trypsin treatment (31) or heating of the toxin molecule (molecular weight of 63,000) under reducing conditions (32) causes it to split into two fragments. Fragment A (molecular weight of 24,000 daltons) is responsible for the enzymatic activity of the toxin, which consists of the catalysis of the hydrolysis of NAD to nicotinamide and adenine diphosphoribose (ADPR) and the attachment of the ADPR to the eucaryotic elongation factor 2. This renders the factor inactive and protein synthesis is thereby stopped. The cell-free protein-synthesizing extracts may come from any tissue, whether sensitive to diphtheria toxin or not, and they will still be inhibited. Bacterial systems are not inhibited by toxin because the comparable elongation factor (factor G) for procaryotic cells does not react with ADPR. In this case, therefore, the toxin is active in the animal and man and not in the bacteria because of a difference in the structure of elongation factors.

Isolated fragment A inhibits protein synthesis in a cell-free system, but has no effect on whole cells. Penetration into cells seems to be the function of the B fragment. The idea is that the total toxin attaches to the sensitive cell via receptors on the cell surface. This is followed by a proteolytic "nicking" and reduction of the toxin molecule. Fragment A now enters the cell, causing the formation of ADPR which inhibits protein synthesis.

A mutant of the β phage has been isolated (33) that produced a nontoxic protein, immunologically related to the toxin. This protein, with a molecular weight of 45,000 (rather than 63,000), can be nicked with trypsin and reduced with thiol, yielding two fragments, one of which is the toxic material A. Other mutants have been isolated (34,35,36) that are deficient in parts of either fragment A or B, and an active toxin may be reconstituted *in vitro* by mixing the appropriate nontoxic fragments derived from these mutants.

Other toxins may act in a similar manner (37). Exotoxin A (*Pseudomonas aeruginosa*) also kills mammalian cells by forming ADPR although the toxin is quite different chemically from diphtheria toxin. The action of *E. coli* enterotoxin and cholera toxin, while different from diphtheria toxin, are themselves closely related and both act by activating adenylate cyclase and thus increase cyclic AMP. Cholera toxin is composed of two parts A and B and part B is associated with the transport into the cell. But A can be split into two fragments of which A_1 is active in the presence of a soluble cytoplasmic protein, NAD and ATP, in stimulating cyclic AMP synthesis. A fragment of the *E. coli* toxin has the same molecular weight as the A_1 fraction of cholera toxin, requires NAD, ATP and a component of the host cytoplasm and is inhibited by antibody against cholera toxin.

It is beyond the scope of this book to consider further aspects of the mechanisms of pathogenicity, virulence, and host-microbe interactions, but it should be clear that this is an important field which now appears to be approachable by the methods of physiology and from which one may expect significant advances in the future.

REFERENCES

Autotrophy

Kelly, D. P. 1971. Autotrophy: concepts of lithotrophic bacteria and their organic metabolism. *Ann. Rev. Microbiol.* *25*:177—210.

McFadden, B. A. 1973. Autotrophic CO_2 assimilation and the evolution of ribulose diphosphate carboxylase. *Bacteriol. Reviews* *37*:289—319.

Peck, H. D., Jr. 1968. Energy-coupling mechanisms in chemolithotrophic bacteria. *Ann. Rev. Microbiol.* *22*: 489—518.

Rittenberg, S. C. 1969. The role of exogenous organic matter in the physiology of chemolithotrophic bacteria. *Advan. Microbial Physiol.* *3*:159—196.

Suzuki, I. 1974. Mechanisms of inorganic oxidation and energy coupling. *Ann. Rev. Microbiol.* *28*:85—101.

Trudinger, P. A. 1969. Assimilatory and dissimilatory metabolism of inorganic sulfur compounds by microorganisms. *Advan. Microbial Physiol.* *3*:111—158.

Endogenous Metabolism

Goldberg, A. L., and J. F. Dice. 1974. Intracellular protein degradation in mammalian and bacterial cells. *Ann. Rev. Biochem.* *43*:835—869.

Nitrogen Fixation

Benemann, J. R., and R. C. Valentine. 1972. The pathways of nitrogen fixation. *Advan. Microbial Physiol* *8*L59—104.

Dixon, R. O. D. 1969. Rhizobia (with particular reference to relationships with host plants). *Ann. Rev. Microbiol.* *23*:137—158.

Shanmugam, K. T., and R. C. Valentine. 1975. Molecular biology of nitrogen fixation. *Science* *187*:919—924.

Stewart, W. D. P. 1973. Nitrogen fixation by photosynthetic microorganisms. *Ann. Rev. Microbiol.* *27*:283–316.

Streicher, S. L., and R. C. Valentine. 1973. Comparative biochemistry of nitrogen fixation. *Ann. Rev. Biochem.* *42*:279–304.

Yates, M. G., and C. W. Jones. 1974. Respiration and nitrogen fixation in *Azotobacter*. *Advan. Microbial Physiol.* *11*:97–136.

Spores

Gould, G. W., and G. J. Dring. 1974. Mechanisms of spore heat resistance. *Advan. Microbial Physiol.* *11*:137–164.

Halvorson, H. O., J. C. Vary, and W. Steinberg. 1966. Developmental changes during the formation and breaking of the dormant state in bacteria. *Ann. Rev. Microbiol.* *20*:169–188.

Holt, S. C., and E. R. Leadbetter. 1969. Comparative ultrastructure of selected aerobic spore-forming bacteria: a freeze-etching study. *Bacteriol. Reviews* *33*:346–378.

Murrell, W. G. 1967. The biochemistry of the bacterial endospore. *Advan. Microbial Physiol.* *1*:133–252.

Pathogenicity and Virulence

Barksdale, L. 1970. *Corynebacterium diphtheriae* and its relatives. *Bacteriol Reviews* *34*:278–422.

Collier, R. J. 1975. Diphtheria toxin: mode of action and structure. *Bacteriol. Reviews* *39*:54–85.

Smith, H. 1968. Biochemical challenge of microbial pathogenicity. *Bacteriol. Reviews* *32*:164–184.

Papers

Autotrophy

1. *J. Bacteriol.* *87*:1243 (1964).
2. *Soil Science* *51*:331 (1941).
3. *J. Bacteriol.* *85*:137 (1963).
4. *J. Bacteriol.* *43*:141 (1942).
5. *J. Bacteriol.* *46*:451 (1943).
6. *J. Bacteriol.* *88*:620 (1964).
7. *J. Bacteriol.* *82*:788 (1961).
8. *J. Bacteriol.* *95*:2182 (1968).
9. *J. Bacteriol.* *108*:612 (1971).
10. *Biochemistry* *2*:194 (1963).
11. *Biochim. Biophys. Acta* *104*:359; *122*:22 (1966).
12. *J. Bacteriol.* *92*:899; *95*:147 (1968).
13. *J. Bacteriol.* *91*:729 (1966).
14. *Zeit. f. Allg. Mikrobiol.* *8*:445 (1968).

15. *J. Gen. Physiol. 26*:103 (1942).
16. *Arch. Biochem. Biophys. 116*:97 (1966).
17. *J. Bacteriol. 109*:1149 (1972).
18. *Can. J. Microbiol. 18*:275 (1972).
19. *Can. J. Microbiol. 18*:1907 (1972).
20. *J. Bacteriol. 95*:844 (1968).
21. *J. Bacteriol. 97*:256 (1969)
22. *J. Bacteriol. 108*:328, 334 (1971)
23. *J. Bacteriol. 94*:972 (1967).
24. *J. Bacteriol. 97*:966 (1969).
25. *J. Bacteriol. 117*:834 (1974).
26. *J. Bacteriol. 93*:597 (1967).

Endogenous Metabolism

27. *Ann. N. Y. Acad. Sci. 102*:515 (1963).
28. *J. Bacteriol. 93*:1467 (1967).

Pathogenicity and Virulence

29. *J. Immunol. 88*:505 (1962).
30. *J. Expt. Med. 123*:723 (1966).
31. *J. Biol. Chem. 246*:1485, 1492, 1496, 1504 (1971).
32. *J. Bacteriol. 115*:277 (1973).
33. *Nature 233*:8 (1971).
34. *Virology 50*:664 (1972).
35. *J. Biol. Chem. 248*:3838, 3851 (1973).
36. *Science 125*:901 (1972).
37. *Science 190*:969 (1975).

Index